Molecular and Integrative Toxicology

Series editor

Rodney R. Dietert

For further volumes:
http://www.springer.com/series/8792

Suzanne M. Snedeker
Editor

Toxicants in Food Packaging and Household Plastics

Exposure and Health Risks to Consumers

Humana Press

Editor
Suzanne M. Snedeker
Independent Consultant
Ithaca, NY
USA

ISSN 2168-4219 ISSN 2168-4235 (electronic)
ISBN 978-1-4471-6499-9 ISBN 978-1-4471-6500-2 (eBook)
DOI 10.1007/978-1-4471-6500-2
Springer London Heidelberg New York Dordrecht

Library of Congress Control Number: 2014943126

Printed on acid-free paper

Humana Press is a brand of Springer
Springer is part of Springer Science+Business Media (www.springer.com)

Preface

The purpose of this book is to provide a comprehensive resource on what is known and what we need to know about toxicants present in food packaging materials and household plastics. This includes identifying human exposure scenarios for consumers, summarizing relevant known and emerging health effects, and recommending additional research needed to address data gaps that will allow for improved risk assessment for human populations. This book is meant to be a resource across disciplines and should be useful to toxicologists, environmental health scientists, food scientists, and regulators in the areas of food safety and environmental health sciences.

This book is unique in several ways:

Format: Each chapter starts with a bulleted list of **Key Take Home Points** and ends with a section on research needed to address data gaps.

Breadth: The toxicants included in the book range from more widely known chemicals such as bisphenol A (Chap. 1), various phthalates (Chap. 2), brominated flame retardants (Chap. 3), perfluorinated compounds (Chap. 7), and the heavy metals lead (Chap. 9), and cadmium (Chap. 10), to chemicals that are just starting to emerge as potential toxicants from food packaging and household plastics, as well as chemicals that will be entering use in the food packaging industry. This includes the alkylphenols nonylphenol and octylphenol that have been identified in food packaging and foodstuff (Chap. 5), chemicals used in UV-cured print inks (benzophenone and 4-methylbenzophenone) that can migrate through porous printed cartonboard and most secondary packaging to foods (Chap. 6), the metal antimony used as a catalyst in the manufacturing of PET-single-use beverage bottles (Chap. 8), methylnaphthalene detected in breakfast cereal box liners (Chap. 10), and nanoparticles (Chap. 4) that will be used in polymer food packaging in future and present emerging toxicological concerns. Several chapters also provide information on the challenges of the use of replacement chemicals, especially for the phthalates (Chap. 2), and the Brominated flame retardant (BFR) (Chap. 3).

International Focus on Exposure: Authors have been encouraged, whenever possible, to include international data on chemical exposure. This includes identification of data gaps where information on a chemical's level in products or biomonitoring data may be limited to one or only a few geographic locations. This is especially important because of the global nature of our food supply and household consumer goods.

Inclusion of Emerging Toxicological Endpoints: Chapters that include Health Effects sections for chemicals have drawn on studies with a wide range of relevant toxicological endpoints, including not only traditional cancer and reproductive endpoints, but when appropriate, emerging research on endocrine disruption, cardiovascular disease, diabetes and obesity, immune function, neurological function and behavior, and transgenerational effects.

Regulatory Approaches and Challenges in the United States and Europe: One of the most unique aspects of this book is the inclusion of information to educate the reader on current approaches and practices used to monitor and evaluate the risk of chemicals that are present as intentional or unintentional substances in food processing and packaging (food contact materials). An overview of the use and functions of food packaging is presented in Chap. 4, Sect. 4.3, and a summary the U.S. Food and Drug Administration's (FDA) approach to address the migration of substances from food contact materials is presented in Sect. 4.5. The final chapter of this book (Chap. 11) provides an overview of several areas, including the use, safety, and exposure to chemicals used in food contact materials, and a comparison of current regulations and risk assessment approaches used by agencies in the U.S. and Europe. Current challenges faced in evaluating chemical risk arising from use of food contact materials are highlighted, including interpreting low-dose effects (non-monotonic dose responses), mixture effects, developmental origins of disease, and transgenerational effects.

New Approaches: This book focuses on a small number of the chemicals used in food packaging and in the manufacturing of household plastics. For thousands of other chemicals, we lack basic toxicological risk information. Realistically, new approaches will be needed, including high-throughput screening, to better identify and assess the toxicological risk of chemicals that are present in household plastics and food packaging materials. Some of these approaches are outlined in Chap. 11, Sect. 11.5.6.

In closing, I would like to thank all the contributing authors for their most precious resource, their time, in developing the concepts and content of their chapters. Their efforts have been outstanding. I thank them for developing carefully thought out and researched chapters that truly make a significant contribution to our understanding of exposure and health risks of toxicants associated with food packaging and household plastics, how they are regulated, and the new avenues that need to be pursued to address what we still need to know about exposure and health effects in human populations.

Suzanne M. Snedeker

Contents

Contributors

Omari J. Bandele Division of Food Contact Notifications, Office of Food Additive Safety, Center for Food Safety and Applied Nutrition, U.S. Food and Drug Administration, College Park, MD, USA

Michelle L. Berger Center for Marine Studies, Marine Environmental Research Institute, Blue Hill, ME, USA

Linda S. Birnbaum National Institute of Environmental Health Sciences, National Institutes of Health, Research Triangle Park, NC, USA; National Cancer Institute, National Institutes of Health, Research Triangle Park, NC, USA

Timothy V. Duncan Division of Food Processing Science and Technology, Office of Food Safety, Center for Food Safety and Applied Nutrition, U.S. Food and Drug Administration, Bedford Park, IL, USA

Jennifer H. Harris Center for Marine Studies, Marine Environmental Research Institute, Blue Hill, ME, USA

Anthony G. Hay Department of Microbiology, Cornell University, Ithaca, NY, USA; Graduate Field of Environmental Toxicology, Cornell University, Ithaca, NY, USA

Paul Honigfort Division of Food Contact Notifications, Office of Food Additive Safety, Center for Food Safety and Applied Nutrition, U.S. Food and Drug Administration, College Park, MD, USA

Piper R. Hunt Division of Toxicology, Office of Applied Research and Safety Assessment, Center for Food Safety and Applied Nutrition, U.S. Food and Drug Administration, Laurel, MD, USA

Allan C. Just Department of Environmental Health, Harvard School of Public Health, Boston, MA, USA

Kurunthachalam Kannan New York State Department of Health, Wadsworth Center, Albany, NY, USA; Department of Environmental Health Sciences, School of Public Health, State University of New York at Albany, Albany, NY, USA

Joseph Laquatra Department of Design and Environmental Analysis, Cornell University, Ithaca, NY, USA

Jane Muncke Food Packaging Forum Foundation, Zurich, Switzerland

Karthik V. Pillai Division of Food Processing Science and Technology, Office of Food Safety, Center for Food Safety and Applied Nutrition, U.S. Food and Drug Administration, Bedford Park, IL, USA

Penelope A. Rice Division of Food Contact Notifications, Office of Food Additive Safety, Center for Food Safety and Applied Nutrition, U.S. Food and Drug Administration, College Park, MD, USA

Kathryn M. Rodgers Silent Spring Institute, Newton, MA, USA

Ruthann A. Rudel Silent Spring Institute, Newton, MA, USA

Thaddeus T. Schug Division of Extramural Research and Training, National Institute of Environmental Health Sciences, Research Triangle Park, NC, USA

Susan D. Shaw Center for Marine Studies, Marine Environmental Research Institute, Blue Hill, ME, USA; Department of Environmental Health Sciences, School of Public Health, State University of New York at Albany, Albany, NY, USA

Suzanne M. Snedeker Independent Consultant, Ithaca, NY, USA

Bikram Subedi New York State Department of Health, Wadsworth Center, Albany, NY, USA

Chapter 1
Human Health Effects of Bisphenol A

Thaddeus T. Schug and Linda S. Birnbaum

Abstract Bisphenol A (BPA) is a high production endocrine disrupting chemical found in numerous consumer products. BPA has been used commercially since 1957 to make hard polycarbonate plastics and epoxy resins used in food-can linings, cash-register receipts, and dental resins. The ubiquity of BPA in our environment results in exposure to this chemical daily in human populations. But controversy remains regarding how much BPA humans actually ingest or otherwise encounter. Many laboratory animal and human studies have linked exposures to BPA, a hormone mimicking chemical, to adverse health effects, including altered behavior and obesity in children, reproductive abnormalities, cardiovascular changes, and various cancers. However, there have been considerable inconsistencies in the outcomes from these studies with respect to the nature of the adverse health effects observed, and questions as to whether the BPA dose at which they occur are within the range of non-occupational human exposures. This chapter reviews the latest research on BPA, focusing on human exposure, discussions of biomonitoring studies and toxicokinetic models, human health effects, and research needs. We also include illustrative examples of animal models that address whether BPA-exposure is associated with changes in certain health endpoints.

T. T. Schug (✉)
Division of Extramural Research and Training, National Institute of Environmental Health Sciences, Research Triangle Park, NC, USA
e-mail: schugt2@niehs.nih.gov

L. S. Birnbaum
National Institute of Environmental Health Sciences, National Institutes of Health, Research Triangle Park, NC, USA
e-mail: birnbaumls@niehs.nih.gov

L. S. Birnbaum
National Cancer Institute, National Institutes of Health, Research Triangle Park, NC, USA

S. M. Snedeker (ed.), *Toxicants in Food Packaging and Household Plastics*,
Molecular and Integrative Toxicology, DOI: 10.1007/978-1-4471-6500-2_1,

Keywords Bisphenol A · Endocrine disruption · Developmental toxicity · Pharmacokinetics · Biomonitoring

1.1 Key Take Home Points

- Bisphenol A exposure is ubiquitous in our environment and it is found in our bodies.
- Bisphenol A has estrogenic and other properties and is considered to be an endocrine disrupting chemical.
- Bisphenol A is a non-persistent chemical and is quickly metabolized by the body.
- Bisphenol A exposure has been tied to many adverse health effects in animal models and humans including: prostate cancer, breast cancer, obesity, diabetes, cardiovascular problems, some neurobehavioral effects, including anxiety, as well as reproductive effects.
- Enhanced biomonitoring studies are needed to better evaluate bisphenol A exposure and health outcomes in humans.

1.2 Introduction

Manufacturers produce more than 8 billion pounds of BPA every year, making it one of the most common industrial chemicals produced worldwide (Rubin 2011). Plastics made with BPA are used in many consumer products, including food and beverage containers, toys, eyeglasses, computers, kitchen appliances, and medical equipment. Epoxy resins containing the chemical are used in dental work and in metal coatings for food cans, pipes, cars, dairy equipment, office equipment, and other metal products. BPA and its derivatives are also used in flame retardants (tetrabromobisphenol A), in engineering applications such as laminates for printed circuit boards, and as color developers in thermal receipt paper (Birnbaum et al. 2012). Some, but not all, plastics that are marked with recycle codes "3" or "7" may be made with BPA.

BPA has been detected in air, soil, water, landfill leachate, and the human body. The chemical has been shown to leach into foods and beverages from some types of food packaging (e.g., polycarbonate containers and epoxy lining of metal cans) and reusable containers (von Goetz et al. 2010). People also may be exposed to BPA through skin contact, inhalation, dental fillings, and occupational exposures. BPA has been found in human serum, milk, saliva, urine, and amniotic fluid (Vandenberg et al. 2009, 2010, 2012).

The ubiquity of BPA in the environment and in the human body has led to concerns about potential adverse health effects. BPA's chemical structure allows it to fit in the estrogen receptor (ER) binding pocket, and BPA is considered to act as an endocrine disruptor. BPA binds to both nuclear and cell membrane ERs; at higher levels, BPA acts as an androgen receptor (AR) antagonist, interacts with the thyroid receptor (Vandenberg et al. 2009) and induces the peroxisome proliferator-activated receptor gamma (PPARγ) (Kwintkiewicz et al. 2010; Wang et al. 2012). Animal and human research has associated BPA with many health problems including infertility, weight gain, behavioral changes, early-onset puberty, prostate and mammary gland cancers, cardiovascular effects, obesity, and diabetes (Birnbaum et al. 2012; Schug et al. 2012).

1.3 Chemical and Biological Properties of Bisphenol A

BPA ($C_{15}H_{16}O_2$; CAS No. 80-05-7) is one of the most common industrial chemicals produced worldwide. BPA was first synthesized by the Russian chemist A.P. Dianin in 1891. The compound consists of two conjoined phenol functional groups, and is synthesized by the condensation of acetone (hence the suffix A in the name) with two equivalents of phenol (Fig. 1.1). BPA is used in polycarbonate plastic to enhance product strength, durability, and transparency. BPA is used in epoxy resins to extend the shelf-life of canned foods as well as in dental composite resins used to fill most cavities (Kingman et al. 2012). BPA also functions as a color developer in carbonless thermal receipt paper.

1.3.1 Endocrine Disrupting Properties of Bisphenol A

By virtue of its binding ability to steroid receptors, BPA was hypothesized to be estrogenic. These properties of BPA were first demonstrated in studies using ovariectomized rats in the 1930s during a search for synthetic estrogens. However, BPA was abandoned for pharmaceutical use when diethylstilbestrol (DES) was determined to be a much more potent estrogen (Vandenberg et al. 2010). Biochemical assays have since shown that BPA does fit within the ER binding pocket and that it binds to both ERα and ERβ, with approximately tenfold higher affinity for ERβ (Gould et al. 1998; Kuiper et al. 1998). However, the binding affinity of BPA for both ER isoforms is nearly 10,000-fold weaker than that of estradiol (EC_{50} 2–7 $\times$ 10^{-7} vs. 1–6 $\times$ 10^{-13} M for estradiol) (Andersen et al. 1999; Kuiper et al. 1998). A recent study suggests that BPA can function as an ER agonist at higher concentrations [$\geq$10 nanomolar (nM)] and as antagonist at lower concentrations ($\leq$10 nM). These paradoxical actions are likely cell-type specific and may

Fig. 1.1 Chemical structures of bisphenol A and estradiol

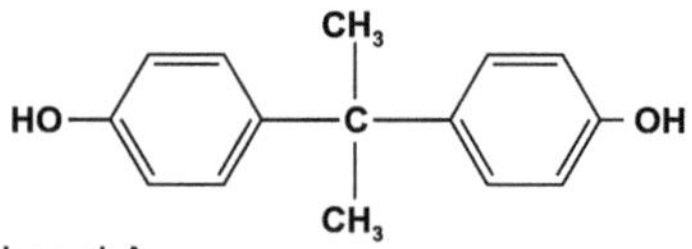

Bisphenol A

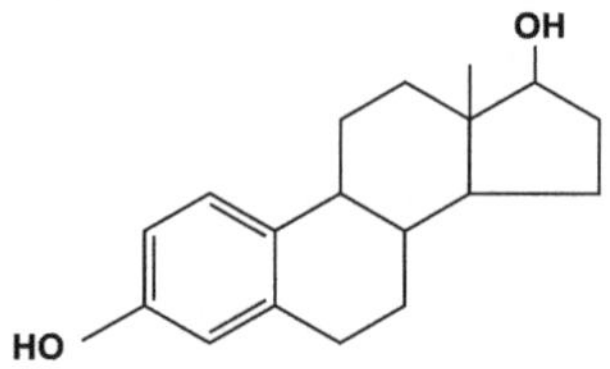

Estradiol

be in part mediated by BPA activation of the AF-2 domain of ERα, not the classical ligand-binding domain (Li et al. 2012).

Studies have shown that BPA also binds to membrane-bound forms of the ER (mER) and with high affinity to a transmembrane ER receptor called G protein-coupled receptor 30 (GPR30) (Watson et al. 2007; Thomas and Dong 2006; Vinas and Watson 2013; Wetherill et al. 2007). In addition to its estrogenic activity, there is mounting evidence that BPA interacts with other nuclear receptors, albeit at higher concentrations. BPA, for example, binds to the thyroid hormone receptor (TR) with lower affinity than the ER (Moriyama et al. 2002). Studies also have shown that BPA binds to the ubiquitous aryl hydrocarbon receptor (AhR) (Pocar et al. 2005), which mediates toxicity through several signaling pathways (Pocar et al. 2005; Barouki et al. 2012). Other evidence suggests that BPA and its derivatives act as obesogens by inducing adipocyte differentiation and adipogenic marker genes in preadipocytes through various mechanisms (Masuno et al. 2005; Chamorro-Garcia et al. 2012).

1.4 Human Exposure, Biomonitoring, and Metabolism of Bisphenol A

1.4.1 Human Exposure to Bisphenol A

Since the 1990s, many studies have been dedicated to determining which consumer products contain BPA, and how much is released from these products into food and beverages under normal conditions of use (von Goetz et al. 2010). In particular, studies have focused on the levels of BPA released from baby bottles, food-contact papers, and epoxy resins used both for dental sealants and the linings of metal food cans (Schug et al. 2012; Carwile et al. 2011). These studies confirm that the majority of canned foods contain measurable levels of BPA. A wide range of concentrations were reported, leading scientists to estimate that current human exposures from canned and bottled goods could be in the nanogram per kilogram (ng/kg) range for children and adults and the low microgram per kilogram (μg/kg) range for bottle-fed infants (von Goetz et al. 2010). However, recent moves by manufacturers in the United States (U.S.) and Europe to remove BPA from baby bottles and infant feeding cups have reduced the risks of exposure to infants. Other recent studies have shown that thermal papers (Geens et al. 2012) and dental procedures (Kingman et al. 2012) can contribute to human exposures.

1.4.2 Biomonitoring of Bisphenol A

The ability to accurately measure exposure to BPA is critical to assessing the chemical's health effects. Measuring BPA in urine is generally considered the most reliable indicator of BPA exposure because it integrates exposure over a recent time period, whereas BPA concentrations in blood are thought to reflect only current exposures due to the chemical's short half-life and evidence that BPA does not bioaccumulate. However, serum measurements are currently the most meaningful way to assess single point in time levels of unconjugated BPA, also known as free BPA, which is the form that is considered to be more biologically active because it can bind to ER and other nuclear receptors (Ye et al. 2011). This is an area that requires further investigation because some posit that rapid processing of BPA in the gastrointestinal tract and the liver during first pass metabolism results in very low levels of circulating parent (and thus biologically active) BPA entering the bloodstream, precluding BPA from causing disease. However, more than 30 reports have shown human blood levels of BPA in the range of 0.1–4.0 ng per milliliter (ml) (Taylor et al. 2011; Vandenberg et al. 2010). Although in some cases consistent results have been reported across studies, these findings have spurred debate in the scientific community about the likelihood that free BPA blood measurements are compromised by contamination (Calafat 2010; Ye et al. 2012). For example, small amounts of BPA may leach into samples from syringes, containers, tubing, or even water used in experiments. In addition, instruments used to measure levels of BPA may be incorrectly calibrated. The National Institute of Environmental Health Sciences (NIEHS) is currently leading a multi-agency effort to improve measurement procedures for detection of BPA in blood (Birnbaum et al. 2012).

While the National Health and Nutrition Examination Survey (NHANES) (Stahlhut et al. 2009; Calafat et al. 2008), a statistically-based sampling of the U.S. population conducted every 2 years, and several birth cohorts, such as the Center for the Health Assessment of Mothers and Children of Salinas (CHAMACOS) study cohort (Castorina et al. 2010), have contributed useful data on BPA exposure and health impacts, the design and methodology of these studies make it difficult to draw conclusions from them about BPA's health effects (Stahlhut et al. 2009). These studies are effective at inventorying exposure data for large populations of people, but they are not geared to assess the impact of BPA exposure on individuals whom may have significant fluctuations in exposures or wide-ranges in susceptibility.

Because of BPA's short half-life, depending on the nature of the BPA exposure (e.g., diet) and the timing of urine collection, hourly variability of BPA urinary concentrations is expected. Since diet, which often changes daily for most adults, is considered the main pathway of BPA exposure, information on the temporal variability of urinary BPA concentrations measured from a 24-hour (h) urine collection and first morning voids is of interest (Calafat 2010). For example, a recent report from the U.S. Centers for Disease Control and Prevention (CDC) showed that the between-day BPA concentrations in spot samples of urine collected from the same person varied substantially (Ye et al. 2011). Such variance highlights the importance of the time of the day of sampling for spot urine collections in study designs to evaluate exposure to BPA using biomonitoring (Calafat 2010). Overall, epidemiological and biomonitoring studies could be enhanced by better exposure assessment strategies, including collecting and pooling multiple spot urine samples, implementing 24-h urine sampling, or identifying new matrices for measuring non-persistent chemical exposure (e.g., meconium, teeth, metabolomics) (Braun et al. 2010a; Arora and Austin 2013).

1.4.3 Pharmacokinetics of Bisphenol A

In contrast to the dozens of studies that have measured BPA and its metabolites in environmentally exposed individuals, only two studies have determined the kinetics of BPA metabolism following direct administration of BPA. In the first kinetics study, healthy adult volunteers were administered 5 mg deuterium-labeled BPA (equivalent to 54–90 μg/kg body weight [bd wt]) and then were monitored for blood and urinary BPA and BPA metabolites levels (Volkel et al. 2002). Unconjugated BPA was always below the limits of detection (LOD) (LOD for urine = 1.37 ng/ml, LOD for plasma = 2.28 ng/ml) and BPA-glucuronide (Fig. 1.2) concentrations fell below the LOD in both urine and blood 24–34 h after BPA administration. The authors concluded that BPA was rapidly metabolized to BPA-glucuronide, and this conjugate is rapidly cleared from blood.

In a second study, subjects were administered 25 μg BPA, and unconjugated and conjugated BPA were measured in urine at 0, 1, 3, 5, and 7 h after exposure (Volkel et al. 2005). The authors concluded that due to the rapid and complete elimination of BPA in humans as BPA-glucuronide in urine (Volkel et al. 2002), that human exposure to BPA is less than 2.3 μg/person per day (based on a volume

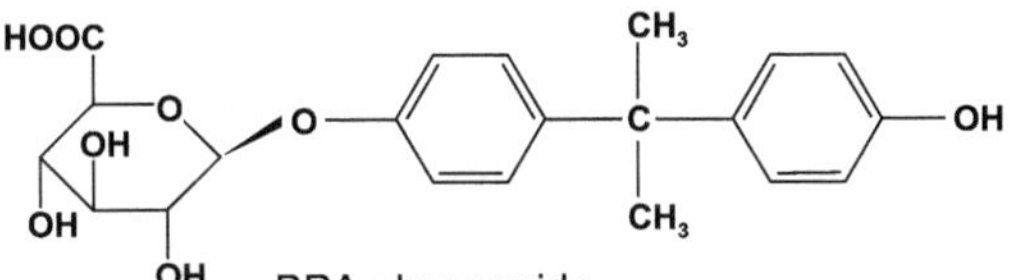

Fig. 1.2 Chemical structure of bisphenol A glucoronide, which is considered less biologically active than aglycone (or free) bisphenol A due to its inability to bind to the estrogen receptor

of 2 liters (L) of urine excreted per day). This value corresponds to a BPA dose of 38 ng/kg bd wt per day for a 60-kg adult (Timms et al. 2005). However, it is worth noting that exposure levels vary significantly between individuals and that the upper bound (95 %) for adults exposure and exposure in children is reported to be over 250 ng/kg bd wt per day (Lakind and Naiman 2011). The rapid elimination of BPA as seen in this and a previous study of human excretion kinetics of BPA also indicates a need for frequent monitoring of BPA in blood and urine, since BPA taken up from food will be rapidly eliminated and will only be present in detectable concentrations for a relatively short period (Dekant and Volkel 2008).

Consumption of canned foods and beverages is thought to be a major route of human exposure to BPA (Shelby 2008; Teeguarden et al. 2011). In a study in which individuals consumed a diet heavy in canned foods and juices, Teeguarden et al. (2011) were able to detect measurable blood levels of total BPA (free and conjugated) in only a small fraction of subjects. Further analysis of these samples revealed undetectable levels of free BPA in serum (LOD = 0.3 ng/ml), despite the high dietary exposure. The authors stipulate that rapid conjugation of BPA after oral ingestion results in serum fractions containing roughly 1 % free BPA compared to 99 % conjugated BPA. The highest level of total serum BPA measured in their study of 1.3 ng/ml would equate to a free BPA concentration of just 0.13 ng/ml, which was below their LOD and below levels claimed in many biomonitoring studies (Teeguarden et al. 2011). These authors argue that measurable free serum BPA levels reported in many human studies is likely due to contamination. Others point out that other routes of exposure, such as dermal contact and sublingual uptake, could account for detection of free BPA in blood (Vandenberg et al. 2013). While trace sample contamination can occur with any matrix, the impact is likely less critical in urine than in blood because of the higher detectable levels of conjugated BPA in urine (Ye et al. 2013). Indeed, recent pharmacokinetic studies conducted at the NIEHS Clinical Research Unit indicate that total BPA concentrations are between 40 and 100-fold higher in urine than blood (Birnbaum unpublished data). Altogether, evidence strongly suggests that use of 24-h urine collections is the optimal method to accurately measure total daily exposure to BPA (Ye et al. 2013; Koch et al. 2012).

Numerous studies have examined BPA metabolism in animal models including non-human primates (Taylor et al. 2011; Prins et al. 2010; Doerge et al. 2010;

Fisher et al. 2011; Patterson et al. 2013; Richter et al. 2007). While there are still uncertainties regarding metabolism of BPA, results of these studies have been especially important in improving the understanding of how BPA is handled once inside the body. This has greatly reduced key uncertainties concerning potential levels of internal exposure in humans. Although newborn and young rodents have significant age dependent differences in metabolic capabilities, resulting in the younger animals not being able to metabolize BPA as well as adult rodents and thus being exposed to higher levels internally, this may not be the case for non-human primates (Doerge et al. 2010). Multiple pharmacokinetic studies in monkeys, supported by preliminary results in adult humans (Birnbaum unpublished data), have now demonstrated that newborn and young primates metabolize BPA at or very near the level of adults (Fisher et al. 2011; Patterson et al. 2013). Additional studies in pregnant Rhesus monkeys have shown both that potential fetal exposure is significantly reduced by the mother's metabolic capabilities and that the fetus (third trimester) can effectively metabolize BPA (Doerge et al. 2010; Patterson et al. 2013). Further supporting these findings, in a recent study of 11 human neonates and one young infant, researchers found detectable levels of conjugated BPA in all urine samples but no free BPA, suggesting that neonates have full capacity to metabolize BPA (Nachman et al. 2013). However, a study using umbilical cord blood from 2nd trimester human pregnancies reported universal detection of free and conjugated BPA in umbilical cord serum samples, with relatively high levels of free BPA in the upper bound exposure groups (Gerona et al. 2013). Further studies in humans and in animal models are needed to address the capacity of the embryo or fetus to metabolize BPA.

BPA is commonly used in thermal receipt papers, and cashiers handle them more frequently than the general population. Accordingly, the NIEHS/National Toxicology Program (NTP) is conducting an exposure study to measure BPA, BPA conjugates, and bisphenol S (a common replacement for BPA) in cashiers' blood and urine samples before and after their work shifts. The study will yield insights about the contribution of thermal receipt papers to BPA exposure, although it will not determine the route of exposure (dermal absorption vs. oral ingestion) (Liao and Kannan 2011).

The NTP and the NIEHS Clinical Research Unit have developed a protocol to investigate BPA metabolism and excretion in humans after oral and dermal exposures. Adult volunteers will be administered a low dose of deuterated BPA (d-BPA) to support development of a refined physiologically-based pharmacokinetic model. Participants will receive 100 μg of d-BPA per kg bd wt by oral administration; blood (starting at 10 min) and urine samples will be collected at regular intervals for up to 5 days following dosing to measure d-BPA and d-BPA conjugates. Use of isotopically-labeled BPA will be used to rule out possible contamination during sample collection, storage, and analysis by parent BPA and provide a limit of quantification much lower than previous studies in humans (Volkel et al. 2002). Preliminary results from a pilot phase of the study show rapid elimination and low systemic bioavailability of d-BPA from food, with 90 % excretion by 12 h and 99 % by 48 h (Birnbaum unpublished data). These preliminary results confirm and extend previous human studies using gelatin capsule dosing (Volkel et al. 2002) and dietary studies (Teeguarden et al. 2011) using canned food.

1.5 Health Studies of Bisphenol A Exposures

Normal human growth, development, and homeostasis are dependent on numerous potent hormonal messengers that act on evolutionarily conserved receptors (McLachlan 2001; Palanza et al. 1999). Exposure to endocrine disrupting compounds, like BPA, may interact with hormonally mediated pathways that include the ER, TR, and AR, and, in turn increase the risk of disease (Vandenberg et al. 2009). In recent years, hundreds of experimental studies in animals have reported associations between low-dose BPA exposures (doses below the regulatory no observed adverse effect level [NOAEL]) and adverse health effects. These studies give plausibility to the notion that BPA exposure may be associated with some adverse health outcomes in humans. A growing number of human studies have reported associations between BPA exposure and disease (Table 1.1). Early life exposure is of particular concern given the unique susceptibility of the fetus and young child to environmental toxicant exposures; however chronic low-level exposures may also increase the risk of some adult health outcomes (Mendola et al. 2002). Since human BPA exposure is widespread, even small adverse effects of BPA could have large public health implications (Bellinger 2004).

1.5.1 Historical Context of Health Studies on Bisphenol A

Hundreds of studies were published on the health effects of BPA between the mid-1990s and the mid-2000s. Many showed some form of toxicity, but critical data gaps and uncertainties led to discussion about how the research should be interpreted. In response to increasing concerns about BPA toxicity, NIEHS organized a workshop to examine the body of evidence related to BPA. The resulting report, known as the *Chapel Hill Consensus Statement*, published in 2007 along with five review articles (Wetherill et al. 2007; Vandenberg et al. 2007; Crain et al. 2007; Keri et al. 2007; Richter et al. 2007; vom Saal et al. 2007), concluded that human exposure to BPA is widespread and that the adverse health effects observed in animal studies raised significant concerns about the potential for similar effects in humans (vom Saal et al. 2007). This report also outlined research gaps and needs.

Table 1.1 Recent epidemiological studies on the health effects of bisphenol A exposure

Endpoint	Cohort	Number	Matrix	Study results	Reference
Cancer					
Cancer (general)	NHANES adults	1,455	Spot urine	No association between urinary levels of BPA and cancer incidence	Lang et al. (2008)
Breast	Korean women	167	Blood	No association between blood levels of BPA and breast cancer risk	Yang et al. (2009)
Male reproduction					
	Men from fertility clinic	167	Spot urine	Increased BPA levels in urine associated with altered hormone levels	Meeker et al. (2010b)
	Men from study for future families	375	Spot urine	BPA exposure associated with modest reduction in markers of free testosterone	Mendiola et al. (2010)
	In CHIANTI adults	715	24 h. urine	BPA exposure associated with increased testosterone levels in men	Galloway et al. (2010)
Sexual function	Men in workplace	427	Spot urine	Higher BPA levels associated with decreased sexual function in males	Li et al. (2010)
Female reproduction					
Polycystic ovarian syndrome (PCOS)	PCOS patients and controls	171	Blood	Higher BPA levels in women with PCOS compared to controls	Kandaraki et al. (2011)
Hormone levels	Women undergoing IVF	84	Spot urine	BPA exposure associated with a decrease in peak serum estradiol levels prior to oocyte retrieval	Mok-Lin et al. (2010)
		44	Spot urine		Bloom et al. (2011a)
		74	Spot urine		Ehrlich et al. (2012b)

(continued)

Table 1.1 (continued)

Endpoint	Cohort	Number	Matrix	Study results	Reference
Egg quality	Sub-fertile women	69	Blood	No association between serum BPA concentrations and oocyte maturation.	Fujimoto et al. (2011)
Oocytes retrieved	Women undergoing IVF	147	Spot urine	Higher urinary BPA concentration associated with decreased numbers of retrieved oocytes	Mok-Lin et al. (2010)
	Women undergoing IVF	174	Spot urine	Urinary BPA concentration associated with a decreased trend of blastocyst formation	Ehrlich et al. (2012a)
Embryo quality	Couples undergoing IVF	54	Blood	Low quality embryos associated with unconjugated BPA in males	Bloom et al. (2011b)
Endometriosis	Women	69	Blood	BPA concentrations associated with the occurrence of endometriosis	Cobellis et al. 2009
Implantation	Women undergoing IVF	137	Spot urine	BPA concentrations associated with increased odds of implantation failure	Ehrlich et al. (2012a)
Pubertal development					
Breast development	9 years old girls	192	Spot urine	Higher urinary BPA concentrations not associated with advanced breast or pubic hair development	Wolff et al. (2008)
	8 years old girls	1,151	Spot urine	BPA exposure associated with modest effects on pubertal development in girls	Wolff et al. (2010)
Metabolic and Cardio.					
General obesity	NHANES adults	2,747	Spot urine	Higher BPA exposure associated with general and central obesity in the general adult population	Carwile and Michels (2011)

(continued)

Table 1.1 (continued)

Endpoint	Cohort	Number	Matrix	Study results	Reference
Childhood obesity	NHANES children	2,838	Spot urine	Higher urinary BPA concentrations significantly associated with obesity in children and adolescents	Trasande et al. (2012)
	Children	1,326	Spot urine	Higher urinary BPA concentration associated with a two-fold increase risk of weight gain in girls	Li et al. (2013)
Cardiovascular disease and diabetes	NHANES adults	1,455	Spot urine	Higher urinary BPA concentrations associated with cardiovascular disease and diabetes	Lang et al. (2008)
Heart disease			Spot urine	Higher urinary concentrations of BPA consistently associated with reported heart disease	Melzer et al. (2010)
Growth					
Birth weight	Pregnant women	40	Blood	No association between serum BPA concentrations and birth weight	Padmanabhan et al. (2008)
Pre-term birth	Pregnant women	60	Spot urine	Positive association between urinary BPA concentrations and pre-term birth	Cantonwine et al. (2010)
Neurodevelopment					
Anxiety	Children	249	Spot urine	No associations between childhood urinary BPA concentrations and anxious/depressive behaviors	Braun et al. (2009)
	Children	244	Spot urine	No associations between childhood urinary BPA concentrations and anxious/depressive behaviors	Braun et al. (2011)

(continued)

Table 1.1 (continued)

Endpoint	Cohort	Number	Matrix	Study results	Reference
	Children	534	BPA not measured	Greater surface-years of exposure to bis-GMA containing dental composites associated with more self-reported anxious and depressive behaviors	Maserejian et al. (2012b)
	Children	198	Spot urine	No associations between childhood urinary BPA concentrations and anxious/ depressive behaviors	Perera et al. (2012)
	Children	1,089	Spot urine	BPA concentrations in children positively associated with parent-reported anxious and depressive problem scores in children	Hong et al. (2013)
Attention deficit hyperactivity disorder (ADHD)	Children	534	BPA not measured	No associations between bis-GMA containing amalgams and ADHD	Maserejian et al. (2012a)
	Children	249	Spot urine	No association between parent-reported ADHD behaviors and BPA concentrations during the first 3 years of life	Braun et al. (2009)
	Children	198	Spot urine	Child urinary BPA concentrations at 3 years of age associated with parent-reported higher emotional reactivity scores	Perera et al. (2012)
	Children	1,089	Spot urine	Positive relationship between concurrent child BPA concentrations and parent-reported attention problems	Hong et al. (2013)
Social behavior	Children	249	Spot urine	Gestational BPA concentrations positively associated with aggression and hyperactivity	Braun et al. (2009)

Around the same time, the NTP Center for the Evaluation of Risks to Human Reproduction (CERHR) (Chapin et al. 2008) convened an expert panel to examine BPA research related to human reproduction and development. Based largely on the panel's assessment, the NTP reported *negligible concern* for reproductive effects in non-occupationally exposed adults and *minimal concern* for occupationally exposed workers, but identified *some concern* for effects on the brain, behavior, and prostate gland in fetuses, infants, and children at current levels of human BPA exposure (NTP 2008).

In response to these findings, the U.S. Food and Drug Administration (FDA), which has regulatory authority over many consumer and medical products containing BPA, issued a statement expressing its agreement with the NTP's conclusion for some concern about the effects of BPA on the brain, behavior, and prostate gland of fetuses, infants, and children (FDA 2010). FDA also identified substantial uncertainties in BPA research findings and their implications for human health. The statement called for further research to address these uncertainties.

The Chapel Hill consensus statement, the NTP-CERHR monograph, and FDA's statements have helped to focus the field by identifying gaps in BPA research and highlighting concerns that need to be addressed to move the field forward. Key sources of uncertainty that have been identified include: absent or inconsistent data on dose response including low-dose effects and non-monotonic dose response behaviors; pharmacokinetics across species and the lifespan; gender differences; routes and extent of human exposures; sensitive windows of exposure; mechanism(s); and specific disease endpoints. In addition, these reports cited difficulties extrapolating data from animals to humans and comparing results from studies compliant with good laboratory practices (GLP) with academic investigator-initiated studies (Birnbaum et al. 2012).

1.5.2 Human Health Studies of Bisphenol A Exposure

Table 1.1 summarizes human epidemiological studies that have evaluated various health endpoints in relation to exposure to BPA, including cancer, male and female reproduction, obesity and heart disease, developmental effects, and neurological development. Additional analysis and comment on these studies is provided below.

1.5.2.1 Cancer Outcomes

The ubiquitous presence of BPA in the environment together with an increase in incidence of endocrine-related cancers, such as breast, uterine, and prostate cancers, has led to research evaluating the role of BPA in human carcinogenesis.

In 2008, Lang et al. reported the levels of urinary BPA in samples collected by the NHANES of U.S. adults whom had also completed a health survey (Lang et al. 2008). No association between elevated BPA concentrations and general cancer incidence was evident. Since the most common form of breast cancer is sensitive to estrogen (ER-positive cancers), several studies have investigated the potential impact of BPA specifically on breast cancer in humans and mammary cancer in animal models. Yang et al. analyzed 167 blood samples collected between 1994 and 1997 from Korean women, and measured total BPA levels in blood by high performance liquid chromatography with fluorescence detection (HPLC/FD). Blood BPA levels were not found to be associated with increased breast cancer risk (Yang et al. 2009). However, given the unknown latency period between exposure and cancer detection, this study design was likely to be insufficient in order to identify an association.

Numerous studies in rodents have reported that early-life BPA exposure can cause development alterations that in turn increase the risk of cancer later in life. For example, several studies have shown effects of maternal administration of low doses of BPA on development of the prostate in male mouse fetuses, as well as a permanent increase in prostate size and number of ARs in adulthood (Nagel et al. 1997; vom Saal et al. 1997; Gupta 2000). More recent studies have demonstrated that treatment of neonatal rats with low doses of BPA (10 μg/kg/day) either orally or by subcutaneous injection, results in early stage prostate cancer when males are treated with estrogen in adulthood to mimic the increase in estrogen that occurs during aging in men (Ho et al. 2006).

There is extensive literature showing that exposure to BPA alters rodent mammary gland differentiation and subsequent adult function and increases the risk of developing neoplastic lesions. For example, studies in mice and rats report developmental exposure to environmentally relevant levels of BPA during gestation and lactation may induce changes in mammary gland development leading to neoplasms (Acevedo et al. 2013; Markey et al. 2001). Others have shown that BPA exposure leads to development of cystic ovaries, adenomyosis, leiomyomas, atypical hyperplasia, and stomal polyps (Newbold et al. 2007). In summary, more research is needed to bridge the data gaps between the carcinogenic properties of BPA found in animals to the associations reported in humans. This will require prospective epidemiological studies that account for the long latency between chemical exposure and cancer formation.

1.5.2.2 Male Reproductive Outcomes

Several studies have examined relationships between BPA and reproductive endpoints in men exposed to BPA. Outcomes in these studies include sexual function and semen quality. Several epidemiological studies observed associations between urinary BPA concentrations and serum reproductive hormones, but the direction of the relationship between BPA and male hormones has been inconsistent (Hanaoka et al. 2002; Meeker et al. 2010a; Mendiola et al. 2010; Galloway et al. 2010). Three studies reported consistent associations between increasing urinary BPA concentrations and one or more measures of reduced semen quality (Meeker et al. 2010b; Mendiola et al. 2010). An additional study found that occupational BPA exposure and urinary BPA concentrations were associated with decreased sexual desire, erectile function, and ejaculation difficulty in Chinese men (Li et al. 2010). These findings suggest that recent BPA exposure may be associated with lower semen quality and increased sexual dysfunction in men. However, due to the studies cross sectional designs and incomplete assessment of potential co-exposures, the clinical relevance of these findings remains uncertain (Schug et al. 2012).

1.5.2.3 Female Reproductive Outcomes

Human studies also have reported associations between BPA exposures and female reproductive abnormalities, including effects on the ovary, uterus, and oocyte quality. Studies of BPA effects on ovarian morphology are limited and have focused on polycystic ovarian syndrome (PCOS). One case control study (71 women with PCOS and 100 women without) reported a positive association between serum BPA levels and PCOS (Kandaraki et al. 2011). In three studies of women undergoing in vitro fertilization (IVF), BPA exposure was associated with a decrease in peak serum estradiol levels prior to oocyte retrieval, which may have negative effects on embryo quality (Ehrlich et al. 2012b; Mok-Lin et al. 2010; Bloom et al. 2011a).

A few human studies have evaluated serum or urinary BPA levels in relation to egg quality. A small prospective study of 58 sub-fertile women and 37 male partners undergoing intracytoplasmic sperm injection (ICSI) or conventional IVF did not find an association between serum BPA concentrations and oocyte maturation (Fujimoto et al. 2011). In two publications from the same prospective cohort of 84 women (Mok-Lin et al. 2010) and 147 women (Ehrlich et al. 2012b) undergoing IVF, increasing urinary BPA concentration was associated with decreased numbers of retrieved oocytes. These results suggest that BPA is associated with impaired oocyte yield during IVF and could affect oocyte maturation and fertilization, adversely affecting the success of IVF treatment.

Few studies have examined the association of BPA on blastocyst and embryo development. In one prospective preconception cohort study of 174 women, total urinary BPA concentration was associated with a decreased trend of blastocyst formation (Ehrlich et al. 2012a). Further, a prospective cohort study involving sub-fertile male and female partners undergoing IVF treatment found that increasing urinary BPA concentrations in the male, but not female partner, decreased the odds of a high embryo cell count and embryo fragmentation score, suggesting the embryos were of low quality for IVF treatments (Bloom et al. 2011b).

The results of one case control study of 69 women suggests that serum BPA levels may be associated with the occurrence of endometriosis (Cobellis et al. 2009) but the study was limited by a small sample size and the selected methods for measurement of serum BPA. Future studies of endometriosis risk should evaluate BPA exposure using 24-h urine samples. In a study of 137 women undergoing IVF, higher quartiles of urinary BPA concentrations were associated with increased odds of implantation failure (Ehrlich et al. 2012a).

1.5.2.4 Pubertal Development Outcomes

Early onset of puberty is known to increase the risk of breast cancer later in life (Apter 1996; Vanderloo et al. 2007). Investigators have started to explore whether higher levels BPA exposure can increase the incidence of precocious puberty in young girls (Wolff et al. 2008, 2010), and whether early BPA exposure in animal models can affect puberty onset, mammary development, and subsequent risk of mammary neoplasia (Soto et al. 2013). A cross-sectional study conducted in New York City examined the association between concurrent urinary BPA concentrations with pubic hair and breast development in 192 9-year old girls (Wolff et al. 2008). Higher urinary BPA concentrations were not associated with advanced breast or pubic hair development. However, body mass index (BMI) modified the association between BPA and breast development, whereas higher urinary BPA concentrations were more associated with delayed breast development among girls with lower BMI.

A multi-center prospective cohort study of 1,151 6 to 8-year old female children from Cincinnati, OH; San Francisco, CA; and New York, NY, examined the relationship between urinary BPA concentrations and pubertal development 1 year later (Wolff et al. 2010). Compared to the lowest quintile of urinary BPA concentrations, the odds of advanced breast development were similar but slightly below the null for the top four quintiles, with no discernible dose-response pattern. Compared to the first quintile of urinary BPA concentrations, the odds of advanced pubic hair development (stage 2+) were slightly increased in the second and third quintiles of urinary BPA concentrations and null in the fourth and fifth quintiles. The authors did not report whether BMI modified the association between urinary BPA concentrations and pubertal development. These studies suggest that in utero or childhood exposure to BPA may have modest positive associations with altered pubertal timing in girls and that in utero may be the most sensitive window to BPA exposure.

1.5.2.5 Metabolic and Cardiovascular Outcomes

Several laboratory animal studies have demonstrated that BPA exposure can disrupt multiple metabolic mechanisms (Masuno et al. 2002; Sakurai et al. 2004) suggesting the possibility that BPA exposure may increase body mass in environmentally relevant doses and therefore contribute to the obesity epidemic in humans. In 2011 the Office of Health Assessment and Translation of the NTP organized a workshop to evaluate the concept that environmental chemicals such as BPA may be contributing factors to the epidemics of diabetes and obesity (Thayer et al. 2012). The group concluded that animal and in vitro studies are suggestive of an effect of BPA on glucose homeostasis, insulin release, cellular signaling in pancreatic β cells, and adipogenesis, while the existing human data on BPA and diabetes were insufficient to draw meaningful conclusions (Thayer et al. 2012).

Three recent cross-sectional studies have found an association between urinary BPA concentration and increased risk of obesity in humans. Carwile et al. reported that BPA exposure, as measured by urinary BPA levels, is associated with general and central obesity in the general adult population of the U.S. (Carwile and Michels 2011). Two other studies reported associations between concurrent BPA exposure (as measured by spot urine samples) and weight gain in children (Li et al. 2013; Trasande et al. 2012). Other studies have demonstrated associations between urinary BPA concentrations and adult diabetes, cardiovascular diagnoses, and abnormalities in liver function (Lang et al. 2008; Melzer et al. 2010). The interpretation of all these results is limited by their cross-sectional design. Cardiovascular disease (CVD) and metabolic disorders have long latency periods and contemporaneous urinary BPA concentrations may be unreliable reflections of BPA exposures during the relevant etiologic windows for the development of CVD and metabolic diseases. It has been suggested that other factors may be responsible for observed associations since obese individuals are at increased risk for CVD and other metabolic disorders, and may consume more packaged and processed foods that contain BPA (Sharpe and Drake 2013). In summary, there is a clear need for long duration prospective human epidemiological studies that include more robust biomonitoring, such as multiple 24-h urine sampling, to determine whether exposure to BPA has metabolic effects in humans.

1.5.2.6 Fetal and Childhood Growth Outcomes

Several studies have examined associations between BPA exposure and infant/childhood growth (Padmanabhan et al. 2008; Cantonwine et al. 2010). One small cross-sectional study reported no association between serum BPA concentrations and birth weight or gestational age; however, numerical results were not presented (Padmanabhan et al. 2008). A pilot case control nested in a cohort study involving

60 pregnant women found a positive association between urine BPA concentration and pre-term birth (Cantonwine et al. 2010).

1.5.2.7 Neurodevelopmental Outcomes

Steroid hormones play a critical role in brain organization of the neuroendocrine circuitry that coordinates sex-specific physiology and behavior in vertebrates. Thus, endocrine disrupting effects may be irreversible and result in atypical behavior (Wolstenholme et al. 2011; Patisaul and Adewale 2009; Wright et al. 2010). Two epidemiological studies have examined the relationship between childhood BPA exposure and cognitive abilities. Maserejian and colleagues examined childhood BPA exposure from bisphenol-A-diglycidyl-dimethacrylate (bis-GMA) containing composite tooth fillings and cognition in the New England Child Amalgam Trial (Maserejian et al. 2012a). They found no evidence of associations between total surface-years of bis-GMA composite filling exposure and cognitive abilities, memory, and visual-spatial abilities (Maserejian et al. 2012a). A study in Korean children reported that current urinary BPA concentrations were associated with higher teacher-reported learning difficulty scores in children 8 to 11 years of age (Hong et al. 2013).

Several studies have examined the association between gestational or childhood BPA exposure measures and anxiety or depression (Braun et al. 2009, 2011; Hong et al. 2013; Maserejian et al. 2012b; Perera et al. 2012). Three out of five recent studies examining childhood BPA exposures reported increased anxiety among children with higher childhood BPA exposures. Despite similar levels of exposure, neither Perera et al. (2012) nor Braun et al. (2011) reported associations between childhood urinary BPA concentrations and anxious/depressive behaviors, however, they did report other neurobehavioral effects. Maserejian et al. (2012b) and colleagues reported that greater surface-years of exposure to bis-GMA containing dental composites were associated with more self-reported anxious and depressive behaviors. A cross-sectional study from Korea also reported that concurrent child BPA concentrations were positively associated with parent-reported anxious and depressive problem scores in children age 8 to 11 years (Hong et al. 2013).

The rising incidence of some neurobehavioral disorders, like autism spectrum disorders (ASDs) and attention-deficit/hyperactivity disorder (ADHD), has generated suspicions that environmental chemicals, including BPA, may be contributing to these diseases (Boyle et al. 2011). Maserejian and colleagues observed no associations between bis-GMA containing amalgams and ADHD problems (Maserejian et al. 2012a). Braun and colleagues reported no association between parent-reported ADHD behaviors and urine BPA concentrations during the first 3 years of life (Braun et al. 2009). In contrast, Perera et al. reported that child urinary BPA concentrations at 3 years of age were associated with parent-reported higher emotional reactivity scores in children, but the association did not differ by child sex (Perera et al. 2012). Finally, Hong et al. reported a positive relationship between concurrent child BPA urinary concentrations and parent-reported attention problems (Hong et al. 2013).

In a prospective birth cohort of 249 mothers and infants from Cincinnati, OH, Braun et al. (2009) examined the association between prenatal urinary BPA concentrations and childhood social behavior at 2 years of age. Gestational BPA concentrations were positively associated with aggression and hyperactivity in children, but this association was stronger in girls than in boys. The magnitude of the observed associations was similar to those seen for other environmental toxicants and neurodevelopment (e.g., environmental lead and IQ/behavior).

1.5.3 Limitations of Human Studies

Several limitations emerge from this review of the human studies on BPA exposure and health outcomes. Since few studies include similar health endpoints, future studies should utilize comparable health endpoints and measurement instruments to facilitate systematic reviews. Also, there is the potential for misclassification of BPA exposure. Because of BPA's short biological half-life, urinary BPA concentrations usually reflect exposure over the past 6–12 h. Therefore, a single spot urine sample may not accurately classify long-term or episodic exposure over weeks, months or years (Braun et al. 2010b). Additional large prospective cohort studies are needed to confirm and validate findings from cross sectional human studies as well as findings in laboratory animals.

1.5.4 Conclusions of the Human Studies

To date, most human studies of BPA have utilized cross-sectional designs that offer suggestive results, but cannot address the temporality of exposure and disease. For many diseases and disorders (e.g., asthma, cancer, neurobehavioral disorders, obesity, or early-onset puberty), gestational, childhood, or pubertal BPA exposure may be more relevant to the development of disease than concurrent BPA exposures. Despite these limitations, results from cross-sectional results can be used to guide the design of stronger and more powerful studies. The limitation of BPA exposure misclassification can be addressed in prospective cohort studies by collecting multiple 24-h urine samples during the etiologic relevant window prior to disease onset (Schug et al. 2012). It is also important to note that BPA levels in human populations can vary considerably, both within and between individuals. Accordingly, studies should examine the associations between upper bound levels of BPA exposure in adults, children, and infants to adverse health effects.

1.6 Regulation of Bisphenol A

The Food and Agriculture Organization of the United Nations (FAO) and World Health Organization (WHO) have estimated that the average human exposure to BPA is between 0.4–1.4 μg/kg per day (WHO 2010). The FDA currently estimates average dietary exposure is 0.2–0.4 μg/kg per day for infants and 0.1–0.2 μg/kg per day for children and adults (FDA 2012). In a 2010 statement that was subsequently updated in 2012, FDA, in agreement with the NTP panel's 2008 findings, expressed *some concern* about the effects of BPA on the brain, behavior, and prostate gland of fetuses, infants, and children and encouraged consumers and industry to take reasonable steps to reduce human exposure to BPA, particularly among infants (FDA 2012). In 2012, FDA amended its food additive regulations to no longer provide for the use of polycarbonate resins in baby bottles and "sippy" cups in response to a food additive petition submitted by the American Chemistry Council proposing that FDA amend its regulations because manufacturers no longer use polycarbonate resins in these products (FDA 2013).

FDA has called for further studies to clarify substantial uncertainties in BPA research, including: the routes of exposure employed, the lack of consistency among some of the measured endpoints or results between studies, the relevance of some animal models to human health, differences in the metabolism (and detoxification) of and responses to BPA both at different ages and in different species, and limited or absent dose response information for some studies (FDA 2012).

In 2010 the European Food and Safety Authority (EFSA) received a request from the European Commission to review scientific arguments in support of the Danish government's decision to ban the use of BPA in food contact materials for infants aged from 0 to 3 years. The Danish risk assessment was based mainly on the rat study by Stump et al. (2010) looking at possible neurodevelopmental effects of BPA at a range of dose levels. Following a comprehensive review of recent scientific literature and studies on the toxicity of BPA at low doses, EFSA's Food Contact Materials, Enzymes, Flavourings and Processing Aids Panel (CEF) Panel concluded they could not identify any new evidence that would lead them to revise the current European guidelines of tolerable daily intake (TDI) for BPA of 0.05 milligram (mg) per kg body weight (bd wt) (EFSA 2010). The Panel also stated that the data currently available do not provide convincing evidence of neurobehavioral toxicity of BPA (EFSA 2010). However, in 2011 the European Union banned BPA from infant feeding bottles, citing uncertainty deriving from new studies, which showed that BPA may have effects on development, immune response, and tumor promotion. In a statement released on July 25, 2013, EFSA cited that new data warranted lowering exposure estimates for infants and toddlers (aged 6 months to 3 years). Based on 24 h urine sampling, average exposure from the diet is estimated to amount to 0.375 μg/kg per day whereas for the population above 18 years of age (including women of child-bearing age) the figure is up to 0.132 μg/kg per day (EFSA 2013). These values are in line with current FDA exposure estimates.

1.7 Addressing the Data Gaps

To work toward a comprehensive, integrated assessment of the health effects of BPA, the NIEHS brought existing BPA grantees together with the American Recovery and Reinvestment Act (ARRA) into a BPA Grantee Consortium in 2009. This consortium includes over 40 researchers from academia, the NTP and NIEHS intramural investigators. Consortium members have gathered three times in person and met approximately once per month via conference call from late 2009 to the present. The BPA Grantee Consortium has several specific activities and areas that focus on resolving discrepancies, and producing results that will be more interpretable across studies. Consortium members have worked to establish consistency in the models, approaches, doses, and end points used across the spectrum of BPA research (Birnbaum et al. 2012).

Another factor, contributing to uncertainty in BPA research, stems from inconsistencies in the characterization of low-dose effects. Studies in both animals and humans have indicated effects of BPA both at low (nanomolar or ng/kg per day or lower) and high (micromolar or μg/kg per day or higher) doses, often with fewer effects at mid-level doses (Vandenberg et al. 2012). Such nonmonotonic effects curves may reflect BPA activity in different systems, with low doses causing effects in the endocrine system and high doses potentially causing effects in other organ systems. Other studies have reported effects at very low doses but have not included higher doses, making the dose–response relationship difficult to characterize (Birnbaum et al. 2012).

Given the body of diverse and often difficult to interpret evidence on the health effects of BPA, the NTP and NIEHS determined in 2010 that a new guideline-compliant study conducted in accordance with good laboratory practice (GLP) would be helpful to reconcile uncertainties on the toxicity of BPA and offer risk assessors and risk managers a more comprehensive body of animal research to inform decision making. GLP practices ensure the uniformity, consistency, reliability, reproducibility, quality, and integrity of chemical toxicity tests. In addition, FDA's ongoing review of BPA offered an opportunity to test a new research model based on enhancing the links between academic and guideline-compliant research. These factors led to the genesis of the Consortium Linking Academic and Regulatory Insights on BPA Toxicity (CLARITY-BPA) research program (Schug et al. 2013).

The intense discussions and scientific research on BPA have slowly contributed to a process of improving our strategies on testing environmental chemicals.

Future studies on environmental contaminants may benefit from an early focus on identifying data gaps and on collaborative efforts to confront controversies (or prevent them before they arise). Fractured, uncoordinated research efforts can leave significant unanswered questions, impeding progress and making it difficult for risk assessors and regulators to interpret findings. In the future, perhaps earlier investments to identify needs and coordinate research efforts will save time and money, as well as improve our ability to protect human health.

1.7.1 Disclosure Statement

This book chapter is in part a work product supported by the intramural research program at the National Cancer Institute (NCI), National Institutes of Health (NIH) and by the NIEHS/NIH.

References

Acevedo N, Davis B, Schaeberle CM et al (2013) Perinatally administered bisphenol A acts as a mammary gland carcinogen in rats. Environ Health Perspect 121(9):1040–1046

Andersen HR, Andersson A-M, Arnold SF et al (1999) Comparison of short-term estrogenicity tests for identification of hormone-disrupting chemicals. Environ Health Perspect 107(Suppl 1):89–108

Apter D (1996) Hormonal events during female puberty in relation to breast cancer risk. Eur J Cancer Prev 5(6):476–482

Arora M, Austin C (2013) Teeth as a biomarker of past chemical exposure. Curr Opin Pediatr 25(2):261–267

Barouki R, Aggerbeck M, Aggerbeck L et al (2012) The aryl hydrocarbon receptor system. Drug Metab Drug Interact 27(1):3–8

Bellinger DC (2004) What is an adverse effect? A possible resolution of clinical and epidemiological perspectives on neurobehavioral toxicity. Environ Res 95(3):394–405

Birnbaum LS, Bucher JR, Collman GW et al (2012) Consortium-based science: the NIEHS's multipronged, collaborative approach to assessing the health effects of bisphenol A. Environ Health Perspect 120(12):1640–1644

Bloom MS, Kim D, Vom Saal FS et al (2011a) Bisphenol A exposure reduces the estradiol response to gonadotropin stimulation during in vitro fertilization. Fertil Steril 96(3):672–677

Bloom MS, Vom Saal FS, Kim D et al (2011b) Serum unconjugated bisphenol A concentrations in men may influence embryo quality indicators during in vitro fertilization. Environ Toxicol Pharmacol 32(2):319–323

Boyle CA, Boulet S, Schieve LA et al (2011) Trends in the prevalence of developmental disabilities in US children, 1997–2008. Pediatrics 127(6):1034–1042

Braun JM, Yolton K, Dietrich KN et al (2009) Prenatal bisphenol A exposure and early childhood behavior. Environ Health Perspect 117(12):1945–1952

Braun JM, Daniels JL, Poole C et al (2010a) A prospective cohort study of biomarkers of prenatal tobacco smoke exposure: the correlation between serum and meconium and their association with infant birth weight. Environ Health 9(53):1–11

Braun JM, Kalkbrenner AE, Calafat AM et al (2010b) Variability and predictors of urinary bisphenol A concentrations during pregnancy. Environ Health Perspect 119(9):131–137

Braun JM, Kalkbrenner AE, Calafat AM et al (2011) Impact of early-life bisphenol A exposure on behavior and executive function in children. Pediatrics 128(5):873–882

Calafat AM (2010) BPA biomonitoring and biomarker studies. In: FAO/WHO expert meeting on bisphenol A (BPA), Ottawa, Canada, 1–5 Nov 2010

Calafat AM, Ye X, Wong LY et al (2008) Exposure of the U.S. population to bisphenol A and 4-tertiary-octylphenol: 2003–2004. Environ Health Perspect 116(1):39–44

Cantonwine D, Meeker JD, Hu H et al (2010) Bisphenol a exposure in Mexico City and risk of prematurity: a pilot nested case control study. Environ Health 9(62):2–7

Carwile JL, Michels KB (2011) Urinary bisphenol A and obesity: NHANES 2003–2006. Environ Res 111(6):825–830

Carwile JL, Ye X, Zhou X et al (2011) Canned soup consumption and urinary bisphenol A: a randomized crossover trial. JAMA 306(20):2218–2220

Castorina R, Bradman A, Fenster L et al (2010) Comparison of current-use pesticide and other toxicant urinary metabolite levels among pregnant women in the CHAMACOS cohort and NHANES. Environ Health Perspect 118(6):856–863

Chamorro-Garcia R, Kirchner S, Li X et al (2012) Bisphenol A diglycidyl ether induces adipogenic differentiation of multipotent stromal stem cells through a peroxisome proliferator-activated receptor gamma-independent mechanism. Environ Health Perspect 120(7):984–989

Chapin RE, Adams J, Boekelheide K et al (2008) NTP-CERHR expert panel report on the reproductive and developmental toxicity of bisphenol A. Birth Defects Res B Dev Reprod Toxicol 83(3):157–395

Cobellis L, Colacurci N, Trabucco E et al (2009) Measurement of bisphenol A and bisphenol B levels in human blood sera from healthy and endometriotic women. Biomed Chromatogr 23(11):1186–1190

Crain DA, Eriksen M, Iguchi T et al (2007) An ecological assessment of bisphenol-A: evidence from comparative biology. Reprod Toxicol 24(2):225–239

Dekant W, Volkel W (2008) Human exposure to bisphenol A by biomonitoring: methods, results and assessment of environmental exposures. Toxicol Appl Pharmacol 228(1):114–134

Doerge DR, Twaddle NC, Woodling KA et al (2010) Pharmacokinetics of bisphenol A in neonatal and adult Rhesus monkeys. Toxicol Appl Pharmacol 248(1):1–11

Ehrlich S, Williams PL, Missmer SA et al (2012a) Urinary bisphenol A concentrations and implantation failure among women undergoing in vitro fertilization. Environ Health Perspect 120(7):978–983

Ehrlich S, Williams PL, Missmer SA et al (2012b) Urinary bisphenol A concentrations and early reproductive health outcomes among women undergoing IVF. Hum Reprod 27(12):3583–3592

European Food Safety Authority (EFSA) (2010) Scientific opinion on bisphenol A: evaluation of a study investigating its neurodevelopmental toxicity, review of recent scientific literature on its toxicity and advice on the Danish risk assessment of bisphenol A. EFSA J 8(9):1829. http://www.efsa.europa.eu/en/efsajournal/pub/1829.htm. Accessed 4 Feb 2014

European Food Safety Authority (EFSA) (2013) Press release on human BPA exposure estimates. http://www.fda.gov/newsevents/publichealthfocus/ucm064437.htm. Accessed 14 Aug 2013

Fisher JW, Twaddle NC, Vanlandingham M et al (2011) Pharmacokinetic modeling: prediction and evaluation of route dependent dosimetry of bisphenol A in monkeys with extrapolation to humans. Toxicol Appl Pharmacol 257(1):122–136

Food and Drug Administration (FDA) (2010) Update on bisphenol A for use in food contact applications. http://www.fda.gov/NewsEvents/PublicHealthFocus/ucm197739.htm. Accessed 13 Mar 2012

Food and Drug Administration (FDA) (2012) Bisphenol A (BPA): use in food contact application. http://www.fda.gov/NewsEvents/PublicHealthFocus/ucm064437.htm. Accessed 21 Sept 2012

Food and Drug Administration (FDA) (2013) Indirect food additives: adhesives and components of coatings. https://www.federalregister.gov/articles/2013/07/12/2013-16684/indirect-food-additives-adhesives-and-components-of-coatings. Accessed 1 Aug 2013 2013

Fujimoto VY, Kim D, vom Saal FS et al (2011) Serum unconjugated bisphenol A concentrations in women may adversely influence oocyte quality during in vitro fertilization. Fertil Steril 95(5):1816–1819

Galloway T, Cipelli R, Guralnik J et al (2010) Daily bisphenol A excretion and associations with sex hormone concentrations: results from the InCHIANTI adult population study. Environ Health Perspect 118(11):1603–1608

Geens T, Aerts D, Berthot C et al (2012) A review of dietary and non-dietary exposure to bisphenol-A. Food Chem Toxicol 50(10):3725–3740

Gerona RR, Woodruff TJ, Dickenson CA et al (2013) BPA, BPA glucuronide, and BPA sulfate in mid-gestation umbilical cord serum in a northern California cohort. Environ Sci Technol 13:1–34

Gould JC, Leonard LS, Maness SC et al (1998) Bisphenol A interacts with the estrogen receptor α in a distinct manner from estradiol. Mol Cell Endocrinol 142:203–214

Gupta C (2000) Reproductive malformation of the male offspring following maternal exposure to estrogenic chemicals. Proc Soc Exp Biol Med 224(2):61–68

Hanaoka T, Kawamura N, Hara K et al (2002) Urinary bisphenol A and plasma hormone concentrations in male workers exposed to bisphenol A diglycidyl ether and mixed organic solvents. Occup Environ Med 59(9):625–628

Ho SM, Tang WY, Belmonte de Frausto J et al (2006) Developmental exposure to estradiol and bisphenol A increases susceptibility to prostate carcinogenesis and epigenetically regulates phosphodiesterase type 4 variant 4. Cancer Res 66(11):5624–5632

Hong SB, Hong YC, Kim JW et al (2013) Bisphenol A in relation to behavior and learning of school-age children. J Child Psychol Psychiatry 8:890–899

Kandaraki E, Chatzigeorgiou A, Livadas S et al (2011) Endocrine disruptors and polycystic ovary syndrome (PCOS): elevated serum levels of bisphenol A in women with PCOS. J Clin Endocrinol Metab 96(3):480–484

Keri RA, Ho SM, Hunt PA et al (2007) An evaluation of evidence for the carcinogenic activity of bisphenol A. Reprod Toxicol 24(2):240–252

Kingman A, Hyman J, Masten SA et al (2012) Bisphenol A and other compounds in human saliva and urine associated with the placement of composite restorations. J Am Dent Assoc 143(12):1292–1302

Koch HM, Kolossa-Gehring M, Schroter-Kermani C et al (2012) Bisphenol A in 24 h urine and plasma samples of the German environmental specimen bank from 1995 to 2009: a retrospective exposure evaluation. J Expo Sci Environ Epidemiol 22(6):610–616

Kuiper GG, Lemmen JG, Carlsson B et al (1998) Interaction of estrogenic chemicals and phytoestrogens with estrogen receptor beta. Endocrinology 139:4252–4263

Kwintkiewicz J, Nishi Y, Yanase T et al (2010) Peroxisome proliferator-activated receptor-gamma mediates bisphenol A inhibition of FSH-stimulated IGF-1, aromatase, and estradiol in human granulosa cells. Environ Health Perspect 118(3):400–406

Lakind JS, Naiman DQ (2011) Daily intake of bisphenol A and potential sources of exposure: 2005–2006 National Health and Nutrition Examination Survey. J Expo Sci Environ Epidemiol 21(3):272–279

Lang IA, Galloway TS, Scarlett A et al (2008) Association of urinary bisphenol A concentration with medical disorders and laboratory abnormalities in adults. JAMA 300(11):1303–1310

Li DK, Zhou Z, Miao M et al (2010) Relationship between urine bisphenol-A (BPA) level and declining male sexual function. J Androl 1(13):500–506

Li Y, Burns KA, Arao Y et al (2012) Differential estrogenic actions of endocrine-disrupting chemicals bisphenol A, bisphenol AF, and zearalenone through estrogen receptor in vitro. Environ Health Perspect 120(7):1029–1035

Li DK, Miao M, Zhou Z et al (2013) Urine bisphenol-A level in relation to obesity and overweight in school-age children. PLoS ONE 8(6):1–6

Liao C, Kannan K (2011) Widespread occurrence of bisphenol A in paper and paper products: implications for human exposure. Environ Sci Technol 45(21):9372–9379

Markey CM, Luque EH, Munoz De Toro M et al (2001) In utero exposure to bisphenol A alters the development and tissue organization of the mouse mammary gland. Biol Reprod 65(4):1215–1223

Maserejian NN, Trachtenberg FL, Hauser R et al (2012a) Dental composite restorations and neuropsychological development in children: treatment level analysis from a randomized clinical trial. Neurotoxicology 33(5):1291–1297

Maserejian NN, Trachtenberg FL, Hauser R et al (2012b) Dental composite restorations and psychosocial function in children. Pediatrics 130(2):328–338

Masuno H, Kidani T, Sekiya K et al (2002) Bisphenol A in combination with insulin can accelerate the conversion of 3T3-L1 fibroblasts to adipocytes. J Lipid Res 43(5):676–684

Masuno H, Iwanami J, Kidani T et al (2005) Bisphenol A accelerates terminal differentiation of 3T3-L1 cells into adipocytes through the phosphatidylinositol 3-kinase pathway. Toxicol Sci 84(2):319–327

McLachlan JA (2001) Environmental signaling: what embryos and evolution teach us about endocrine disrupting chemicals. Endocr Rev 22(3):319–341

Meeker JD, Calafat AM, Hauser R (2010a) Urinary bisphenol A concentrations in relation to serum thyroid and reproductive hormone levels in men from an infertility clinic. Environ Sci Technol 44(4):1458–1463

Meeker JD, Ehrlich S, Toth TL et al (2010b) Semen quality and sperm DNA damage in relation to urinary bisphenol A among men from an infertility clinic. Reprod Toxicol 4:532–539

Melzer D, Rice NE, Lewis C et al (2010) Association of urinary bisphenol A concentration with heart disease: evidence from NHANES 2003/06. PLoS ONE 5(1):1–9

Mendiola J, Jorgensen N, Andersson AM et al (2010) Are environmental levels of bisphenol A associated with reproductive function in fertile men? Environ Health Perspect 118(9):1286–1291

Mendola P, Selevan SG, Gutter S et al (2002) Environmental factors associated with a spectrum of neurodevelopmental deficits. Ment Retard Dev Disabil Res Rev 8(3):188–197

Mok-Lin E, Ehrlich S, Williams PL et al (2010) Urinary bisphenol A concentrations and ovarian response among women undergoing IVF. Int J Androl 33(2):385–393

Moriyama K, Tagami T, Akamizu T et al (2002) Thyroid hormone action is disrupted by bisphenol A as an antagonist. J Clin Endocrinol Metab 87:5185–5190

Nachman RM, Fox SD, Golden WC et al (2013) Urinary free bisphenol A and bisphenol A-glucuronide concentrations in newborns. J Pediatr 162(4):870–872

Nagel SC, vom Saal FS, Thayer KA et al (1997) Relative binding affinity-serum modified access (RBA-SMA) assay predicts the relative in vivo bioactivity of the xenoestrogens bisphenol A and octylphenol. Environ Health Perspect 105(1):70–76

National Toxicology Program (NTP) (2008) NTP-CERHR monograph on the potential human reproductive and developmental effects of bisphenol A. National Toxicology Program, U.S. Department of Health and Human Services. http://oehha.ca.gov/prop65/CRNR_notices/state_listing/data_callin/pdf/NTP_CERHR_0908_bisphenolA.pdf. Accessed 3 July 2013

Newbold RR, Jefferson WN, Padilla-Banks E (2007) Long-term adverse effects of neonatal exposure to bisphenol A on the murine female reproductive tract. Reprod Toxicol 24(2):253–258

Padmanabhan V, Siefert K, Ransom S et al (2008) Maternal bisphenol-A levels at delivery: a looming problem? J Perinatol 28(4):258–263

Palanza P, Morellini F, Parmigiani S et al (1999) Prenatal exposure to endocrine disrupting chemicals: effects on behavioral development. Neurosci Biobehav Rev 23(7):1011–1027

Patisaul HB, Adewale HB (2009) Long-term effects of environmental endocrine disruptors on reproductive physiology and behavior. Front Behav Neurosci 3:1–10

Patterson TA, Twaddle NC, Roegge CS et al (2013) Concurrent determination of bisphenol A pharmacokinetics in maternal and fetal Rhesus monkeys. Toxicol Appl Pharmacol 267(1):41–48

Perera F, Vishnevetsky J, Herbstman JB et al (2012) Prenatal bisphenol A exposure and child behavior in an inner-city cohort. Environ Health Perspect 120(8):1190–1194

Pocar P, Fischer B, Klonisch T et al (2005) Molecular interactions of the aryl hydrocarbon receptor and its biological and toxicological relevance for reproduction. Reproduction 129(4):379–389

Prins GS, Ye SH, Birch L et al (2010) Serum bisphenol A pharmacokinetics and prostate neoplastic responses following oral and subcutaneous exposures in neonatal Sprague-Dawley rats. Reprod Toxicol 1:1–20

Richter C, Birnbaum LS, Farabollini F et al (2007) In vivo effects of bisphenol A in laboratory rodent studies. Reprod Toxicol 24(2):199–224

Rubin BS (2011) Bisphenol A: an endocrine disruptor with widespread exposure and multiple effects. J Steroid Biochem Mol Biol 127(1–2):27–34

Sakurai K, Kawazuma M, Adachi T et al (2004) Bisphenol A affects glucose transport in mouse 3T3-F442A adipocytes. Br J Pharmacol 141(2):209–214

Schug TT, Vogel SA, Vandenberg LN et al (2012) Bisphenol A. In: Schecter A (ed) Dioxins and health: including other persistent organic pollutants and endocrine disruptors, 3rd edn. Wiley, Hoboken, pp 381–414

Schug TT, Heindel JJ, Camacho L et al (2013) A new approach to synergize academic and guideline-compliant research: the CLARITY-BPA research program. Reprod Toxicol 5(13):00121–00124

Sharpe RM, Drake AJ (2013) Obesogens and obesity—an alternative view? Obesity 20(10):20373–20378

Shelby MD (2008) NTP-CERHR monograph on the potential human reproductive and developmental effects of bisphenol A. NTP CERHR MON (22):v, vii-ix, 1–64

Soto AM, Brisken C, Schaeberle C et al (2013) Does cancer start in the womb? Altered mammary gland development and predisposition to breast cancer due to in utero exposure to endocrine disruptors. J Mammary Gland Biol Neoplasia 24:199–208

Stahlhut RW, Welshons WV, Swan SH (2009) Bisphenol A data in NHANES suggest longer than expected half-life, substantial nonfood exposure, or both. Environ Health Perspect 117(5):784–789

Stump DG, Beck MJ, Radovsky A et al (2010) Developmental neurotoxicity study of dietary bisphenol A in Sprague-Dawley rats. Toxicol Sci 115(1):167–182

Taylor JA, Vom Saal FS, Welshons WV et al (2011) Similarity of bisphenol A pharmacokinetics in Rhesus monkeys and mice: relevance for human exposure. Environ Health Perspect 119(4):422–430

Teeguarden JG, Calafat AM, Ye X et al (2011) Twenty-four hour human urine and serum profiles of bisphenol A during high-dietary exposure. Toxicol Sci 123(1):48–57

Thayer KA, Heindel JJ, Bucher JR et al (2012) Role of environmental chemicals in diabetes and obesity: A National Toxicology Program workshop review. Environ Health Perspect 120(6):779–789

Thomas P, Dong J (2006) Binding and activation of the seven-transmembrane estrogen receptor GPR30 by environmental estrogens: a potential novel mechanism of endocrine disruption. J Steroid Biochem Mol Biol 102:175–179

Timms BG, Howdeshell KL, Barton L et al (2005) Estrogenic chemicals in plastic and oral contraceptives disrupt development of the fetal mouse prostate and urethra. Proc Natl Acad Sci USA 102(19):7014–7019

Trasande L, Attina TM, Blustein J (2012) Association between urinary bisphenol A concentration and obesity prevalence in children and adolescents. JAMA 308(11):1113–1121

Vandenberg LN, Hauser R, Marcus M et al (2007) Human exposure to bisphenol A (BPA). Reprod Toxicol 24(2):139–177

Vandenberg LN, Maffini MV, Sonnenschein C et al (2009) Bisphenol-A and the great divide: a review of controversies in the field of endocrine disruption. Endocr Rev 30(1):75–95

Vandenberg LN, Chahoud I, Heindel JJ et al (2010) Urinary, circulating, and tissue biomonitoring studies indicate widespread exposure to bisphenol A. Environ Health Perspect 118(8):1055–1070

Vandenberg LN, Colborn T, Hayes TB et al (2012) Hormones and endocrine-disrupting chemicals: low-dose effects and nonmonotonic dose responses. Endocr Rev 33:1–78

Vandenberg LN, Hunt PA, Myers JP et al (2013) Human exposures to bisphenol A: mismatches between data and assumptions. Rev Environ Health 28(1):37–58

Vanderloo MJ, Bruckers LM, Janssesn JP (2007) Effects of lifestyle on the onset of puberty as determinant for breast cancer. Eur J Cancer Prev 16(1):17–25

Vinas R, Watson CS (2013) Bisphenol S disrupts estradiol-induced nongenomic signaling in a rat pituitary cell line: effects on cell functions. Environ Health Perspect 121(3):352–358

Volkel W, Colnot T, Csanady GA et al (2002) Metabolism and kinetics of bisphenol A in humans at low doses following oral administration. Chem Res Toxicol 15(10):1281–1287

Volkel W, Bittner N, Dekant W (2005) Quantitation of bisphenol A and bisphenol A glucuronide in biological samples by high performance liquid chromatography-tandem mass spectrometry. Drug Metab Dispos 33:1748–1757

vom Saal FS, Timms BG, Montano MM et al (1997) Prostate enlargement in mice due to fetal exposure to low doses of estradiol or diethylstilbestrol and opposite effects at high doses. Proc Nat Acad Sci 94(5):2056–2061

vom Saal FS, Akingbemi BT, Belcher SM et al (2007) Chapel Hill bisphenol A expert panel consensus statement: Integration of mechanisms, effects in animals and potential to impact human health at current levels of exposure. Reprod Toxicol 24:131–138

von Goetz N, Wormuth M, Scheringer M et al (2010) Bisphenol A: how the most relevant exposure sources contribute to total consumer exposure. Risk Anal 30(3):473–487

Wang J, Sun B, Hou M et al (2012) The environmental obesogen bisphenol A promotes adipogenesis by increasing the amount of 11beta-hydroxysteroid dehydrogenase type 1 in the adipose tissue of children. Int J Obes 23(10):173–179

Watson CS, Bulayeva NN, Wozniak AL et al (2007) Xenoestrogens are potent activators of nongenomic estrogenic responses. Steroids 72:124–134

Wetherill YB, Akingbemi BT, Kanno J et al (2007) In vitro molecular mechanisms of bisphenol A action. Reprod Toxicol 24(2):178–198

Wolff MS, Britton JA, Boguski L et al (2008) Environmental exposures and puberty in inner-city girls. Environ Res 107(3):393–400

Wolff MS, Teitelbaum SL, Pinney SM et al (2010) Investigation of relationships between urinary biomarkers of phytoestrogens, phthalates, and phenols and pubertal stages in girls. Environ Health Perspect 118(7):1039–1046

Wolstenholme JT, Rissman EF, Connelly JJ (2011) The role of bisphenol A in shaping the brain, epigenome and behavior. Horm Behav 59(3):296–305

World Health Organization (WHO) (2010) Toxicological and health aspects of bisphenol A. In: Joint FAO/WHO Expert meeting 2–5 November 2010 and Stakeholder meeting on Bisphenol A 1 November 2010, Ottawa, Canada, 2010. World Health Organization, Geneva, Switzerland

Wright CL, Schwarz JS, Dean SL et al (2010) Cellular mechanisms of estradiol-mediated sexual differentiation of the brain. Trends Endocrinol Metab 21(9):553–561

Yang M, Ryu JH, Jeon R et al (2009) Effects of bisphenol A on breast cancer and its risk factors. Arch Toxicol 83(3):281–285

Ye X, Zhou X, Needham LL et al (2011) In vitro oxidation of bisphenol A: Is bisphenol A catechol a suitable biomarker for human exposure to bisphenol A? Anal Bioanal Chem 399(3):1071–1079

Ye X, Zhou X, Wong LY et al (2012) Concentrations of bisphenol A and seven other phenols in pooled sera from 3–11 year old children: 2001–2002 National Health and Nutrition Examination Survey. Environ Sci Technol 46(22):12664–12671

Ye X, Zhou X, Hennings R et al (2013) Potential external contamination with bisphenol A and other ubiquitous organic environmental chemicals during biomonitoring analysis: an elusive laboratory challenge. Environ Health Perspect 121(3):283–286

Chapter 2
Phthalates in Food Packaging, Consumer Products, and Indoor Environments

Kathryn M. Rodgers, Ruthann A. Rudel and Allan C. Just

Abstract Phthalates are a diverse group of chemicals, including five with production volumes of over 1 million pounds per year in the United States (U.S.). They are used as plasticizers in a variety of plastics including polyvinyl chloride (PVC), medical devices (e.g., intravenous bags and tubing), food contact materials (FCMs), toys, and household goods, and as solvents in fragranced personal care and household products. Although not all phthalates have been evaluated for their toxic effects, many that are in widespread use have displayed endocrine disrupting properties on the developing reproductive system, especially in males, in laboratory, animal, and human studies. Widespread exposure to phthalates has been documented in the U.S. and in European countries, with some examples of unusually high exposures in certain populations, such as neonates with intravenous interventions in hospital settings. For the majority of the population, the primary route of exposure to the endocrine disrupting phthalates produced in the highest volume, bis(2-ethylhexyl) phthalate (DEHP) and diisononyl phthalate (DINP), is through diet. DEHP is used in food packaging, and also has been found to contaminate food sources directly. Some newer phthalates that have been introduced as alternatives to phthalates with known health concerns are also endocrine disruptors, while others have not been evaluated. Regulatory agencies are considering ways to define phthalates and assess their risk as a group based on chemical structure.

K. M. Rodgers · R. A. Rudel (✉)
Silent Spring Institute, Newton, MA, USA
e-mail: rudel@silentspring.org

K. M. Rodgers
e-mail: rodgers@silentspring.org

A. C. Just
Department of Environmental Health, Harvard School of Public Health, Boston, MA, USA
e-mail: acjust@hsph.harvard.edu

S. M. Snedeker (ed.), *Toxicants in Food Packaging and Household Plastics*,
Molecular and Integrative Toxicology, DOI: 10.1007/978-1-4471-6500-2_2,

Keywords Phthalates · Testosterone · Male reproduction · Female reproduction · Reproductive toxicity · Sperm count · Plastics · Vinyl · Polyvinyl chloride · PVC · Food · Food packaging · Food contact material · Consumer products · Children's toys · Medical equipment · Neonatal intensive care unit · Intravenous bags · DEP · DBP · DnBP · DiBP · BBzP · DEHP · DINP

2.1 Key Take Home Points

- Phthalates are ubiquitous in the modern indoor environment, and they have a variety of uses and diverse exposures.
- Food packaging and processing, vinyl household products, and vinyl flooring are the major contributors of exposure to bis(2-ethylhexyl) phthalate (DEHP) and several other phthalates that are used as plasticizers, while fragranced products are an important source of diethyl phthalate (DEP) exposure.
- Population-level variations in exposures are influenced by factors such as geography, dietary habits, gender, and personal care product use.
- Some phthalates, such as DEHP, are endocrine disruptors that affect the developing male reproductive system by interrupting the production of fetal testosterone and the protein involved in testicular descent, insulin-like 3 (insl3), while other phthalates, such as DEP, are not endocrine disruptors.
- Because phthalates vary in their potency, and some do not show endocrine disruption, all phthalates cannot be grouped together or be grouped based on molecular weight.
- Since mixtures of phthalates act in a dose-additive manner, risks should be considered cumulatively.

2.2 Introduction

Phthalates are a group of compounds commonly used in plastics and personal care products. Although they are often referred to collectively by the single term "phthalates," there are many types of phthalates in widespread commercial use with different predominant uses, physical and chemical characteristics, and toxicologic profiles. Examples of phthalates, U.S. production volumes, endocrine disrupting properties, and major uses are given in Table 2.1. DEHP and butylbenzyl phthalate (BBzP) are often used as plasticizers in flexible vinyl materials, such as items ranging from food containers to vinyl flooring to medical devices. Lower molecular weight phthalates such as dibutyl phthalate (DBP) and diethyl phthalate (DEP) are used in personal care products, pesticides, glues, and paints as solvents, and in time-released pharmaceuticals (Meeker and Ferguson 2012).

Phthalates are weakly bound to their incorporating substrate, and are thus easily leached from products, enabling them to enter the environment. They are high-production volume chemicals present in food packaging, foods, and a variety of consumer products, and they are some of the most abundant chemicals measured

Table 2.1 Selected high production volume phthalates

Chemical name (abbreviation) CAS no.	U.S. production volume 2012 (United States Environmental Protection Agency 2012a) (million lbs/year)	Anti-androgenic endocrine disruptor	Uses and products
Diethyl phthalate (DEP) CAS: 84-66-2	10 < 50	No (Gray et al. 2000)	*Solvent* fragrance, soaps and detergents, adhesives and sealants, pharmaceuticals *Plasticizer* rubber and plastic products (e.g., toys, artificial turf, food packaging)
Dibutyl phthalate (DBP) Di-*n*-butyl phthalate (DnBP) CAS: 84-74-2 Diisobutyl phthalate (DiBP) CAS: 84-69-5	10 < 50	Yes (Gray et al. 2000)	*Solvent* adhesives, binding agents, paint, and coating manufacturing *Plasticizer* plastic and vinyl products (e.g., shower curtains, rain coats)
Butylbenzyl phthalate (BBzP) CAS: 85-68-7	50 < 100	Yes (Gray et al. 2000)	*Solvent* adhesives and glues *Plasticizer* plastic manufacturing, automotive products, vinyl floors
Bis(2-ethylhexyl) phthalate (DEHP) CAS: 117-81-7	100 < 500	Yes (Gray et al. 2000)	*Plasticizer* Vinyl flooring, wallpaper, toys, electronics, plumbing, shoes
Diisononyl phthalate (DINP) CAS: 68515-48-0 and 28553-12-0	100 < 500	Yes (Gray et al. 2000)	*Plasticizer* rubber and PVC plastic, toys, food packaging and other plastic packaging material as substitute for DEHP

in indoor air and house dust (Schecter et al. 2013; Dodson et al. 2012; Rudel et al. 2003). Not surprisingly, greater than 98 % of the U.S. population is considered exposed to some widely used phthalates, although urinary concentrations of different phthalate metabolites typically span orders of magnitude within a population (Silva et al. 2004; Aylward et al. 2011). The variability in concentrations of phthalates in environmental media and biologic samples arises primarily from variation in the sources and concentrations, individuals' behaviors, and the chemicals' short biological half-lives.

Some phthalates are considered to be endocrine disruptors, thereby disrupting normal hormonal signaling and functioning in the body. Consistent laboratory evidence shows that DEHP, DBP, BBzP, and DINP are anti-androgenic; they adversely affect the developing male reproductive system through inhibition of testosterone and insl3 synthesis during fetal development, and there are supporting human associations for these observations (Wilson et al. 2004; Swan et al. 2005). Adverse effects on male and female reproduction have also been demonstrated in animal studies (Gray et al. 2006a). In humans there is also some evidence of effects on metabolism, neurological development, asthma, and allergy (Meeker and Ferguson 2012). While DEP is commonly found in personal care products, particularly as a solvent for fragrance compounds, it has not been demonstrated to show anti-androgenic endocrine disrupting activity.

2.3 Human Exposure and Biomonitoring

Widespread human exposure to phthalates has been documented by independent studies and representative samples in the U.S., as well as European countries, Mexico, Taiwan, and other populations (Meeker and Ferguson 2012). Phthalates have been measured in indoor and outdoor air and house dust, foods and food packaging, consumer products, and other media (Rudel et al. 2003; Dodson et al. 2012; Schecter et al. 2013; Rudel et al. 2010, 2011).

2.3.1 Exposure Biomarkers

Human exposure to phthalates is most commonly measured by chemical analysis of urine to detect monoester and oxidative metabolites (Silva et al. 2007). Table 2.2 lists the common metabolites for biomonitoring major phthalates. Phthalates are ubiquitous in the environment and also in sample collection and laboratory equipment, so great care is required to produce valid measurements. The most reliable exposure biomarkers are monoesters and oxidative metabolites in urine. Measurements of the parent compound are vulnerable to contamination from laboratory or sampling equipment. Additionally, hydrolytic enzymes found in blood, breast milk, and other biologic matrices like amniotic fluid and meconium, but not

Table 2.2 Selected urinary phthalate metabolites and median levels in the U.S. general population (Centers for Disease Control and Prevention 2009)

Parent (abbreviation)	Metabolite(s) (abbreviation)	U.S. population levels NHANES 2009–2010 median (μg/L)
Dimethyl phthalate (DMP)	Mono-methyl phthalate (MMeP)	0.94
Diethyl phthalate (DEP)	Monoethyl phthalate (MEP)	54.90
Butylbenzyl phthalate (BBzP) Dibutyl phthalate (DBP)	Monobutyl phthalate (MBuP)	15.90
	Mono-isobutyl phthalate (MiBP)	8.30
Dicyclohexyl phthalate (DCHP)	Mono-cyclohexyl phthalate (MCHP)	<0.40
Butylbenzyl phthalate (BBzP)	Monobenzyl phthalate (MBzP)	6.50
Bis(2-ethylhexyl) phthalate (DEHP)	Mono-(2-ethyl-5-hydroxyhexyl) phthalate (MEHHP)	12.90
	Mono-(2-ethyl-5-oxohexyl) phthalate (MEOHP)	8.02
	Mono-2-ethylhexyl phthalate (MEHP)	1.59
Di-*n*-octyl phthalate (DnOP)	Mono-(3-carboxypropyl) phthalate (MCPP)	3.02
Di-isononyl phthalate (DINP)	Mono-iso-nonyl phthalate (MiNP)	<0.77

urine, can convert these contaminants into the monoesters after sample collection (Calafat et al. 2006). Measurements of oxidative metabolites in blood are thought to be reliable because they are not formed as a result of sample contamination.

Urinary levels of phthalate metabolites are considered the best measure of exposure as phthalates are rapidly excreted within hours, primarily in the urine (Janjua et al. 2008). The measurement of phthalates in urine is a valuable biomarker for epidemiologic studies of health effects because it integrates exposure across multiple routes: ingestion, inhalation, and dermal absorption. The contribution of each of the routes of exposure appears to vary between phthalates, with some phthalates primarily coming from dietary sources, while others come from personal care products (Wormuth et al. 2006; Fromme et al. 2007). Because of the short half-lives of phthalates and because urine levels vary throughout the day, average exposure levels are best estimated using 24-hour (h) urine samples, possibly from multiple days.

Urinary concentrations of phthalate metabolites are available for many populations, geographies, and age groups, and these provide useful information on relative exposure levels among these groups. Pharmacokinetic models and other techniques have been used to back-calculate exposure intake levels that correspond to the urine levels (Lorber and Calafat 2012; Aylward et al. 2009; Koch et al. 2011). Based on these models, researchers have estimated that the typical intake for DEHP in the U.S. 2001 general population was in the range of 0.6–2.2 microgram/kg per day (μg/kg per day), corresponding to median urine concentrations of MEHP, 5OH-MEHP, and 5oxo-MEHP of 4.1, 20.1, and 14.0 μg per liter (L), respectively (Lorber and Calafat 2012). Others have used these relationships and a cumulative assessment of exposure to three common phthalates (DEHP, DnBP and DiBP)

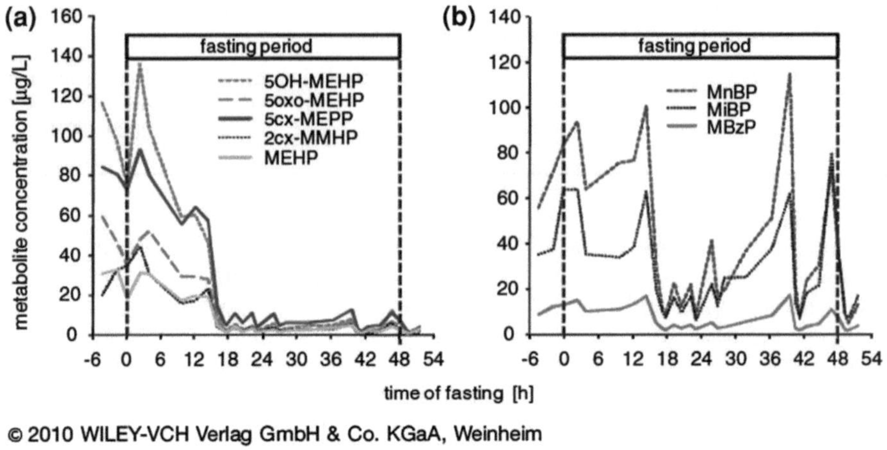

Fig. 2.1 Excretion of several phthalate metabolites during 48-h fasting. *Credit* figure reproduced with permission from John Wiley and Sons, Copyright © 2010 WILEY-VCH Verlag GmbH & Co. KGaA, Weinheim; original figure appeared in: Wittassek et al. (2011) as Fig. 1

and showed that 24 % of German children exceeded the health-based guidance value, the Tolerable Daily Intake (TDI) (Koch et al. 2011).

Because phthalates are not listed on ingredient labels in most product categories in the U.S., there is no way to know which phthalates may be commonly found in everyday products (Dodson et al. 2012). Since formulations change regularly, substitutes may enter the market without the public knowing. As a result, scientists may not always know which phthalates to measure in biomonitoring studies.

2.3.2 Diet as a Primary Route of Exposure

Several studies provide evidence that exposure to DEHP and its substitutes is primarily from diet. In two similar studies, volunteers participated in a 48-h fast. DEHP, DINP, DnBP, DiBP, and BBzP metabolites were measured in urine before and during the fasting period (Koch et al. 2013; Wittassek et al. 2011). Within 24 h, DEHP and DINP metabolites were reduced to levels five to ten times lower than pre-fasting measurements. DINP is a widely used substitute for DEHP. However, no significant changes were seen for the urinary excretion of DnBP, DiBP, or BBzP metabolites during the fasting period (Fig. 2.1), suggesting their exposures were from non-food sources. In a similar attempt to reduce dietary exposure sources, another intervention study removed food packaging from five families' diets (Rudel et al. 2011). Switching from a conventional diet to a diet of whole foods and beverages that had limited contact with plastic, aluminum, or canned packaging resulted in an average decrease of urinary DEHP metabolites levels by over 50 % during the 3-day intervention period (Fig. 2.2). Reductions

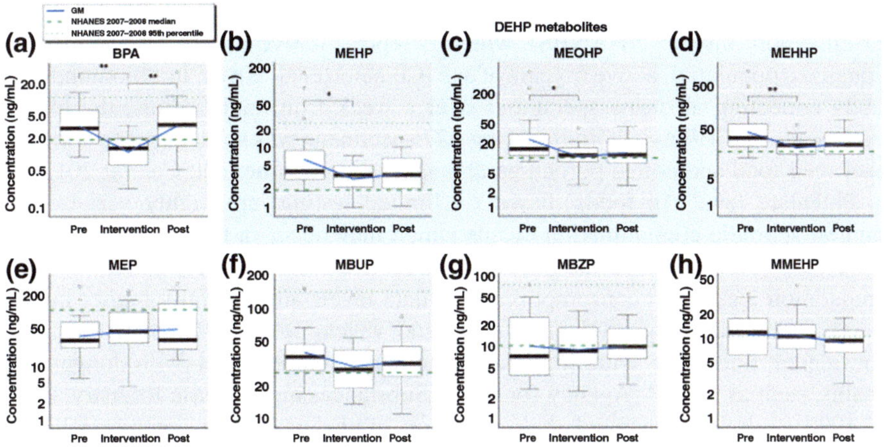

Fig. 2.2 Change in phthalate excretion during a dietary intervention to reduce food packaging. *Credit* figure reproduced with permission from Environmental Health Perspectives; original figure appeared in: Rudel et al. (2011) as Fig. 2

were even more pronounced for people with the highest pre-intervention exposures, which resulted in more than a 90 % reduction of urinary DEHP. Levels of the urinary metabolites of DMP, DEP, DBP, and BBzP did not significantly change during the intervention period, again suggesting that exposure to these phthalates is largely from non-food related sources.

2.3.3 Foods and Food Packaging

Phthalates, especially DEHP, have been reported at high concentrations in some foods. In a dietary intervention study similar to that discussed in Sect. 2.3.2 (Rudel et al. 2011), researchers substituted food with limited contact with plastic packaging into ten families' diets, and an unexpected spike in DEHP measurements occurred during the dietary intervention period (Sathyanarayana et al. 2013). This was an unexpected finding, and the authors concluded that the prepared meals that were served to the participants were contaminated with DEHP. Milk, cream and coriander used for meal preparation in the study had higher DEHP levels than normally found in those foods, demonstrating food contamination may occur even in the most careful of circumstances. Another study revealed phthalate contamination in Taiwanese sport drinks, which contained high levels of DEHP added illegally as a clouding agent (Yang et al. 2013).

Additional evidence that food and food packaging are important sources of DEHP exposure come from observational human epidemiologic and panel studies. For example, in analyses adjusted for body mass index (BMI), each ounce of poultry consumed in the previous day was associated with an approximately 6 % higher level of urinary DEHP metabolites in the 2003–2004 National Health and Nutrition Examination Survey (NHANES), which is representative of the U.S. non-institutionalized population above 6 years of age (Colacino et al. 2010). In a biomonitoring study collecting all urine specimens over a week from eight adults, the highest excretion of DEHP metabolites among 427 specimens was subsequent to a meal of packaged food and coffee purchased at a gas station (Lorber and Calafat 2012).

Phthalate levels in foods, based on limited testing, are highly variable and suggest sporadic contamination events which may occur in the source or original collection of the food, in the packaging or processing prior to market, or in the preparation leading to consumption. Phthalate levels are generally higher in fatty foods, including dairy products, meats, and vegetable oils. Phthalate esters are considered lipophilic with higher lipophilicity in phthalates with longer side chains, such as DEHP (Agency for Toxic Substances and Disease Registry 1995). A 1990 study demonstrated that collection tubing used in a commercial dairy resulted in DEHP-contaminated milk, and that the level of DEHP was at a higher level in cream (Castle et al. 1990). This suggests that the DEHP migrated from the collection tubing to the fat component of the milk and cream.

Recent testing of 72 foods purchased from a U.S. supermarket demonstrated that various phthalates were detectable in all classes of food with DEHP being the highest of the phthalates tested in most food categories, particularly in pork, dairy products, vegetable oils and grains, although the sample sizes were limited (Schecter et al. 2013). This is largely consistent with concentrations found in European foods, reviewed in a scenario-based model that found that diet had a major influence on DEHP exposure (Wormuth et al. 2006). The sporadic contamination of food was seen in the three U.S. cooking oil samples, which had concentrations of BBzP of 459, 2.20 and 0.35 nanograms per gram (ng/g), with the highest value detected in virgin olive oil from a glass container, although the source of phthalates found in that sample was unknown (Schecter et al. 2013). A study of FCMs by the European Food Safety Authority (EFSA) found that the use of gaskets for metal lids made from PVC on imported foods accounted for several of the highest observed concentrations of phthalates in foods, although conveyor belts, gloves, and tubes for liquids were also contributing sources (Petersen and Jensen 2010). Items used in the handing, processing, and packaging of foods and beverages are all considered to be FCMs. The packaging (polyethylene coated aluminum dishes sealed with polyethylene terephthalate-coated foil) of hot cooked foods led to further increases in both DEHP and DnBP in a study examining Italian school lunches (Cirillo et al. 2011).

It is unclear which food packaging materials contain phthalates and how different foods become contaminated. On the one hand, the American Plastics Council states that:

> phthalates are not used in plastic beverage bottles, nor are they used in plastic food wrap, food containers, or any other type of plastic food packaging sold in the United States (Enneking 2006).

On the other hand, the studies described above show that diet remains an important source of phthalate exposure, and that food packaging is one established contributing source (Schecter et al. 2013; Rudel et al. 2011). Some possible explanations for this discordance include inconsistent compliance with industry claims, particularly in imported foods; introduction of phthalates from recycled content or during manufacturing; use of packaging on food that was not intended for food use; and use of PVC and other phthalate-containing plastics in food processing and handling (Petersen and Jensen 2010; Tsumura et al. 2001; Montuori et al. 2008). Identifying the specific sources of phthalate contamination in food has been difficult because of the variety and vast assortment of FCMs in use, and the large number of manufacturers of foods and food packaging materials. Surveys of phthalates in materials and foods are further limited by the expense of laboratory measurements and the technical challenge of directly measuring phthalate diesters that are prone to laboratory contamination.

2.3.4 Non-dietary Sources

Exposures to phthalates other than DEHP and its substitutes are primarily from non-dietary sources, possibly from consumer goods in the indoor environment and personal care products. A study that paired indoor and outdoor air measurements found that phthalates were the most abundant chemicals in indoor air and house dust, with maximum levels of DBP in air of 1.1 micrograms per cubic meter ($\mu g/m^3$) and levels in house dust approaching 1 mg per gram (g) (Rudel et al. 2003, 2010). There was no correlation between indoor and outdoor air measurements of DEP, DBP, DiBP, or DEHA, an alternate plasticizer, demonstrating that phthalate exposures are primarily from indoor sources. In a cohort in New York City, homes with vinyl floors were correlated with indoor air and urine measurements of BBzP, but not DEHP, which is further support for the idea that DEHP exposure is primarily from food and food packaging (Adibi et al. 2008; Just 2012). In a Swedish study of infants, the urinary metabolites of BBzP, but not DEHP, were higher among those with PVC flooring in their bedrooms (Carlstedt et al. 2013). These studies demonstrate that BBzP used in vinyl migrates out of its substrate, entering indoor air and dust, leading to exposures through inhalation or ingestion of dust. Recent research has shown that exposure via dermal absorption in the indoor environment is a major contributor to the total intake for certain phthalates that predominantly partition to the gas-phase. A Danish study of children aged 3–6 estimated that dermal absorption of DEP, DnBP, and DiBP was the dominant exposure pathway of these phthalates in the indoor environment. More research in the dermal exposure pathway from ambient indoor air is forthcoming (Bekö et al. 2013).

DEHP and BBzP are found in consumer products at high concentrations. DEHP is especially high in vinyl products, up to 40 % by weight (Agency for Toxic Substances and Disease Registry 2002). DEP, also commonly found in the indoor environment, is used as a solvent for fragrance in perfumes, lotions, hair products, and other personal care products. A 2012 study tested 213 commercial products, including "conventional" products and some "alternative," or advertised as healthier, products (Dodson et al. 2012). DEHP was detected at the highest concentration of the phthalates tested: 28 % by weight of a vinyl shower curtain, and 14 % by weight of a vinyl pillow protector; BBzP was also detected in the pillow protector at lower levels. DEHP also was detected in conventional diapers, hand sanitizer, shaving cream, deodorant, lipstick, car air freshener, dryer sheet, nail polish, and sunscreen. DEHP also was found in alternative products, including hand sanitizer, shaving cream, lipstick, and an alternative pillow protector (at very low levels). DEP was one of the most commonly detected chemicals in both conventional products and alternative products. Of the ten phthalates tested for, four were found in both conventional products and alternative products, including DEP, di-*n*-butyl phthalate (DnBP), BBzP, and DEHP. Other types of conventional products with additional phthalates, such as di-*n*-octyl phthalate (DnOP), di-*n*-hexyl phthalate (DnHP), and diisobutyl phthalate (DiBP) included foundation makeup, lipstick, and perfume. Other alternative products that contained DnOP, DINP, and di-cyclohexyl phthalate were laundry detergent, shaving cream, bar soap, baking soda, and borax. Exposure to these phthalates may result from direct contact with the product, such as shaving cream applied directly to skin, and also from inhalation or ingestion of phthalates that migrate out of the product, as in the case of DEHP in vinyl shower curtains. The European Union (EU) also identified DEHP, DnBP, and DiBP in PVC-containing sandals and DEHP in PVC-containing sex toys as potentially significant sources of exposure (European Chemicals Agency 2012). Although there are now some restrictions on phthalates in children's toys in the EU and U.S., high levels have been reported in PVC-based toys, some containing up to 40 % DINP or DEHP (Schettler 2006; Stringer et al. 2000).

Phthalates have also been detected in art supplies, including modeling clay. High concentrations of DnOP, DnHP, BBzP, and DEHP have been reported in Sculpey™ and Fimo® modeling clay, with levels up to 14 % by weight (Schettler 2006). Contact with these concentrated sources can likely be associated with much higher exposure levels than reported in the general population, especially for children who may exhibit greater hand-to-mouth-activity resulting in incidental ingestion of DEHP. Inhalation exposure may also be important when the clay products are baked dry.

A survey of phthalates in personal care products purchased in the U.S. in 2002, with a follow-up published in 2010, found that eight of 25 products that were retested no longer contained phthalates, possibly indicating reformulation following market pressures (Hubinger and Havery 2006; Hubinger 2010). In the 2010 follow-up study, while DEP was still widely found and in high concentrations, the only other phthalate that was found in adult-use cosmetic products was DnBP in 11 of 24 nail products with widely ranging concentrations (<10–62,607 μg/g) (Hubinger 2010). Of the phthalates tested, only DEP was detected in a survey of 24 baby-care products in the U.S. (baby shampoo, body wash, cream, lotion, and oil products) (Hubinger 2010). Similar results, also showing DEP detection, were reported in a survey of personal care products purchased in Canada (Koniecki et al. 2011). These changes may reflect restrictions in the EU on DEHP, DBP, and BBzP in personal care products that were enacted in 2004 (European Union 2004).

Several studies have demonstrated that adult use of personal care products, particularly the use of fragrance (cologne or perfume), the use of multiple products, and use reported in the house before urine collection, is associated with higher levels of urinary DEP metabolites in Israel and the U.S. (Berman et al. 2009; Duty et al. 2005a; Just et al. 2010; Parlett et al. 2013; Braun et al. 2013). In one study, a model adjusting for urinary dilution, age, education, and use of perfume or cologne in the previous 24 h explained 41 % of the population variance in urinary DEP metabolites, with fragrance users having almost three times the concentration of non-users on average (Parlett et al. 2013). In two separate studies, personal care product use in Mexico was associated with higher urinary levels of DEHP and DEP metabolites among adult women, and among both boys and girls 8–13 years of age (Romero-Franco et al. 2011; Lewis et al. 2013). A study measuring phthalate metabolites in infant urine found that reported use of baby care products (lotion, powder, and shampoo) was associated with higher levels of infant exposure to and urinary excretion of DMP, DEP, and DiBP (Sathyanarayana et al. 2008).

Contact with certain phthalate-containing products may lead to unusually high exposure levels. For example, Hauser et al. reported that study participants using phthalate-containing medications had urinary DBP levels 100 times above the NHANES 95th percentile (Hauser et al. 2004). Use of phthalate-containing medical devices can also affect potential exposures. For instance, infants in a neonatal intensive care unit showed elevated urinary levels of DEHP, DBP, and BBzP metabolites, with highest levels observed in conjunction with the most intensive use of medical devices that contained or contacted PVC (Weuve et al. 2006).

2.3.5 Exposure Trends: Demographic and Geographic Patterns

Results from a biobank of samples from German university students show that the daily intake of both DEHP and DnBP, as estimated from urinary excretion patterns of metabolites, decreased substantially over 15 years, with the median DnBP intake falling to less than one-third of 1988 levels by 2003. Over the same period, there was some evidence of increasing exposure to the replacement phthalates DiNP (a DEHP replacement) and DiBP (a DnBP replacement) (Wittassek et al. 2007). It also was noted that the annual median intake of DEHP for German students, estimated from urine samples from this biobank, had an extremely close correlation ($r = 0.97$) with yearly German industrial DEHP production data over the same period (Helm 2007).

Gender and age may also affect use patterns of phthalate-containing products. DEP exposures were higher in females than in males, and higher in adults and adolescents compared to children in one U.S. study, likely reflecting patterns of personal care product use (Silva et al. 2004). DBP, BBzP, and DEHP metabolites were higher among children and among females than in males (Silva et al. 2004). In a study conducted in Mexico City, levels of metabolites of DEP, DnBP, DiBP,

BBzP and DEHP in stored maternal urine collected during the third trimester of pregnancy were compared with paired urine samples from the delivered children when they were aged 8–13 years; children had higher levels than their mothers except for the DEP metabolite which is found in personal care products, particularly perfume (Just et al. 2010; Hubinger and Havery 2006).

Biomonitoring results from NHANES demonstrate that socioeconomic factors are associated with large differences in population exposure to different phthalates. This may be the result of economic and social differences in dietary preferences, housing materials, and consumer product use patterns. In a pooled analysis of female NHANES 2001–2008 participants 20–39 years of age, those in the lowest quartile of overall socioeconomic status (SES) had lower urinary DEHP metabolite levels but an average of 1.83 times the BBzP metabolite concentrations of those in the highest quartile of SES. Non-White ethnicities had higher urinary DBP and DEP metabolite levels compared with Non-Hispanic Whites but lower urine BBzP after accounting for SES (Kobrosly et al. 2012). Whether these phthalate exposure and urinary excretion patterns would apply to other populations and other demographic groups requires further research. Women in a Mennonite community that had a limited exposure to plastic food packaging, cosmetics, and automobiles had lower exposures to DEP, DiBP, and DEHP, measured from urinary excretion than the general population (Martina et al. 2012).

2.4 Health Effects

2.4.1 Reproductive Toxicity: Laboratory Animals

Animal studies provide consistent evidence that certain phthalates target the developing male reproductive system. These effects in animal studies have been termed *Phthalate Syndrome* (PS), and the effects mirror a set of reproductive symptoms seen in human males, termed *Testicular Dysgenesis Syndrome* (TDS). TDS and PS include symptoms that arise from insufficient testosterone production during in utero development, including undescended testes, malformations of the penis, reduced anogenital distance (AGD), decreased sperm motility and mobility, infertility, and testicular cancer (Meeker and Ferguson 2012). Phthalates that have exhibited these effects in animal studies are DiBP, DnBP, BBzP, DEHP, diisoheptyl phthalate (DiHP), and DiNP (Gray et al. 2000; Hannas et al. 2011). Testicular toxicity has also been observed following pubertal exposure, and, at higher doses, in adult males (Mahood et al. 2007; Noriega et al. 2009).

Although the effects of phthalates on the male reproductive system are the most fully described, DEHP has been shown to affect the female reproductive system as well. Female rats treated with DnBP from weaning through pregnancy aborted litters subsequent to decreases in progesterone levels, and these effects are observed at similar doses that induce testicular toxicity in adult male rats (Gray et al. 2006a). Female marmosets (primates) showed effects on ovarian and uterine weight and blood levels of estradiol following 65 weeks of pubertal exposure to DEHP (Tomonari et al. 2006). In a study that examined generational affects, female pups of mice orally exposed to high doses of MEHP during pregnancy had shorter life expectancy, prolonged estrus cycles, and hyperplasia of mammary tissue (Moyer and Hixon 2012). At extremely high doses, adult female rats orally exposed to DEHP have lowered circulating estradiol, prolonged estrus cycles, and experience anovulation (Lovekamp-Swan and Davis 2003).

Phthalates are thought to cause their toxic effects on male reproductive development by interrupting testosterone and insl3 production in the testes during the sensitive in utero masculinization programing window. In male mammals, testosterone and insl3 are necessary to change the default phenotype from female to male during fetal development. Therefore, the fetus is at the most sensitive developmental stage for the reproductive health effects of anti-androgenic chemicals. The specific developmental period of greatest sensitivity to these effects is from 15.5 to 17.5 days gestation in the rat and 8–20 weeks gestation in the human (Scott et al. 2009). Rodent studies with DEHP and DiBP show that the inhibition of testosterone production in Leydig cells, which produce testosterone in the testes, may occur through down-regulation of key genes, including StAR, HMC-CoA synthase, and SRB1 (involved in cholesterol uptake), and the enzymes CYP11A, 3B-Hsd, and CYP17 (involved in steroid biosynthesis). Because cholesterol is the first building block of testosterone synthesis, down-regulation of its uptake and transport may result in less testosterone production. CYP11A is the rate-limiting step of testosterone biosynthesis, and CYP17 is regarded as the qualitative regulator of steroidogenesis because it is necessary in more than half of the conversions from cholesterol to testosterone (Scott et al. 2009). During pubertal development, Sertoli cells are also a target of phthalate toxicity. Some phthalates, such as DEHP, impair the production of spermatozoa by interrupting the Sertoli cells' ability to nourish the germ cells (Hauser et al. 2006). Insl3, which is a protein involved in controlling testicular descent, is adversely affected by DEHP's monoester metabolite, MEHP (Gray et al. 2006b). One study found that DINP reduced immunostaining intensity in Leydig cells for StAR, P45scc, and CYP17 in similar manner as DEHP (Boberg 2007). Scientists expect harm to the male reproductive system observed in animal models to be relevant to humans because similar effects have been observed in many species of mammals, including rats, mice, hamsters, ferrets, and guinea pigs (Voss et al. 2005; Hannas et al. 2012; Hotchkiss et al. 2008).

Differences in phthalate structure determine how they are metabolized and their toxic effects. MBuP and MBzP, metabolites of DBP and BBzP, have been identified as active metabolites toxic to reproductive endpoints, including sperm concentration in adult males (Hauser et al. 2006). Direct measures of parent phthalates or phthalate monoester metabolites in blood are difficult to interpret because of the high likelihood of contamination from sampling or from lab materials given that esterase activity after sample collection can generate monoester metabolites in samples contaminated after collection (Calafat et al. 2013).

Phthalates differ in their ability to affect androgen pathways, with some showing anti-androgenic effects and others showing no effect (see Table 2.1).

These effects may partially be due to the potency of the phthalate. Dr. Earl Gray, a reproductive biologist and toxicologist at U.S. Environmental Protection Agency (USEPA), has conducted experiments to assign relative potencies to different phthalates. DMP, DEP, BBzP, DEHP, dioctyl tere-phthalate (DOTP), and DINP were orally administered to pregnant rats perinatally, and their offspring were assessed for androgen-sensitive endpoints. While administration of DEP, DMP, and DOTP did not exhibit adverse effects on the male reproductive system, DEHP and BBzP produced male malformations, including reduced AGD and nipple retention, with similar potencies and frequency. At the same dose, DINP produced these malformations with approximately ten times less prevalence (Gray et al. 2000). DnBP and DiBP have also shown to exhibit similar potencies to DEHP and BBzP, while dipentyl phthalate (DPP) was three times more potent than DEHP (Howdeshell et al. 2008). Examples of human reproductive toxicity with exposure to various phthalates are provided in Sect. 2.4.4.

2.4.2 Epigenetic Changes

There is some evidence that phthalates can cause other types of toxicity, including epigenetic changes. Epigenetic effects are changes to genetic expression, which can occur through the way DNA is packaged, such as histone modification and methylation patterns. Changes to genetic expression may have adverse effects, and heritable epigenetic changes may alter DNA expression across several generations. A study with human breast cell line MCF10A and human breast cancer cell line MCF7 found that exposure to BBzP led to demethylation of a promoter region of estrogen receptor alpha (ERα) (Kang and Lee 2005). ERα is a cellular receptor for the hormone estrogen, and it is necessary for normal estrogen function in both men and women. In another study, DnBP was associated with hypomethylation in the c-myc proto-oncogene in liver cells (Ge et al. 2002). Prenatal exposure to DEHP in mice was found to induce PS, increase DNA methylation of testes, and increase methyltransferase expression (Wu et al. 2010).

2.4.3 Carcinogenicity and Genotoxicity

In addition to having adverse effects on the male reproductive system, DEHP has also been found to produce liver tumors in both male and female Fischer-344 rats and B6C3F1 mice. Animals fed DEHP in a 2-year cancer bioassay had significantly higher incidence of malignant liver tumors (hepatocellular carcinomas), compared to control groups (Integrated Risk Information System 1997). In a study with male Sprague-Dawley rats fed DEHP in their diet for 3 years, benign Leydig cell tumors in the testes occurred with nearly twice the incidence than the control animals (Voss et al. 2005). Testicular tumors are a symptom of TDS and the finding is consistent with the testicular damage associated with phthalates. A 2-year study with Fischer-

344 rats fed DEHP in their diet resulted in significantly higher incidence of benign pancreatic tumors in the male rats in the highest dose group (David et al. 2000). DEHP has been rated as *reasonably anticipated to be a human carcinogen* by the National Toxicology Program (NTP) based on evidence from laboratory animal cancer bioassays (National Toxicology Program 2011). DINP has structural and mechanistic similarities to DEHP and was recently identified as a carcinogen based on several rodent studies (Office of Environmental Health Hazard Assessment California Environmental Protection Agency 2013). Like DEHP, DINP also has been found to cause increased hepatocellular carcinomas in male and female Fischer-344 rats and B6C3F1 mice. Male and female Fischer-344 rats also experienced significantly increased incidence of mononuclear cell leukemia following chronic DINP exposure. Increased incidence of renal tubular cell carcinoma was observed in male Fischer-344 rats, although a statistically significant association was observed in only one study. Incidence of pancreatic islet cell tumors, Leydig cell tumors, and uterine tumors all increased in Sprague-Dawley rats following DINP exposure, although findings did not reach statistical significance.

Additional genotoxic effects in vivo also have been observed, such as DNA damage, chromosomal aberrations, and tumor promotion (International Agency for Research on Cancer 2012). The mono-ester metabolite of DEHP, MEHP, activates peroxisome proliferation activated receptor alpha (PPARα) in the liver, which is thought to lead to liver tumors, and it has been hypothesized that rats are more sensitive to this liver effect than humans because of species differences in PPARα (Melnick 2001; Rusyn and Corton 2012). DINP also activates PPARα, as well as PPAR gamma (γ). However, a 2-year dietary DEHP study with PPARα-null mice found significantly increased incidence of malignant liver tumors in the PPARα-null mice compared to wild-type mice, demonstrating the liver effects are not entirely PPARα-dependent. The authors hypothesize DEHP exposure may lead to the formation of reactive oxygen species, causing DNA damage in liver tissue, contributing to tumor effects through other pathways (Ito et al. 2007).

2.4.4 Reproductive Toxicity: Human Epidemiology

Epidemiologic studies measuring phthalate exposure largely show consistency with animal models' evidence of harm to the male reproductive system (Meeker and Ferguson 2012). One longitudinal cohort study in the U.S. described decreased AGD in male children whose mothers had higher levels of MEP, MBP, MBzP, and MiBP in their urine (Swan et al. 2005). AGD is a testosterone-dependent endpoint of genital development. Males have longer AGDs due to increased fetal testosterone levels as compared to females, who have shorter AGDs due to lower testosterone levels. This finding is consistent with evidence from animal models for exposures to the anti-androgenic phthalates BBzP and DBP, although DEP does not produce anti-androgenic effects in animal models. These results suggest that phthalates are interfering with fetal testosterone levels. A Danish-Finish cohort study examined the levels of phthalate monoester metabolites in breast milk of mothers of 3-month old male infants (Main et al. 2005). Higher breast milk levels of MBP, the active metabolite of DBP, were statistically significantly negatively correlated with free serum testosterone levels in male offspring. Breast milk levels of MiNP, the primary metabolite of DINP, were significantly positively correlated with serum levels of luteinizing hormone (LH) in the boys. Increased LH is an indirect indicator of anti-androgenic effects because its production is stimulated in the presence of low testosterone levels. MMP (a metabolite of DMP) and MBP (a metabolite of DBP and BBzP), measured in mother's breast milk, were both significantly positively correlated with LH:free testosterone ratio in male offspring. This finding may indicate that impaired testosterone production has been compensated by increased levels of LH (Scott et al. 2009). Of these associations, DBP and DINP, but not DMP, are supported by anti-androgenic activity in animal models. The authors found no statistically significant difference between lactational phthalate exposure and the presence or absence of cryptorchidism (undescended testes). Another study found higher risk of pubertal gynecomastia (breast tissue development) in boys aged 11–15 years in Turkey, who had higher plasma concentrations of DEHP and MEHP (Durmaz et al. 2010). Additional studies with adult men have shown associations with phthalate exposures and reduction in serum free testosterone and increased DNA damage in sperm (Duty et al. 2005b; Hauser et al. 2006; Pan et al. 2006).

Although few epidemiological studies have focused on female reproductive endpoints, one study that examined MEHP levels in women near the time of conception found that increased urinary levels of MEHP were associated with an increased risk of miscarriage, compared to women with lower MEHP levels, consistent with a study in female rats (Toft et al. 2012; Gray et al. 2006a).

Gender-specific behavioral endpoints also have been assessed in relation to phthalate exposure. Sexually dimorphic play behavior was observed among children as measured by a validated questionnaire; elevations in mid-pregnancy urinary concentrations of MnBP, MiBP, MEOHP, and MEHP in the mothers were significantly positively correlated with decreased masculine play in their male children (Swan et al. 2010). All of these phthalates show anti-androgenic activity in animals, and this study supports the idea that androgen production may also alter sexual differentiation in the brain during fetal development. Confounding is an important consideration and limitation of epidemiological studies, and it is discussed in Sect. 2.4.8.

2.4.5 Epidemiological Associations with Evidence of Neurotoxicity

Neurological effects associated with phthalate exposure also have been reported in human studies. Although a mechanism of action has not been demonstrated, it is hypothesized that disruption of maternal thyroxine levels may contribute to the downstream clinical effects of thyroid dysfunction (Meeker et al. 2007).

Measurements of maternal urinary phthalate measurements in the third trimester, specifically low molecular weight phthalates, including MBP, MEP, MiBP, and MMP, were associated with symptoms of attention deficit hyperactive disorder in follow up of their children at ages 4 through 9. Statistically significant poorer results in emotional control and Global Executive Composite scale, a measure of executive function or regulation of cognitive processes, were seen among boys with increasing third trimester measurements of MEP, MBP, and MMP (Engel et al. 2010).

2.4.6 Epidemiologic Associations with Metabolic Effects

Metabolic health effects as a result of phthalate exposure are thought to occur via the PPAR pathway and through endocrine disruption of the thyroid hormones. The sodium/iodide symporter, necessary for normal thyroid function, can be down-regulated by some phthalates in animal models (Breous et al. 2005). The PPARs are nuclear receptor proteins that are involved in metabolism, cellular differentiation, and development. Two of these receptors, PPARα and PPARγ, have been shown to be activated by MEHP, MBzP, and MBuP. PPARγ is primarily found in adipose tissue and controls adipogenesis, or fat cell creation (Hurst and Waxman 2003). Because of the phthalates' interaction with fat cells and the endocrine system, there is concern that these chemicals may have health effects related to metabolism, such as obesity and diabetes. In a study based on NHANES data, children in the 2003–2008 surveys that had higher levels of urinary DEHP metabolites had elevated systolic blood pressure. This cross-sectional analysis adjusted for diet, BMI, and demographic characteristics (Trasande et al. 2013). Decreased levels of thyroid hormones in adult males and females has been observed in association with MEHP and MBP exposure (Meeker and Ferguson 2012). Decreased insulin resistance and increased waist circumference in adult males has been observed in association with BBzP, DEHP, and DEP exposures estimated from urinary excretion (Meeker and Ferguson 2012).

2.4.7 Epidemiologic Associations with Immune System and Respiratory Effects

Phthalate exposure has been associated with respiratory symptoms in both male and female children and adults. Individual studies have used a variety of endpoints and identified links with several different phthalates. BBzP in house dust was associated with childhood eczema, a persistent skin rash that may be related to the development of an allergic phenotype, in a cross-sectional Swedish case control study (Bornehag et al. 2004). Similarly, prenatal concentrations of the urinary metabolite of BBzP were positively associated with a report of early eczema in an urban cohort of children from New York City (Just et al. 2012b). Although lacking direct exposure measures, PVC materials in the home such as plastic wall materials or PVC flooring (potential sources of BBzP and/or DEHP) were associated with wheezing and asthma symptoms in several cross-sectional and prospective studies of children (Jaakkola et al. 1999, 2000; Larsson et al. 2010). Increased urinary metabolites of DEP and DBP were associated with decreased lung function but only among males in a cross-sectional sample of 240 U.S. adults (Hoppin et al. 2004). Metabolites of DEP and BBzP, both believed to have substantial contributions from inhalation, were associated with higher airway inflammation as measured by fractional exhaled nitric oxide among children 5–9 years old. There were stronger associations for BBzP among children with report of recent wheeze who may be more susceptible to inflammatory effects of pollutants (Just et al. 2012a). In a study of Norwegian children at age 10, the urinary metabolites mono(carboxyoctyl) phthalate and mono(carboxynonyl) phthalate, but not the metabolites of other common phthalates, were associated with current asthma in a cross-sectional design (Bertelsen et al. 2013). Given the etiologic period of allergy and asthma that develops over a period of months or years, more longitudinal studies are needed to assess whether exposure to different types of phthalates could contribute to incident allergy or asthma.

2.4.8 Limitations of Epidemiological Studies

Confounding is important to consider in epidemiological studies. Exposure studies have demonstrated that people with higher levels of DEP use many more fragranced and personal care products, and these products contain many different chemicals, including endocrine disruptors such as parabens and some fragrances (Duty et al. 2005a; Dodson et al. 2012). Thus, MEP in urine may be a proxy for a complex set of exposures that could be important for the health endpoints in studies. Similarly, DEHP in urine was reduced in people eating a fresh food diet with limited packaged food. DEHP metabolite levels may be a proxy for people who eat a less healthy diet of more processed and packaged foods. Therefore, health outcomes associated with the phthalate levels may be due to nutritional factors, or other chemicals that may migrate from packaging to foods or beverages (Rudel et al. 2011). Additionally, cross-sectional analyses, such as those using NHANES, must be interpreted with caution when used to draw associations between short-lived exposure and complex chronic diseases (LaKind et al. 2012).

2.5 Risk Assessment and Cumulative Effects

Phthalates act in a dose-additive manner; exposure to multiple anti-androgenic phthalates will result in additive risk to reproductive harm, so risks must be evaluated considering cumulative exposure (Hotchkiss et al. 2010; Koch et al. 2011). Phthalates also act additively with other chemicals that disrupt fetal testosterone synthesis by different mechanisms (Howdeshell et al. 2008). These findings prompted the National Research Council and others to report that phthalates, along with other compounds with similar actions, should be considered for their cumulative effects on the male reproductive system (National Research Council 2008; Koch et al. 2011).

A recent study in Germany with 111 school children aged 5–6 conducted a cumulative risk assessment, using an approach that combined the ratios of daily exposure and TDI for each phthalate measured in the study. Roughly a quarter of the children exceeded the cumulative TDI for DEHP, DnBP and DiBP, the three phthalates with the most evidence of male reproductive toxicity (Koch et al. 2011). This study's findings indicate that although these phthalates are regulated in the EU in children's toys, childcare goods, and cosmetics, exposures are currently above acceptable levels for male reproductive health.

2.6 Regulations and Policies

2.6.1 United States

In the U.S., the Food and Drug Administration (FDA) has issued guidance for restricting the use of DBP and DEHP in prescription and nonprescription products (Food and Drug Administration 2012). This does not address phthalate use in drug delivery systems, packaging, or medical equipment. The FDA states that there is no health risk posed by phthalates used in cosmetics, and labeling of phthalate ingredients is not required if they are used in fragrances or in professional salon products (Food and Drug Administration 2013). Three phthalates, DEHP, DnBP, and BBzP are banned in the U.S. in children's toys and some child care articles intended to facilitate feeding, sucking, or teething under the *Consumer Product Safety Improvement Act* (United States Congress 2008). Three additional phthalates, DINP, DIDP, DnOP are temporarily banned from children's toys that can be placed in a child's mouth, or children's toys smaller than 5 centimeters (cm). The ban does not apply to inaccessible parts of a toy (United States Congress 2008). The Consumer Product Safety Commission's (CPSC) Chronic Hazard Advisory Panel is conducting a hazard assessment on these phthalates to determine if their ban will be lifted or remain in place (United States Environmental Protection Agency 2012b).

The USEPA developed an Action Plan for phthalates based upon their toxicity, widespread use, and human exposure. Eight phthalates are included in the Action Plan: DBP, DIBP, BBzP, di-*n*-pentyl phthalate (DnPP), DEHP, DnOP, DINP, and DIDP. The USEPA will coordinate with the CPSC and FDA on regulatory action to address the manufacturing, use, sale, and distribution of these compounds in the U.S. (United States Environmental Protection Agency 2012b). To date, a Significant New Use Rule has been proposed for DnPP, which requires manufacturers or processors of the chemical to obtain USEPA approval (United States Environmental Protection Agency 2013). Levels of DEHP in drinking water are regulated by the *Clean Drinking Water Act*, with a maximum contaminant level (MCL) of 0.006 mg/L for DEHP. DEHP and DBP are listed as hazardous pollutants under the *Clean Air Act* (United States Environmental Protection Agency 2012b).

California's Safe Drinking Water and Toxic Enforcement Act of 1986, more commonly known as Proposition 65, requires products that contain chemicals identified as carcinogens or toxic to development or reproduction by the state of California's Office of Environmental Health Hazard Assessment to be labeled as such in the state. DEHP is labeled as a carcinogen and male developmental toxicant. DINP is recognized as a carcinogen, and DBP, DIDP, BBzP, and DnHP are developmental toxicants under this act (State of California Environmental Protection Agency Office of Environmental Health Hazard Assessment 1986). This rule can provide an incentive for companies to reformulate products with safer alternatives.

Several phthalates are listed among the risks to public health associated with PVC materials, and the American Public Health Association, which represents a broad array of public health professionals, urges federal and local governments to replace PVC when possible in medical care settings, schools, public housing, and building materials (American Public Health Association 2011).

Screening phthalates for anti-androgenic activity has been proposed by the USEPA's *Endocrine Disruptor Screening Program* (EDSP). DnBP, BBzP, and DEHP are included in the first group of 67 chemicals to be screened as part of the EDSP. Because phthalates' anti-androgenic activity is not a result of direct action of the chemical on the androgen receptor, receptor-based screening assays will not detect the anti-androgenic activity of phthalates or evidence of other indirect endocrine disruption (Ankley and Gray 2013).

2.6.2 European Union

Food, food packaging, and pharmaceuticals have been shown to increase human exposure to phthalates. To address this, the EFSA restricted the use of DBP, DIDP, BBzP, DEHP, and DINP in FCMs (Petersen and Jensen 2010). In the EU, DEHP is allowed in food production facilities, such as in conveyor belts, provided it does next exceed the substance migration limit of 1.5 mg/kg of food, although it is prohibited from single-use lips or caps (European Union 2007). DEHP, DBP, and BBzP are banned from children's toys and child care articles intended to be placed in the mouth by children under age 3, and DINP, DIDP, and DnOP are banned from toys and childcare articles that can be placed in the mouth by children (European Parliament Council 2005). DEHP, DBP, and BBzP are prohibited from use in cosmetics (European Union 2004). Regulations in the EU require human health and environmental testing and data sharing for chemicals within its legislative chemical framework under *Registration, Evaluation, Authorization and Restriction of Chemicals* (REACH). Substances of Very High Concern are evaluated for restriction, or authorization for certain uses only. DnPP, bis(2-methoxyethyl) phthalate,

DiBP, BBzP, DEHP, and DnBP are on the candidate list for REACH authorization as toxic for reproduction (European Chemicals Agency 2013).

The European Medicines Agency recently released a guidance document recommending a reduction in the content of DEP and DBP in medicines in order to protect the safety of all patient populations, in recognition that phthalate exposure through medicines may contribute to the overall burden of phthalate exposure (European Medicines Agency 2013).

2.7 Alternatives

A variety of chemicals are commonly used as alternatives to phthalate plasticizers. Di(isononyl) cyclohexane 1,2-dicarboxylate (DINCH), di(ethylhexyl) adipate (DEHA) and *O*-acetyl tributyl citrate (ATBC) have been reported as substitutes that are commonly used in DEHP-free products in medical settings, in children's products, and in plastic cling wrap for food storage (Health and Consumer Protection 2007; United States Environmental Protection Agency 2012b; Lowell Center for Sustainable Production 2011). Some alternatives have been evaluated for health effects, and there is evidence of both reproductive and non-reproductive health endpoints associated with these chemicals. In a 2-year study, DINCH administration was associated with increased thyroid weight in both male and female rats, and a 90-day repeated dose study in rats resulted in increased liver and testes weight (Lowell Center for Sustainable Production 2011; Health and Consumer Protection 2007). Studies have shown that DEHA leaches from its plastic polymer to a greater degree than DEHP. There is limited evidence of DEHA's carcinogenicity from two rodent feeding studies. There was no evidence of carcinogenicity in rats, while liver carcinomas and adenomas were observed in DEHA-treated male and female mice, leading to its Group 3 classification by IARC (*Group 3: not classifiable as to its carcinogenicity in humans*) (Van Vliet et al. 2011; World Health Organization 2000). ATBC is a citric acid-derived plasticizer with high potential to leach from plastic and health effects noted in T cells, increased liver weight, and reduced body weight rat pups following in utero exposure (Health and Consumer Protection 2007). Biomonitoring data on these alternatives are limited, although German urine samples show an increased in detectable levels of DINCH metabolites from 7 % in 2006 to 98 % of urine samples in 2012 (Schutze et al. 2014).

2.8 Summary

While phthalates are a diverse group of compounds that are used in many applications, they have toxicological differences, and they cannot all be considered as one group. Humans are primarily exposed to DEHP through their diet, although consumer products and building materials in the indoor environment are also significant contributors to exposure to DEP, DBP, BBzP, DEHP, and DINP. Since phthalates are so commonly used, most people experience constant exposure, although levels vary throughout the day due to the compounds' short half-lives. Because of this rapid clearance from the body, studies have indicated it is possible to reduce exposure levels quickly by eating fresh whole foods and avoiding foods or beverages stored in plastic or cans. The best understood toxic effect of some phthalates is on the developing male reproductive system. Their toxicological action on this system is known to be additive, and thus phthalates and other compounds that exert similar effects should be considered cumulatively in a risk assessment. Other toxic endpoints that have been studied include fertility, cancer, epigenetic changes, neurotoxicity, metabolic changes, and immune and respiratory effects. While some phthalates have been restricted in FCMs and children's products due to health concerns, suitable alternatives have not undergone rigorous evaluation for health effects.

2.9 Data Gaps

Risk assessments indicate that current exposure levels for hormonally active phthalates are above a level of concern even for the general population, and there are subpopulations with much higher exposure, for example from medical uses (Koch et al. 2011). Therefore, the most urgent data needs are to identify ways to reduce exposure to these active phthalates and to evaluate potential health effects of substitutes prior to putting them into use. Specific data gaps include:

- Knowledge of major exposure pathways for hormonally active phthalates and identification of opportunities for intervention to reduce exposures. Currently there is limited disclosure of product ingredients and this makes it difficult to identify important pathways or intervene. It is also important to identify phthalate uses that lead to highly exposed subpopulations.
- Sensitive toxicological screening tests that recognize the specific mechanisms of action of phthalates for use in high-throughput chemical testing programs. The initial priority is to be able to screen for the developmental inhibition of testosterone synthesis, but tests are also needed for other biological pathways that may be important for phthalates or their substitutes.
- More comprehensive health effect studies of chemicals used in consumer products and food processing and packaging, especially chemicals being introduced as substitutes for phthalates. Systematic assessments of potential health effects of this family of compounds and better exposure models are needed as a basis for reformulation and regulation.

References

Adibi JJ, Whyatt RM, Williams PL et al (2008) Characterization of phthalate exposure among pregnant women assessed by repeat air and urine samples. Environ Health Perspect 116(4):467–473

Agency for Toxic Substances and Disease Registry (1995) Toxicological profile for diethyl phthalate. U.S. Department of Health and Human Services, Atlanta

Agency for Toxic Substances and Disease Registry (2002) Toxicological profile for di(2-ethylhexyl) phthalate. U.S. Department of Health and Human Services, Atlanta

American Public Health Association (2011) Reducing PVC in facilities with vulnerable populations. http://www.apha.org/advocacy/policy/policysearch/default.htm?id=1419. Accessed 15 Nov 2013

Ankley GT, Gray LE (2013) Cross-species conservation of endocrine pathways: a critical analysis of tier 1 fish and rat screening assays with 12 model chemicals. Environ Toxicol Chem 32(5):1084–1087

Aylward LL, Hays SM, Gagne M et al (2009) Derivation of biomonitoring equivalents for di(2-ethylhexyl)phthalate (CAS No. 117-81-7). Regul Toxicol Pharmacol 55(3):249–258

Aylward LL, Lorber M, Hays SM (2011) Urinary DEHP metabolites and fasting time in NHANES. J Expo Sci Environ Epidemiol 21(6):615–624

Bekö G, Weschler CJ, Langer S et al (2013) Children's phthalate intakes and resultant cumulative exposures estimated from urine compared with estimates from dust ingestion, inhalation and dermal absorption in their homes and daycare centers. PLoS ONE 8(4):e62442

Berman T, Hochner-Celnikier D, Calafat AM et al (2009) Phthalate exposure among pregnant women in Jerusalem, Israel: results of a pilot study. Environ Int 35(2):353–357

Bertelsen RJ, Carlsen KC, Calafat AM et al (2013) Urinary biomarkers for phthalates associated with asthma in Norwegian children. Environ Health Perspect 121(2):251–256

Boberg J (2007) Endocrine disrupters affecting male rat reproductive development—focus on phthalates and the fetal testis. Technical University of Denmark. http://orbit.dtu.dk/fedora/objects/orbit:79926/datastreams/file_3195804/content. Accessed 15 Nov 2013

Bornehag CG, Sundell J, Weschler CJ et al (2004) The association between asthma and allergic symptoms in children and phthalates in house dust: a nested case-control study. Environ Health Perspect 112(14):1393–1397

Braun JM, Sathyanarayana S, Hauser R (2013) Phthalate exposure and children's health. Curr Opin Pediatr 25(2):247–254

Breous E, Wenzel A, Loos U (2005) The promoter of the human sodium/iodide symporter responds to certain phthalate plasticisers. Mol Cell Endocrinol 244(1–2):75–78

Calafat AM, Ye X, Silva MJ et al (2006) Human exposure assessment to environmental chemicals using biomonitoring. Int J Androl 29(1):166–171 (discussion 181-165)

Calafat AM, Koch HM, Swan SH et al (2013) Misuse of blood serum to assess exposure to bisphenol A and phthalates. Breast Cancer Res 15(5):403

Carlstedt F, Jonsson BA, Bornehag CG (2013) PVC flooring is related to human uptake of phthalates in infants. Indoor Air 23(1):32–39

Castle L, Gilbert J, Eklund T (1990) Migration of plasticizer from poly(vinyl chloride) milk tubing. Food Addit Contam 7(5):591–596

Centers for Disease Control and Prevention (2009) Fourth national report on human exposure to environmental chemicals. U.S. Department of Health and Human Services. http://www.cdc.gov/exposurereport/pdf/FourthReport.pdf. Accessed 15 Nov 2013

Cirillo T, Fasano E, Castaldi E et al (2011) Children's exposure to di(2-ethylhexyl)phthalate and dibutylphthalate plasticizers from school meals. J Agric Food Chem 59(19):10532–10538

Colacino JA, Harris TR, Schecter A (2010) Dietary intake is associated with phthalate body burden in a nationally representative sample. Environ Health Perspect 118(7):998–1003

David RM, Moore MR, Finney DC et al (2000) Chronic toxicity of di(2-ethylhexyl)phthalate in mice. Toxicol Sci 58(2):377–385

Dodson RE, Nishioka M, Standley LJ et al (2012) Endocrine disruptors and asthma-associated chemicals in consumer products. Environ Health Perspect 120(7):935–943
Durmaz E, Ozmert EN, Erkekoglu P et al (2010) Plasma phthalate levels in pubertal gynecomastia. Pediatrics 125(1):e122–e129
Duty SM, Ackerman RM, Calafat AM et al (2005a) Personal care product use predicts urinary concentrations of some phthalate monoesters. Environ Health Perspect 113(11):1530–1535
Duty SM, Calafat AM, Silva MJ et al (2005b) Phthalate exposure and reproductive hormones in adult men. Hum Reprod 20(3):604–610
Engel SM, Miodovnik A, Canfield RL et al (2010) Prenatal phthalate exposure is associated with childhood behavior and executive functioning. Environ Health Perspect 118(4):565–571
Enneking PA (2006) Phthalates not in plastic food packaging. Environ Health Perspect 114(2):A89–A90
European Chemicals Agency (2012) Opinion on an Annex XV dossier proposing restrictions on four phthalates. http://echa.europa.eu/documents/10162/77cf7d29-ba63-4901-aded-59cf75536e06. Accessed 17 Dec 2013
European Chemicals Agency (2013) Candidate list of substances of very high concern for authorisation. http://echa.europa.eu/candidate-list-table. Accessed 15 Nov 2013
European Medicines Agency (2013) Guideline on the use of phthalates as excipients in human medicinal products http://www.ema.europa.eu/docs/en_GB/document_library/Scientific_guideline/2013/05/WC500143140.pdf. Accessed 6 Dec 2013
European Parliament Council (2005) Directive 2005/84/EC of the European Parliament and of the Council of 14 December 2005 amending for the 22nd time Council Directive 76/769/EEC on the approximation of the laws, regulations and administrative provisions of the member states relating to restrictions on the marketing and use of certain dangerous substances and preparations (phthalates in toys and childcare articles). http://eur-lex.europa.eu/LexUriServ/LexUriServ.do?uri=OJ:L:2005:344:0040:0043:en:PDF. Accessed 6 Dec 2013
European Union (2004) Commission Directive 2004/93/EC. Official J Eur Union. http://www.dehp-facts.com/upload/documents/webpage/Cosmetics.pdf. Accessed 12 Dec 2013
European Union (2007) Commission Directive 2007/19/EC. http://www.dehp-facts.com/upload/documents/webpage/foodcontact%20leg.pdf. Accessed 12 Dec 2013
Food and Drug Administration (2012) Guidance for industry limiting the use of certain phthalates as excipients in CDER-regulated products. U.S. Department of Health and Human Services. http://www.fda.gov/downloads/Drugs/GuidanceComplianceRegulatoryInformation/Guidances/UCM294086.pdf. Accessed 15 Nov 2013
Food and Drug Administration (2013) Phthalates and cosmetic products. U.S. Department of Health and Human Services. http://www.fda.gov/cosmetics/productandingredientsafety/selectedcosmeticingredients/ucm128250.htm. Accessed 6 Dec 2013
Fromme H, Gruber L, Schlummer M et al (2007) Intake of phthalates and di(2-ethylhexyl)adipate: results of the integrated exposure assessment survey based on duplicate diet samples and biomonitoring data. Environ Int 33(8):1012–1020
Ge R, Tao L, Kramer PM et al (2002) Effect of peroxisome proliferators on the methylation and protein level of the c-myc protooncogene in B6C3F1 mice liver. J Biochem Mol Toxicol 16(1):41–47
Gray LE Jr, Laskey J, Ostby J (2006a) Chronic di-*n*-butyl phthalate exposure in rats reduces fertility and alters ovarian function during pregnancy in female Long Evans hooded rats. Toxicol Sci 93(1):189–195
Gray LE Jr, Wilson VS, Stoker T et al (2006b) Adverse effects of environmental antiandrogens and androgens on reproductive development in mammals. Int J Androl 29 (1):96–104 (discussion 105-108)
Gray LE Jr, Ostby J, Furr J et al (2000) Perinatal exposure to the phthalates DEHP, BBP, and DINP, but not DEP, DMP, or DOTP, alters sexual differentiation of the male rat. Toxicol Sci 58(2):350–365
Hannas BR, Lambright CS, Furr J et al (2011) Dose-response assessment of fetal testosterone production and gene expression levels in rat testes following in utero exposure to diethylhexyl

phthalate, diisobutyl phthalate, diisoheptyl phthalate, and diisononyl phthalate. Toxicol Sci 123(1):206–216
Hannas BR, Lambright CS, Furr J et al (2012) Genomic biomarkers of phthalate-induced male reproductive developmental toxicity: a targeted RT-PCR array approach for defining relative potency. Toxicol Sci 125(2):544–557
Hauser R, Meeker JD, Park S et al (2004) Temporal variability of urinary phthalate metabolite levels in men of reproductive age. Environ Health Perspect 112(17):1734–1740
Hauser R, Meeker JD, Duty S et al (2006) Altered semen quality in relation to urinary concentrations of phthalate monoester and oxidative metabolites. Epidemiology 17(6):682–691
Health and Consumer Protection (2007) Preliminary report on the safety of medical devices containing DEHP plasticized PVC or other plasticizers on neonates and other groups possibly at risk. European Commission. http://ec.europa.eu/health/ph_risk/committees/04_scenihr/docs/scenihr_o_008.pdf. Accessed 15 Nov 2013
Helm D (2007) Correlation between production amounts of DEHP and daily intake. Sci Total Environ 388(1–3):389–391
Hoppin JA, Ulmer R, London SJ (2004) Phthalate exposure and pulmonary function. Environ Health Perspect 112(5):571–574
Hotchkiss A, Ankley GT, Wilson VS et al (2008) Of mice and men (and mosquitofish): antiandrogens and androgens in the environment. Bioscience 58(11):1037–1050
Hotchkiss AK, Rider CV, Furr J et al (2010) In utero exposure to an AR antagonist plus an inhibitor of fetal testosterone synthesis induces cumulative effects on F1 male rats. Reprod Toxicol 30(2):261–270
Howdeshell KL, Wilson VS, Furr J et al (2008) A mixture of five phthalate esters inhibits fetal testicular testosterone production in the Sprague-Dawley rat in a cumulative, dose-additive manner. Toxicol Sci 105(1):153–165
Hubinger JC (2010) A survey of phthalate esters in consumer cosmetic products. J Cosmet Sci 61(6):457–465
Hubinger JC, Havery DC (2006) Analysis of consumer cosmetic products for phthalate esters. J Cosmet Sci 57(2):127–137
Hurst CH, Waxman DJ (2003) Activation of PPARalpha and PPARgamma by environmental phthalate monoesters. Toxicol Sci 74(2):297–308
Integrated Risk Information System (1997) Di(2-ethylhexyl)phthalate (DEHP) summary. United States Environmental Protection Agency. http://www.epa.gov/iris/subst/0014.htm. Accessed 18 Nov 2013
International Agency for Research on Cancer (2012) IARC monographs on the evaluation of carcinogenic risk to humans: di(2-ethylhexyl)phthalate, vol 101. International Agency for Research on Cancer, Lyon, France. http://monographs.iarc.fr/ENG/Monographs/vol101/mono101-006.pdf. Accessed 12 Dec 2013
Ito Y, Yamanoshita O, Kurata Y et al (2007) Induction of peroxisome proliferator-activated receptor alpha (PPARalpha)-related enzymes by di(2-ethylhexyl) phthalate (DEHP) treatment in mice and rats, but not marmosets. Arch Toxicol 81(3):219–226
Jaakkola JJ, Oie L, Nafstad P et al (1999) Interior surface materials in the home and the development of bronchial obstruction in young children in Oslo. Norway. Am J Public Health 89(2):188–192
Jaakkola JJ, Verkasalo PK, Jaakkola N (2000) Plastic wall materials in the home and respiratory health in young children. Am J Public Health 90(5):797–799
Janjua NR, Frederiksen H, Skakkebaek NE et al (2008) Urinary excretion of phthalates and paraben after repeated whole-body topical application in humans. Int J Androl 31(2):118–130
Just AC (2012) Exposure to phthalate mixtures and inner-city pediatric allergic disease and airway inflammation. Ph.D. dissertation, Columbia University http://academiccommons.columbia.edu/catalog/ac%3A156937. Accessed 15 Nov 2013

Just AC, Adibi JJ, Rundle AG et al (2010) Urinary and air phthalate concentrations and self-reported use of personal care products among minority pregnant women in New York City. J Expo Sci Environ Epidemiol 20(7):625–633

Just AC, Whyatt RM, Miller RL et al (2012a) Children's urinary phthalate metabolites and fractional exhaled nitric oxide in an urban cohort. Am J Respir Crit Care Med 86(9):830–837

Just AC, Whyatt RM, Perzanowski MS et al (2012b) Prenatal exposure to butylbenzyl phthalate and early eczema in an urban cohort. Environ Health Perspect 120(10):1475–1480

Kang SC, Lee BM (2005) DNA methylation of estrogen receptor alpha gene by phthalates. J Toxicol Environ Health A 68(23–24):1995–2003

Kobrosly RW, Parlett LE, Stahlhut RW et al (2012) Socioeconomic factors and phthalate metabolite concentrations among United States women of reproductive age. Environ Res 115:11–17

Koch HM, Wittassek M, Bruning T et al (2011) Exposure to phthalates in 5–6 years old primary school starters in Germany—a human biomonitoring study and a cumulative risk assessment. Int J Hyg Environ Health 214(3):188–195

Koch HM, Lorber M, Christensen KL et al (2013) Identifying sources of phthalate exposure with human biomonitoring: results of a 48 h fasting study with urine collection and personal activity patterns. Int J Hyg Environ Health 216(6):672–681

Koniecki D, Wang R, Moody RP et al (2011) Phthalates in cosmetic and personal care products: concentrations and possible dermal exposure. Environ Res 111(3):329–336

LaKind JS, Goodman M, Naiman DQ (2012) Use of NHANES data to link chemical exposures to chronic diseases: a cautionary tale. PLoS One 7(12):e51086

Larsson M, Hagerhed-Engman L, Kolarik B et al (2010) PVC—as flooring material—and its association with incident asthma in a Swedish child cohort study. Indoor Air 20(6):494–501

Lewis RC, Meeker JD, Peterson KE et al (2013) Predictors of urinary bisphenol A and phthalate metabolite concentrations in Mexican children. Chemosphere 93(10):2390–2398

Lorber M, Calafat AM (2012) Dose reconstruction of di(2-ethylhexyl) phthalate using a simple pharmacokinetic model. Environ Health Perspect 120(12):1705–1710

Lovekamp-Swan T, Davis BJ (2003) Mechanisms of phthalate ester toxicity in the female reproductive system. Environ Health Perspect 111(2):139–145

Lowell Center for Sustainable Production (2011) Phthalates and their alternatives. University of Massachusetts Lowell. http://www.sustainableproduction.org/downloads/PhthalateAlternatives-January2011.pdf. Accessed 15 Nov 2013

Mahood IK, Scott HM, Brown R (2007) In utero exposure to di(n-butyl) phthalate and testicular dysgenesis: comparison of fetal and adult end points and their dose sensitivity. Environ Health Perspect 115(Suppl 1):55–61

Main KM, Mortensen GK, Kaleva MM et al (2005) Human breast milk contamination with phthalates and alterations of endogenous reproductive hormones in 3 month old infants. Environ Health Perspect 114(2):270–276

Martina CA, Weiss B, Swan SH (2012) Lifestyle behaviors associated with exposures to endocrine disruptors. Neurotoxicology 33(6):1427–1433

Meeker JD, Ferguson KK (2012) Dioxins and health: including other persistent organic pollutants and endocrine disruptors. In: Phthalates: human exposure and related health effects. Hoboken, Wiley (Chapter 13)

Meeker JD, Calafat AM, Hauser R (2007) Di(2-ethylhexyl) phthalate metabolites may alter thyroid hormone levels in men. Environ Health Perspect 115(7):1029–1034

Melnick RL (2001) Is peroxisome proliferation an obligatory precursor step in the carcinogenicity of di(2-ethylhexyl)phthalate (DEHP)? Environ Health Perspect 109(5):437–442

Montuori P, Jover E, Morgantini M et al (2008) Assessing human exposure to phthalic acid and phthalate esters from mineral water stored in polyethylene terephthalate and glass bottles. Food Addit Contam Part A Chem Anal Control Expo Risk Assess 25(4):511–518

Moyer B, Hixon ML (2012) Reproductive effects in F1 adult females exposed in utero to moderate to high doses of mono-2-ethylhexylphthalate (MEHP). Reprod Toxicol 34(1):43–50
National Research Council (2008) Phthalates and cumulative risk assessment the task ahead. National Academy of Sciences. http://cfpub.epa.gov/ncea/cfm/recordisplay.cfm?deid=202508# Download. Accessed 15 Nov 2013
National Toxicology Program (2011) Twelfth report on carcinogens. U.S. Department of Health and Human Services. http://ntp.niehs.nih.gov/ntp/roc/twelfth/roc12.pdf. Accessed 15 Nov 2013
Noriega NC, Howdeshell KL, Furr J et al (2009) Pubertal administration of DEHP delays puberty, suppresses testosterone production, and inhibits reproductive tract development in male Sprague-Dawley and Long-Evans rats. Toxicol Sci 111(1):163–178
Office of Environmental Health Hazard Assessment California Environmental Protection Agency (2013) Evidence on the carcinogenicity of diisononyl phthalate (DINP). http://oehha.ca.gov/prop65/hazard_ident/pdf_zip/DINP_HID100413.pdf. Accessed 31 Dec 2013
Pan G, Hanaoka T, Yoshimura M et al (2006) Decreased serum free testosterone in workers exposed to high levels of di-n-butyl phthalate (DBP) and di-2-ethylhexyl phthalate (DEHP): a cross-sectional study in China. Environ Health Perspect 114(11):1643–1648
Parlett LE, Calafat AM, Swan SH (2013) Women's exposure to phthalates in relation to use of personal care products. J Expo Sci Environ Epidemiol 23(2):197–206
Petersen JH, Jensen LK (2010) Phthalates and food-contact materials: enforcing the 2008 European Union plastics legislation. Food Addit Contam Part A Chem Anal Control Expo Risk Assess 27(11):1608–1616
Romero-Franco M, Hernandez-Ramirez RU, Calafat AM et al (2011) Personal care product use and urinary levels of phthalate metabolites in Mexican women. Environ Int 37(5):867–871
Rudel RA, Camann DE, Spengler JD et al (2003) Phthalates, alkylphenols, pesticides, polybrominated diphenyl ethers, and other endocrine-disrupting compounds in indoor air and dust. Environ Sci Technol 37(20):4543–4553
Rudel RA, Dodson RE, Perovich LJ et al (2010) Semivolatile endocrine-disrupting compounds in paired indoor and outdoor air in two northern California communities. Environ Sci Technol 44(17):6583–6590
Rudel RA, Gray JM, Engel CL et al (2011) Food packaging and bisphenol A and bis(2-ethyhexyl) phthalate exposure: findings from a dietary intervention. Environ Health Perspect 119(7):914–920
Rusyn I, Corton JC (2012) Mechanistic considerations for human relevance of cancer hazard of di(2-ethylhexyl) phthalate. Mutat Res 750(2):141–158
Sathyanarayana S, Karr CJ, Lozano P et al (2008) Baby care products: possible sources of infant phthalate exposure. Pediatrics 121(2):e260–e268
Sathyanarayana S, Alcedo G, Saelens BE et al (2013) Unexpected results in a randomized dietary trial to reduce phthalate and bisphenol A exposures. J Expo Sci Environ Epidemiol 23(4):378–384
Schecter A, Lorber M, Guo Y et al (2013) Phthalate concentrations and dietary exposure from food purchased in New York State. Environ Health Perspect 121(4):473–479
Schettler T (2006) Human exposure to phthalates via consumer products. Int J Androl 29(1):134–139
Schutze A, Kolossa-Gehring M, Apel P et al (2014) Entering markets and bodies: increasing levels of the novel plasticizer Hexamoll DINCH in 24 h urine samples from the German Environmental Specimen Bank. Int J Hyg Environ Health 217(2–3):421–428
Scott HM, Mason JI, Sharpe RM (2009) Steroidogenesis in the fetal testis and its susceptibility to disruption by exogenous compounds. Endocr Rev 30(7):883–925
Silva MJ, Barr DB, Reidy JA et al (2004) Urinary levels of seven phthalate metabolites in the U.S. population from the National Health and Nutrition Examination Survey (NHANES) 1999–2000. Environ Health Perspect 112(3):331–338

Silva MJ, Samandar E, Preau JL Jr et al (2007) Quantification of 22 phthalate metabolites in human urine. J Chromatogr B Analyt Technol Biomed Life Sci 860(1):106–112

State of California Environmental Protection Agency Office of Environmental Health Hazard Assessment (1986) Safe Drinking Water and Toxic Enforcement Act of 1986. California, United States

Stringer R, Labunska I, Santillo D et al (2000) Concentrations of phthalate esters and identification of other additives in PVC children's toys. Environ Sci Pollut Res Int 7(1):27–36

Swan SH, Main KM, Liu F et al (2005) Decrease in anogenital distance among male infants with prenatal phthalate exposure. Environ Health Perspect 113(8):1056–1061

Swan SH, Liu F, Hines M et al (2010) Prenatal phthalate exposure and reduced masculine play in boys. Int J Androl 33(2):259–269

Toft G, Jonsson BA, Lindh CH et al (2012) Association between pregnancy loss and urinary phthalate levels around the time of conception. Environ Health Perspect 120(3):458–463

Tomonari Y, Kurata Y, David RM et al (2006) Effect of di(2-ethylhexyl) phthalate (DEHP) on genital organs from juvenile common marmosets: I. Morphological and biochemical investigation in 65-week toxicity study. J Toxicol Environ Health A 69(17):1651–1672

Trasande L, Sathyanarayana S, Spanier AJ et al (2013) Urinary phthalates are associated with higher blood pressure in childhood. J Pediatr 163(3):747–753 (e741)

Tsumura Y, Ishimitsu S, Kaihara A et al (2001) Di(2-ethylhexyl) phthalate contamination of retail packed lunches caused by PVC gloves used in the preparation of foods. Food Addit Contam 18(6):569–579

United States Congress (2008) Consumer product safety improvement act of 2008 http://www.cpsc.gov//PageFiles/113865/cpsia.pdf. Accessed 15 Nov 2013

United States Environmental Protection Agency (2012a) Chemical data access tool. http://java.epa.gov/oppt_chemical_search/. Accessed 15 Nov 2013

United States Environmental Protection Agency (2012b) Phthalates action plan. http://www.epa.gov/oppt/existingchemicals/pubs/actionplans/phthalates_ap_2009_1230_final.pdf. Accessed 15 Nov 2013

United States Environmental Protection Agency (2013) Benzidine-based chemical substances; di-*n*-pentyl phthalate (DnPP); and alkanes, C12–13, chloro; proposed Significant New Use Rules. Federal Register. http://www.gpo.gov/fdsys/pkg/FR-2012-03-28/pdf/2012-7208.pdf. Accessed 15 Nov 2013

Van Vliet ED, Reitano EM, Chhabra JS et al (2011) A review of alternatives to di (2-ethylhexyl) phthalate-containing medical devices in the neonatal intensive care unit. J Perinatol 31(8):551–560

Voss C, Zerban H, Bannasch P et al (2005) Lifelong exposure to di-(2-ethylhexyl)-phthalate induces tumors in liver and testes of Sprague-Dawley rats. Toxicology 206(3):359–371

Weuve J, Sanchez BN, Calafat AM et al (2006) Exposure to phthalates in neonatal intensive care unit infants: urinary concentrations of monoesters and oxidative metabolites. Environ Health Perspect 114(9):1424–1431

Wilson VS, Lambright C, Furr J et al (2004) Phthalate ester-induced gubernacular lesions are associated with reduced insl3 gene expression in the fetal rat testis. Toxicol Lett 146(3):207–215

Wittassek M, Wiesmuller GA, Koch HM et al (2007) Internal phthalate exposure over the last two decades—a retrospective human biomonitoring study. Int J Hyg Environ Health 210(3–4):319–333

Wittassek M, Koch HM, Angerer J et al (2011) Assessing exposure to phthalates—the human biomonitoring approach. Mol Nutr Food Res 55(1):7–31. doi:10.1002/mnfr.201000121

World Health Organization (2000) IARC monographs on the evaluation of carcinogenic risks to humans, vol 77. International Agency for Research on Cancer, Lyon, France

Wormuth M, Scheringer M, Vollenweider M et al (2006) What are the sources of exposure to eight frequently used phthalic acid esters in Europeans? Risk Anal 26(3):803–824

Wu S, Zhu J, Li Y et al (2010) Dynamic effect of di-2-(ethylhexyl) phthalate on testicular toxicity: epigenetic changes and their impact on gene expression. Int J Toxicol 29(2):193–200
Yang J, Hauser R, Goldman RH (2013) Taiwan food scandal: the illegal use of phthalates as a clouding agent and their contribution to maternal exposure. Food Chem Toxicol 58:362–368

Chapter 3
Brominated Flame Retardants and Their Replacements in Food Packaging and Household Products: Uses, Human Exposure, and Health Effects

Susan D. Shaw, Jennifer H. Harris, Michelle L. Berger, Bikram Subedi and Kurunthachalam Kannan

Abstract Since the 1970s, the brominated flame retardants (BFRs), polybrominated diphenyl ethers (PBDEs), hexabromocyclododecanes (HBCDs), and tetrabromobisphenol A (TBBPA) have been used as additive or reactive flame retardants (FRs) in household products including foam furniture, baby products, mattresses, textiles, electronics, food packaging, and housing insulation to meet flammability standards. Many of these chemicals are now recognized as global contaminants and are associated with adverse health effects including endocrine disruption, reproductive toxicity, developmental neurotoxicity, and cancer. Additive BFRs migrate out of products and accumulate in dust of indoor environments. Dust ingestion is the primary route of human exposure to BFRs, and studies link body burdens of adults and children to indoor dust concentrations. Because of health concerns, PentaBDE and OctaBDE mixtures were phased out of production in 2005, and DecaBDE and HBCDs are scheduled for global elimination by 2013. However, large amounts of in-use and discarded household products will continue to release these BFRs into the

S. D. Shaw (✉) · J. H. Harris · M. L. Berger
Center for Marine Studies, Marine Environmental Research Institute,
Blue Hill, ME, USA
e-mail: sshaw@meriresearch.org

J. H. Harris
e-mail: jharris@meriresearch.org

M. L. Berger
e-mail: mberger@meriresearch.org

S. D. Shaw · K. Kannan
Department of Environmental Health Sciences, School of Public Health,
State University of New York at Albany, Albany, NY, USA
e-mail: kkannan@wadsworth.org

B. Subedi · K. Kannan
Wadsworth Center, New York State Department of Health, Albany, NY, USA
e-mail: bx24@wadsworth.org

S. M. Snedeker (ed.), *Toxicants in Food Packaging and Household Plastics*,
Molecular and Integrative Toxicology, DOI: 10.1007/978-1-4471-6500-2_3,

environment and TBBPA is still in high-volume use worldwide. Following the PBDE phase-out, many replacement FRs with known or suspected toxicity have been detected at increasing levels in household products and dust. Exposure of infants and children to FRs in baby products and house dust is of special concern. The continued use of FRs with known toxicities and introduction of untested replacement chemicals highlights the need to update flammability standards and modernize chemical policies to require disclosure and safety testing of FR chemicals prior to sale.

Keywords Flame retardants · PBDEs · HBCDs · TBBPA · BFRs · Halogenated flame retardants · Brominated flame retardants · Replacement flame retardants · House dust · Polyurethane foam · Foam furniture · Plastics used in TVs and computers · Electronics · Food packaging · Food contact materials · Children's products · Carpets and textiles

3.1 Key Take Home Points

- For almost 40 years, the brominated flame retardants (BFRs) polybrominated diphenyl ethers (PBDEs), hexabromocyclododecanes (HBCDs), and tetrabromobisphenol-A (TBBPA) have been widely used as additive or reactive flame retardants (FRs) in household products and food packaging to meet flammability standards. These chemicals are now recognized as global contaminants and are associated with adverse health effects in animals and humans.
- Additive BFRs escape from products and accumulate in household dust. Dust ingestion is the primary route of human exposure, especially for children who ingest contaminated dust via hand-to-mouth activity.
- Because of health concerns, PentaBDE and OctaBDE mixtures were phased out of production in 2005. DecaBDE and HBCD are scheduled for global elimination in 2013, but there are no restrictions on production and use of TBBPA.
- Following the PBDE phase-out, many untested replacement chemicals have been detected in furniture and indoor dust, highlighting the need to revise flammability standards and modernize chemical policies to require disclosure and safety testing of chemicals prior to sale.
- A global strategy is urgently needed for the identification and safe disposal of flame retardant-treated household products and electronic waste.

3.2 Introduction

Acronyms for the flame retardants, different types of plastics, biological terminology, and other abbreviations used in this chapter are listed in Table 3.1.

Table 3.1 Acronyms for flame retardant, plastics and biomedical terminology

	Abbreviation	Term
Flame retardant terms	BDE-209	Decabromodiphenyl oxide
	BFR	Brominated flame retardant
	BTBPE	1,2-Bis(2,4,6-tribromophenoxy)ethane
	DBDPE	Decabromodiphenyl ethane
	decaBDE	Decabromodiphenyl ether
	FM	Firemaster 550
	FR	Flame retardant
	HBCD	Hexabromocyclododecane
	hexaBDE	Hexabromodiphenyl ether
	nonaBDE	Nonabromodiphenyl ether
	octaBDE	Octabromodiphenyl ether
	PBDE	Polybrominated diphenyl ether
	PBDF	Polybrominated dibenzofuran
	pentaBDE	Pentabromodiphenyl ether
	TBBPA	Tetrabromobisphenol A
	TBBPA-DBPE	Tetrabromobisphenol A-bis(2,3 dibromopropylether)
	TDCPP	Tris(1,3-dichloro-2-propyl)phosphate
Plastic terms	ABS	Acrylonitrile butadiene styrene
	EPS	Expandable polystyrene
	HIPS	High Impact Polystyrene
	PBT	Polybutylene terephthalate
	PE	Polyethylene
	PET	Polyethylene terephthalate
	PP	Polypropylene
	PS	Polystyrene
	PUF	Polyurethane foam
	TPE	Thermoplastic elastomer
	XPS	Extruded polystyrene
Electronics terms	CRT	Cathode ray tube
	e-waste	Electronic waste
	LCD	Liquid crystal display
	WEEE	Waste electrical and electronic equipment
Biomedical terms	FSH	Follicle-stimulating hormone
	LH	Luteinizing hormone
	NHL	Non-Hodgkin's lymphoma
	SHBG	Sex-hormone binding globulin
	TSH	Thyroid stimulating hormone
	T3	Triiodothyronine
	T4	Thyroxine
Other terms	NTP	National Toxicology Program
	POP	Persistent Organic Pollutant

Since the 1970s, the BFRs, PBDEs, HBCDs, and TBBPA have been used as additive or reactive FRs in household products to meet flammability standards (Alaee et al. 2003). Additive FRs are not chemically bound but rather are physically blended with polymers, thus, can escape from products and enter the air and dust of the indoor environment (Allen et al. 2008).

Indoor dust is a primary route of human exposure to BFRs (Lorber 2008; Harrad et al. 2010; Covaci et al. 2009); several studies have linked indoor dust concentrations of BFRs with human body burdens (Wu et al. 2007; Johnson et al. 2010, 2013; Stapleton et al. 2012; Roosens et al. 2009). BFRs are associated with a wide range of adverse health effects, including endocrine disruption, reproductive/developmental toxicity, immunotoxicity, and neurotoxicity (Birnbaum and Staskal 2004; Shaw et al. 2010; Covaci et al. 2009).

Since their introduction, PBDEs, HBCDs, and TBBPA have become widespread global contaminants (Shaw and Kannan 2009; Hites 2004; Covaci et al. 2006; Tanabe et al. 2008; Covaci et al. 2009). BFRs enter the environment through multiple pathways, such as air emission during manufacturing, combustion, leaching from discarded products in landfills, or recycling at the end of the product's life.

3.2.1 Major Uses

PBDEs are brominated aromatic compounds that have been produced since the 1960s as three commercial mixtures (Penta-, Octa-, and DecaBDE) that vary in degree of bromination (Fig. 3.1). PentaBDE was primarily used as an additive in polyurethane foam (PUF) at levels up to 30 % of the weight of the foam in residential furniture, baby products, and automobile and aircraft interiors to meet California's flammability standard, *Technical Bulletin 117* (Alaee et al. 2003; Birnbaum and Staskal 2004; Shaw et al. 2010). OctaBDE was mainly used in thermoplastic resins, such as the hard plastic components of televisions and computers, with the most prevalent use in acrylonitrile-butadiene-styrene (ABS) plastic, which can contain OctaBDE up to 12 % by weight (European Union 2003a). DecaBDE (97 % BDE-209),[1] the most widely used PBDE in all markets, is added to various plastic polymers such as polyvinyl chloride, polycarbonates, and high-impact polystyrene (HIPS) used in housings of electrical and electronic equipment, as well as back-coating for textiles (commercial furniture, automobile fabrics, and carpets) (BSEF 2012a; Alaee et al. 2003).

HBCDs are primarily used as additives in expandable polystyrene (EPS) and extruded polystyrene (XPS) foam used to thermally insulate buildings. HBCDs also are added to the back-coating of textiles of upholstered furniture as well as in HIPS housing of electrical and electronic equipment and appliances (Alaee et al. 2003). Although γ-HBCD is predominant in the commercial HBCD formulations

[1] Because DecaBDE is predominantly composed of the congener BDE-209, the terms DecaBDE and BDE-209 are used interchangeably in the literature and in this chapter.

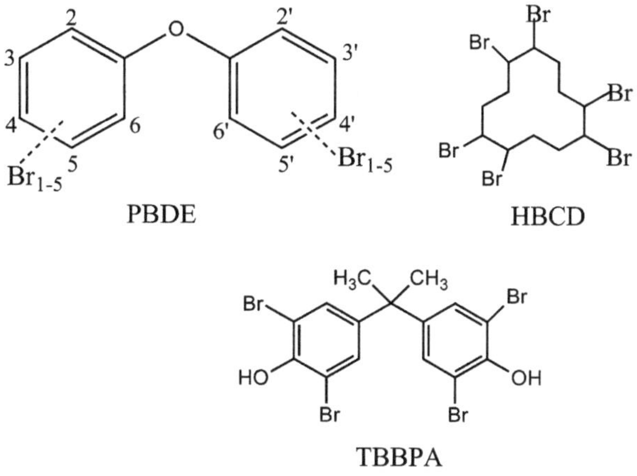

Fig. 3.1 Chemical structures of PBDEs, HBCDs and TBBPA

and most abiotic samples, the persistent α-isomer predominates in biotic and human samples (Covaci et al. 2006).

The highest-volume FR in use worldwide, TBBPA, is used mainly as a reactive flame retardant in epoxy resins for printed circuit boards in computers and telecommunications equipment (Covaci et al. 2009). In reactive applications, TBBPA is covalently bound into the polymer in products and is not expected to migrate from the product. However, TBBPA is also used as an additive to circuit boards for low-energy applications such as remote controls and video recorders as well as the plastic housing for electrical and electronic equipment such as computer monitors and printers (Covaci et al. 2009). Both additive- and reactive-treated products have been shown to release TBBPA and metabolites into the environment (Birnbaum and Staskal 2004).

Contamination patterns of BFRs reflect dramatic differences in worldwide consumption. For example, North America used 95 % or more of the PentaBDE produced globally (Hale et al. 2003), and PentaBDE levels in the North American population and biota are the highest in the world (Shaw and Kannan 2009; Hites 2004). In Asia, the major BFR in use is TBBPA followed by PBDEs (BSEF 2012b). After TBBPA, HBCD is the second highest-volume BFR used in Europe (Alaee et al. 2003). In recent years the global demand for HBCDs has increased, especially in Asia, and temporal studies show that α-HBCD levels are increasing in biota and humans (Covaci et al. 2006; Johnson-Restrepo et al. 2008; Tanabe et al. 2008; Stapleton et al. 2006).

3.2.2 *Flame Retardant Regulatory Restrictions*

Because of health and environmental concerns, the Penta- and OctaBDE mixtures were phased out of production and use in many countries in 2005 (Shaw et al. 2010). In 2009, the Stockholm Convention listed PentaBDE and OctaBDE as persistent organic pollutants (POPs) scheduled for global elimination (Stockholm Convention 2009). The U.S. Environmental Protection Agency (USEPA) negotiated a phase-out of DecaBDE by U.S. producers as of 2013 (USEPA 2009).

Following the PBDE phase-out, HBCD usage has increased as a replacement FR in textiles, plastic enclosures and other products (BSEF 2009). In May 2013, the Stockholm Convention listed HBCD for global elimination, with specific exemptions for use in building materials (UNEP/POPS/COP.6/17) UNEP (2013). The U.S. is currently considering legislation to eliminate the use of HBCD in housing foam insulation (Skinner 2013).

Currently, there are no restrictions on the production and use of TBBPA and global production has increased over the last decade, especially in Asia (Covaci et al. 2009). In the U.S., TBBPA has received little regulatory attention. In 2003, a European Union directive on handling of waste electrical and electronic equipment (WEEE) (European Union 2003b) called for selective treatment of plastics containing BFRs, including TBBPA (BFR-treated plastics must be selectively removed from WEEE). A similar directive for BFR-treated plastics in WEEE was adopted in China, a major global recipient of electronic waste (e-waste) (Wong et al. 2007).

Despite restrictions, large amounts of foam furniture, baby products, carpets, plastics, and insulation containing BFRs are still sources of exposure in homes and buildings, and must be disposed of after their lifetimes, creating outdoor reservoirs (e.g., landfills, wastewater treatment plants, e-waste recycling facilities, or stockpiles of hazardous wastes) for the future dispersal of PBDEs to the environment (Shaw and Kannan 2009).

3.3 Human Exposure

3.3.1 Exposure Routes

For non-occupationally exposed persons, house dust is the primary route of exposure to BFRs (Lorber 2008; Webster et al. 2013; Abdallah and Harrad 2011; Harrad et al. 2010; Covaci et al. 2009). There is ample evidence of contamination of indoor dust with BFRs. Their concentrations in indoor air substantially exceed those outdoors, and studies link body burdens to indoor dust concentrations (Wu et al. 2007; Johnson et al. 2010, 2013; Stapleton et al. 2012; Roosens et al. 2009).

House dust contributes an average of 82 % of an adult U.S. resident's exposure to PBDEs (Lorber 2008). Although diet may also contribute to PBDE exposure (Wu et al. 2007), dust is a direct exposure pathway through incidental ingestion, inhalation of suspended particles, and dermal absorption, and it is a proxy for exposure from product use (Dodson et al. 2012; Webster et al. 2013). PBDE residues on hands were found to be strong predictors of serum levels in both children (Stapleton et al. 2012) and adults (Watkins et al. 2011), underlining the importance of hand-to-mouth contact as a significant oral exposure route. For HBCDs, inhalation exposure was indicated as a minor pathway, while dust ingestion was the major pathway for most individuals, especially toddlers. Dust exposure in toddlers was comparable to occupationally-exposed adults and tenfold greater than dietary exposure (Abdallah et al. 2008a). Similarly, TBBPA in dust is thought to be an important pathway of exposure, contributing 34 and 90 % of the mean overall exposure for adults and toddlers, respectively (Goosey et al. 2008).

3.3.2 Sources of Brominated Flame Retardants in Household Products

BFRs are additive or reactive components in a variety of polymers, such as PUF and PS foams, HIPS, and epoxy resins, which are then used in the manufacturing of foam furniture, electronics, food packaging, and building materials (Table 3.2) Despite their widespread use, for most products data on actual concentrations of BFRs in the polymers or finished products are not available and constitute a data gap.

Foam Furniture Foam furniture items, such as foam upholstered-PUF couches and chairs, as well as mattresses, pads and pillows, are large sources of PentaBDE and DecaBDE in homes (Table 3.2). Stapleton et al. (2012) found BFRs in 85 % of 102 foam samples from residential couches purchased in the U.S. between 1985 and 2010. In couches purchased prior to 2005, PentaBDE congeners −47, −99, and −100 were the predominant BFRs in foam. High concentrations of Penta-BDE were found in mattress foam (mean 20,230 milligrams per kilogram, mg/kg). PentaBDEs were found in 17 % of the couches tested, indicating that many years after the PBDE phase-out, U.S. residents are still exposed to PentaBDE from furniture foam. Because there is currently no strategy in place for the identification or safe disposal of FR-treated furniture, in-use and discarded furniture will continue to be a source of PentaBDE exposure for the foreseeable future (Shaw and Kannan 2009).

Children's Products and Clothing Children's exposure to BFRs is a particular concern due to their frequent hand-to-mouth behavior and contact with dust on floors and surface materials in homes. Exposure to chemicals in baby products is of even greater concern for infants, who are in intimate daily contact with these products at vulnerable stages of their development. Children's foam-containing products such as car seats, mattresses, pillows, and changing pads are sources of PentaBDEs, while children's clothing may be treated with DecaBDE (Table 3.2). Hard and soft plastic toys are sources of all PBDE commercial mixtures (Penta-, Octa- and DecaBDEs).

Stapleton et al. (2011) measured FRs in PUF from 101 commonly used baby products purchased in the U.S. More than 80 % of the products contained a halogenated flame retardant additive. Five samples contained PentaBDE congeners,

Table 3.2 Brominated flame retardants in household products

	Product	Material	Chemical	Concentration	Reference
Foam furniture	Upholstered furniture (chairs, couches)	Polyurethane foam	PentaBDE[a]	20,230 mg/kg (mean)	Stapleton et al. (2012)
	Mattress cores, topper pads	Polyurethane foam	DecaBDE	NA	WHO (1994)
	Pillows				
Children's products and clothing	Child car seats, Car seat pillows	Polyurethane foam	PentaBDE	32,270 mg/kg in foam (mean)	Stapleton et al. (2011)
	Rocking chairs, Cribs				
	Changing table pads				
	Foam toys	Polyurethane foam	∑PBDEs	1.0 mg/kg (median)	Chen et al. (2009)
	Stuffed toys			0.15 mg/kg (median)	Chen et al. (2009)
	Hard plastic toys	Polystyrene copolymers	DecaBDE	34.3 mg/kg (median)	Chen et al. (2009)
			∑PBDEs	53 mg/kg (median)	Chen et al. (2009)
	Tights, Stockings	Polyamide	DecaBDE	NA	Alaee et al. (2003)
	Swimwear				
	Gymnast pit foam cubes	Polyurethane foam	PentaBDE	5 % by weight	Carignan et al. (2013)
Food contact materials	Microwavable containers, yogurt containers, straws	Polyolefin thermoplastic (polypropylene)	DecaBDE	NA	Mingwu et al. (2010)
	Polyolefin shrink wrap	Polyolefin thermoplastic	DecaBDE	NA	Alaee et al. (2003)
	Saran premium and Cling plus wrap	(polyethylene)			
	Butter wrappers	Paper	∑PBDE	0.7–0.8 mg/kg ww	Schecter et al. (2011)
			DecaBDE	0.6 mg/kg ww	Schecter et al. (2011)
	Hot liquid tubing, fatty food containers	Thermoplastic elastomer	DecaBDE	NA	Albemarle (2010)
	Aluminum beverage cans linings, plastic water bottles	Polybutylene- and Polyethylene-terephthalate	BDE-47	NA	Mingwu et al. (2010)
			DecaBDE	NA	ICL-IP (2012)
	Meat and cheese packaging trays	Polyolefin foam	DecaBDE	NA	Albemarle (2010)
	Styrofoam cups, takeout containers, shipping, and packaging	Expandable polystyrene	HBCD	NA	Innes and Innes (2003)
	Rubber gloves	Polyester	DecaBDE	NA	Albemarle (2010)
	Tooth brush monofilaments	Polybutylene terephthalate	DecaBDE	NA	ICL-IP (2012)

(continued)

Table 3.2 (continued)

	Product	Material	Chemical	Concentration	Reference
Textiles, carpets and coatings	*Back coatings*				
	Carpet-, furniture covering-, and mattress ticking- adhesives	Latex /Adhesives	HBCD	NA	SFRC (2000)
			DecaBDE	NA	SFRC (2000)
	Fibers				
	Drapes	Polyethylene fibers	ΣPBDE	120,000 mg/kg (maximum)	Kajiwara et al. (2009)
			HBCD	43,000 mg/kg (maximum)	Kajiwara et al. (2009)
	Carpets, Mattress ticking, Army tents	Polyethylene fibers	DecaBDE	NA	WHO (1994)
	Rope and Cords	Polyamide, Thermoplastic Polyolefins (polypropylene, polyethylene)	DecaBDE	NA	Alaee et al. (2003)
	Clothing				
	Footwear products	Polyvinyl chloride nitrile	DecaBDE	NA	Albemarle (2010)
	Body armor, sports equipment				
Electronics and associated materials	Wire and cable applications	Polyolefin plastic	DecaBDE	NA	Innes and Innes (2003)
	Electrical and Electronic Equipment	Mixed plastics from household electronics and office electronics	OctaBDE	1.8 % by weight of sample (mean)	Kuhn et al. (2004)
			DecaBDE	4.6 % by weight of sample (mean)	Kuhn et al. (2004)
			HBCD	2.3 % by weight of sample (mean)	Kuhn et al. (2004)
			TBBPA	10.2 % by weight of sample (mean)	Kuhn et al. (2004)
	Electric wire coatings	Rubber	DecaBDE	~6 % by weight of sample	Thuresson et al. (2005)
	Electrical outlets and Insulation boards		∑PBDE	0.2 mg/kg (maximum)	Kajiwara et al. (2011)
			HBCD	23,000 mg/kg (maximum)	Kajiwara et al. (2011)
			TBBPA	15 mg/kg (maximum)	Kajiwara et al. (2011)

(continued)

Table 3.2 (continued)

	Product	Material	Chemical	Concentration	Reference
	Packaging for electronics	Expandable polystyrene	HBCD	NA	Innes and Innes (2003)
	Treated /colored surfaces of wiring	Paints	DecaBDE	NA	Albemarle (2010)
			TBBPA	NA	Albemarle (2010)
Hard plastics in TVs, computers, and electronic waste	TV /PC housings	Wood	PentaBDE	10 mg/kg (mean)	Morf et al. (2005)
			OctaBDE	10 mg/kg (mean)	Morf et al. (2005)
			DecaBDE	20 mg/kg (mean)	Morf et al. (2005)
			HBCD	10 mg/kg (mean)	Morf et al. (2005)
			TBBPA	80 mg/kg (mean)	Morf et al. (2005)
	TV /PC housings	Mixed plastics (HIPS and /or ABS)	PentaBDE	50 mg/kg (mean)	Morf et al. (2005)
			OctaBDE	7,500 mg/kg (mean)	Morf et al. (2005)
			DecaBDE	21,000 mg/kg (n = 1); 4,800 mg/kg (mean)	Choi et al. (2009); Morf et al. (2005)
			HBCD	50 mg/kg (mean)	Morf et al. (2005)
			TBBPA	8.1 mg/kg (n = 1); 23,000 mg/kg (mean)	Choi et al. (2009); Morf et al. (2005)
	TV /PC housings (TV rear cabinet)	Mixed plastic (HIPS and /or ABS)	∑PBDE	1.5–130,000 mg/kg (range)	Vehlow et al. (2000)
			HBCD	<0.03–2.2 mg/kg (range)	Vehlow et al. (2000)

(continued)

Table 3.2 (continued)

	Product	Material	Chemical	Concentration	Reference
Hard plastics in TVs, computers, and electronic waste	TV /PC housings (TV rear cover)	Mixed plastic (HIPS and /or ABS)	PentaBDE	50 mg/kg (mean)	Morf et al. (2005)
			OctaBDE	7,700 mg/kg (mean)	Morf et al. (2005)
			DecaBDE	13,000 mg/kg (mean)	Morf et al. (2005)
			HBCD	1,400 mg/kg (mean)	Morf et al. (2005)
			TBBPA	7,300 mg/kg (mean)	Morf et al. (2005)
	TV /PC housings (TV front cabinet)	Mixed plastic (HIPS and /or ABS)	∑PBDE	0.76 – 150,000 mg/kg (range)	Vehlow et al. (2000)
			HBCD	<0.03–0.38 mg/kg (range)	Vehlow et al. (2000)
	TV casings	Composite plastic samples (HIPS and /or ABS)	DecaBDE	110,000 mg/kg (mean)	Kajiwara et al. (2008)
	TV /Monitor housings	Mixed shredder plastic residues (HIPS and /or ABS)	OctaBDE	0.25–1.4 % wt of sample	Schlummer et al. (2007)
			DecaBDE	0.25–2.1 % wt of sample	Schlummer et al. (2007)
			TBBPA	0.30–11 % wt of sample	Schlummer et al. (2007)
	Cathode-ray tube (monitors, TVs)	Mixed plastics (HIPS and /or ABS)	OctaBDE	2,500 mg/kg; 900 mg/kg (mean)	Wäger et al. (2010, 2012)
			DecaBDE	3,200 mg/kg; 4,400 mg/kg (mean)	Wäger et al. (2010, 2012)
			TBBPA	3,700 mg/kg (mean)	Wäger et al. (2010)
	Liquid-crystal display (TV)	Mixed plastics (HIPS and /or ABS)	∑PBDE	0.002–14 mg/kg (range)	Kajiwara et al. (2011)
			HBCD	<LOQ–0.7 mg/kg (range)	Kajiwara et al. (2011)
			TBBPA	0.007–0.9 mg/kg (range)	Kajiwara et al. (2011)
	Liquid-crystal display (Laptop)	Mixed plastics (HIPS and /or ABS)	∑PBDE	0.007–4.8 mg/kg (range)	Kajiwara et al. (2011)
			HBCD	<LOQ–0.6 mg/kg (range)	Kajiwara et al. (2011)
			TBBPA	0.8–9,500 mg/kg (range)	Kajiwara et al. (2011)
	Small appliances (TVs, VCRs, computers, radios, phones)	Mixed plastic (HIPS and /or ABS)	PentaBDE	140; 130 mg/kg (mean)	Wäger et al. (2010, 2012) respectively
	Circuit boards in small appliances	Body reinforced with plastic (HIPS and /or ABS)	TBBPA	7.9–1,300 mg/kg (range)	Vehlow et al. (2000)

(continued)

Table 3.2 (continued)

	Product	Material	Chemical	Concentration	Reference
Hard plastics in TVs, computers, and electronic waste	Printed circuit boards	Body reinforced with plastic (HIPS and /or ABS)	∑PBDEs	40.5 mg/kg (weighted mean)	Richter et al. (1997b)
			PentaBDE	17 mg/kg (mean)	Morf et al. (2005)
			OctaBDE	10 mg/kg (mean)	Morf et al. (2005)
			DecaBDE	27 mg/kg (mean)	Morf et al. (2005)
			HBCD	10 mg/kg (mean)	Morf et al. (2005)
			TBBPA	14.4 mg/kg (weighted mean)	Richter et al. (1997b)
	Waste Electrical and Electronic Equipment	Fractions of shredder residues including plastics[c] (HIPS and /or ABS)	PentaBDE	153 mg/kg (maximum)	Mark et al. (2006)
			OctaBDE	1,070 mg/kg (maximum)	Mark et al. (2006)
			DecaBDE	1,400 mg/kg (maximum)	Mark et al. (2006)
	Waste electrical and electronic equipment	Printed circuit and plastic shredder residues[d] (HIPS and /or ABS)	∑PBDE	45.5 mg/kg (mean)	Ma et al. (2009)
	Waste electrical and electronic equipment	Mixed shredder plastic residues (HIPS and /or ABS)	OctaBDE	0.08–0.4 % wt of sample	Schlummer et al. (2007)
			DecaBDE	0.1–0.3 % wt of sample	Schlummer et al. (2007)
			TBBPA	0.3–0.96 % wt of sample	Schlummer et al. (2007)
	Waste electrical and electronic equipment	Housings including plastics (HIPS and /or ABS)	OctaBDE	NA	Mingwu et al. (2010)
			DecaBDE	36,605 mg/kg (maximum)	Mingwu et al. (2010)
Building Insulation and construction materials	Insulation foam boards	Expandable /Extruded polystyrene	HBCD	NA	Albemarle (2010)
	Construction white goods, insulation	High density /Flexible polyurethane foam, Rigid polyurethane foam	PentaBDE	NA	Alaee et al. (2003)
			DecaBDE	NA	ICL-IP (2012)
	Fuel and oil hoses	Polyvinyl chloride nitrile	DecaBDE	NA	Albemarle (2010)
	Weather seals, gaskets				
	Facing laminates for insulation panels	Laminates /Films	DecaBDE	NA	BSEF (2006)
	Applications exposed to cold (Drainage pipes)	Polypropylene	DecaBDE	NA	Albemarle (2010); ICL-IP (2012)
	Piping, thermal insulation	Polyethylene and Copolymers	DecaBDE	NA	Albemarle (2010)
	Solar panel	(ethylene vinyl acetate)			
	Hot melt adhesives (laminated wood panels)				
	Refrigeration piping	Elastomers	DecaBDE	NA	Albemarle (2010)
	HVAC components				
	Duct-, Mechanical-systems				

Table 3.2 (continued)

Product

NA = not available; LOQ = level of quantification; wt = weight; ww = wet weight

a. Congener composition of PBDE mixtures: PentaBDE (24–38 % tetraBDEs, 50–60 % pentaBDEs, 4–8 % hexaBDEs); OctaBDE (10–12 % hexaBDEs, 44 % heptaBDEs, 31–35 % octaBDEs, 10–11 % nonaBDEs, <1 % decaBDE); DecaBDE (<3 % nonaBDEs; 97–98 % BDE-209)

b. Items from a recycling and disposal company for commercial electronic appliance waste in Germany

c. Shredder residue at a waste electrical and electronic equipment plant

d. Items from an e-waste recycling site in South China

Credits for information reprinted within Table 3.2

Permission granted by Federal Office for the Environment (FOEN) via Nicole Reichen to reproduce information contained in Table 15 from: Kuhn E, Arnet R, Känzig A, Frey T (2004) Bromierte flammschutzmittel in kunststoffprodukten des Schweizer marktes; Swiss Agency for the Environment, Forests and Landscape: Bern, Switzerland. http://www.bafu.admin.ch/publikationen/publikation/00289/index.html?lang=de

Reprinted from Chemosphere, Vol. 74, issue no. 3, Ki-In Choi, Suk-Hui Lee, and Masahiro Osako (Authors), Leaching of brominated flame retardants from TV housing in the presence of dissolved humic matter, pages 460–466, Copyright 2009, with permission from Elsevier

Reprinted from Chemosphere, Vol. 74, issue no. 11, Natsuko Kajiwara, Minekazu Sueoka, Toshio Ohiwa, and Hidetaka Takigami (Authors) Determination of flame-retardant hexabromocyclododecane diastereomers in textiles, pages 1485-1489, Copyright 2009, with permission from Elsevier

Reprinted from Journal of Hazardous Materials, Vol. 192, issue no. 3, Natsuko Kajiwara, Yukio Noma, and Hidetaka Takigami (Authors), Brominated and organophosphate flame retardants in selected consumer products on the Japanese market in 2008, pages 1250-1259, Copyright 2011, with permission from Elsevier

Reprinted (adapted) with permission from Natsuko Kajiwara, Yukio Noma, and Hidetaka Takigami (2008) Photolysis studies of technical decabromodiphenyl ether (DecaBDE) and ethane (DeBDethane) in plastics under natural sunlight, Environmental Science and Technology 42:4404-4409. Copyright 2008 American Chemical Society

Reprinted (adapted) with permission from Shao Mingwu, Wei Chao, Jia Yongjuan, Dai Xinhua, and Fang Xiang (2010) Determination of selected polybrominated diphenylethers and polybrominated biphenyl in polymers by ultrasonic-assisted extraction and high-performance liquid chromatography-inductively coupled plasma mass spectrometry, Analytical Chemistry 82:5154-5159. Copyright 2010 American Chemical Society

Reprinted (adapted) with permission from Heather M. Stapleton, Susan Klosterhaus, Alex Keller, P. Lee Ferguson, Saskia van Bergn, Ellen Cooper, Thomas F. Webster and Arlene Blum (2011) Identification of flame retardants in polyurethane foam collected from baby products. Environmental Science and Technology 45:5323-5331. Copyright 2011 American Chemical Society

Reprinted (adapted) with permission from Heather M. Stapleton, Smriti Sharma, Gordon Getzinger, P. Lee Ferguson, Michelle Gabriel, Thomas F. Webster and Arlene Blum (2012) Novel and high volume use flame retardants in US couches reflective of the 2005 PentaBDE phase out, Environmental Science and Technology 46:13432-13439. Copyright 2012 American Chemical Society

Reprinted (adapted) with permission from Morf LS, Tremp J, Gloor R, Huber Y, Stengele M, Zennegg M (2005) Brominated flame retardants in waste electrical and electronic equipment: substance flows in a recycling plant. Environmental Science and Technology 39:8691-8699. Copyright 2005 American Chemical Society suggesting that baby products treated with PentaBDE are still in use in the U.S. PentaBDEs were detected at average concentrations of 32,270 mg/kg (approximately 3–4 % by weight).

The U.S. imports 70–80 % of its toys from China, 70 % of which are manufactured in South China (Chen et al. 2009). In a recent study of toys purchased in South China, DecaBDE was the predominant PBDE, accounting for 45–79 % of BFRs in hard plastic, foam, rubber/soft plastic, and stuffed toys (Chen et al. 2009). Hard plastic toys (made with PS copolymers) contained the highest total PBDE concentrations (median 53 mg/kg dry weight, dw), 65 % of which was contributed by DecaBDE (median 34.3 mg/kg). Foam toys contained PBDEs at concentrations four to 160-fold lower than those in hard plastic toys. Stuffed toys contained the lowest PBDE concentrations.

Children may be exposed to BFRs in clothing items such as nylon in tights, stockings and swimwear, which are polyamide products treated with DecaBDE and other FRs (Alaee et al. 2003). Children training for sports competitions may be exposed to BFRs in gymnasiums, as foam cubes placed in open pits designed to train young gymnasts may contain PentaBDE at 5 % by weight in foam (Carignan et al. 2013).

Food Contact Materials Many plastic food contact materials including microwavable containers, packaging materials, and plastic wraps are potential sources of BFR exposure in households (Table 3.2). Polyolefin thermoplastics such as polypropylene (PP) and polyethylene (PE) are common food packaging materials that are flame-retarded with PBDEs, primarily DecaBDE (Alaee et al. 2003). DecaBDE is the major BFR used in PP items such as microwaveable plastic food containers, flexible yogurt containers, bottles and straws (Mingwu et al. 2010). DecaBDE is also added to plastic wraps containing PE, such as polyolefin shrink wrap and Saran Wrap® (International Plastics 2013; Alaee et al. 2003). Butter wrapper paper may be contaminated with PBDE congeners, predominantly DecaBDE (BDE-209) (Schecter et al. 2011).

DecaBDE is the main BFR added to thermoplastic elastomers (TPEs) used in tubing of hot liquids and fatty food containers (Albemarle 2010), and thermoplastic polyesters such as polyethylene terephthalate (PET) and polybutylene terephthalate (PBT), plasticizers incorporated into soft plastic products to improve flexibility (Talsness et al. 2009; ICL-IP 2012). Aluminum beverage can linings and plastic water bottles contain PBT and PET treated with BDE-47 and DecaBDE (Mingwu et al. 2010; ICL-IP 2012). BFRs easily migrate from thermoplastics in products into beverages and foods. A recent study reported that plastic PET bottles leached bromine into drinking water; the primary congener found was BDE-209 (Andra et al. 2012).

Polyolefin foams (also called *rubber foam*) treated with DecaBDE (BDE-209) are added to plastic trays used in packaging of meat and cheese (Saigal Polymers Pvt Ltd 2013; Albemarle 2010). Styrofoam used in packaging for shipments, take-out boxes and coffee cups is made of EPS treated with HBCDs (Innes and Innes 2003). Other food contact items may contain BDE-209 and replacement FRs. BDE-209 is added to rubber gloves made from polyester (Albemarle 2010) and to toothbrush monofilaments containing PBT (ICL-IP 2012). More information is needed on actual concentrations of FRs in food packaging and contact materials.

Textiles, Carpets and Coatings Textiles including curtains, carpets, furniture fabric, and back coatings are made with materials such as cotton, polyester, PP, nylon, PBT, PET and hot melts. These materials and textile products themselves are potential sources of PBDEs and HBCDs in the home (Table 3.2). The major exposures result from mattresses, drapery, tents and upholstered furniture (LCSP 2005). Multiple washings of textiles can cause additive FRs to leach out of the products into washwater (USEPA 2005).

Today in the U.S., HBCD is the major BFR used in textiles, replacing DecaBDE that was extensively used in the past (LCSP 2005; Posner 2004). In Japan, HBCDs are the major BFRs added to textiles, followed by DecaBDE (Kajiwara et al. 2009). Few data are available on estimated breakdown and emission rates of BFRs from household textiles. Kajiwara and Takigami (2010) reported that emission rates of HBCDs from curtains were two orders of magnitude higher than those of PBDEs, indicating that HBCDs more readily leach out from textiles into the indoor environment. HBCD showed more resistance than BDE-209 to photodegradation via sunlight exposure, whereas BDE-209 added to the curtain material photodegraded and contributed to the synthesis of polybrominated dibenzofurans (PBDFs), suggesting that treated products in daily usage have the potential to be sources of PBDFs in indoor air and dust. The same trend was reported earlier for emission of BFRs from TV casings (Kajiwara et al. 2008).

Electronics and Associated Materials Electronics, including wires and cables in household appliances, contain plastic materials that can ignite, and have been the subject of plastics flammability specifications issued by the Underwriters Laboratories, Inc. Despite the use of BFRs in plastics for over 40 years, data on the presence of FRs in plastics are limited to presence of BFRs in the components of wires and cables (Innes and Innes 2003). A survey conducted in Switzerland during 1999–2002 measured PBDEs in 486 plastic products from 366 consumer goods including electrical and electronic equipment (EEE), and reported the mean PBDE, HBCD and TBBPA concentrations in the range of 2–10 % by weight (Kuhn et al. 2004).

Many FRs are contained in electronic products, and high concentrations are detected in air and dust at e-waste recycling sites, potentially exposing workers, and often entire communities, to these chemicals (Ma et al. 2009). Electric wires and cables are flame retarded with rubber containing DecaBDE at a concentration of $\sim$6 % by weight (Thuresson et al. 2005); however, actual measurements of concentrations of BFRs in wires and electric cables are not available in the literature. Kajiwara et al. reported HBCD (up to 23,000 mg/kg) and TBBPA (up to 15 mg/kg) in electric outlets and insulation boards (Kajiwara et al. 2011). No studies have reported on the occurrence of BFRs in batteries.

Hard Plastics in TVs, Computers and Electronic Waste Products The hard plastics used in the manufacturing of TV and computer housings include HIPS and ABS (Riess et al. 2000). While the predominant FR detected in plastic housings of TVs and computers is PentaBDE, TBBPA has also been detected in shredder residues of housings in e-waste (Morf et al. 2005; Vehlow et al. 2000; Schlummer et al. 2007; Reiss et al. 2000; Wäger et al. 2010, 2012; Richter et al. 1997). PBDEs, HBCD, and TBBPA have not only been found in housings of older, cathode ray tube (CRT) monitors (Wäger et al. 2010; Wager et al. 2012), but also in liquid crystal display (LCD) panel TVs (Kajiwara et al. 2011) and in e-waste from small appliances (Richter et al. 1997). Concentrations of PBDEs in e-waste plastics were found up to 1,400 mg/kg (Mark et al. 2006), whereas TBBPA weighted mean concentrations in printed circuit boards from a German e-waste recycling site were 14.4 mg/kg, respectively (Richter et al. 1997). Higher levels of BFRs were recently found in WEEE plastic products including PBDEs (mean 2,500 mg/kg) and TBBPA (mean 3,700 mg/kg) (Wäger et al. 2010, 2012).

Kajiwara et al. (2011) conducted a comprehensive analysis of BFRs in consumer products including multiple components of LCD-TVs and laptop computers from Japan (Kajiwara et al. 2011). DecaBDE was the dominant congener accounting for 73–98 % of the total (Σ) PBDE content in most of the TV and laptop computer components analyzed. The highest ΣPBDE concentrations in LCD-TVs were found in rear and front plastic covers (up to 14 mg/kg), followed by printed circuit boards and LCD panels (up to 0.059 mg/kg) (Table 3.2). In laptops, ΣPBDE concentrations were highest in cooling fans and speakers (4.8 mg/kg), followed by keyboards (0.13 mg/kg) and LCD panels (0.0067 mg/kg). The highest HBCD concentrations were found in printed circuit boards (up to 0.68 mg/kg) followed by front covers (up to 0.54 mg/kg) in LCD-TVs. TBBPA concentrations ranged from 0.007 mg/kg in LCD-panel of LCD-TVs to extremely high levels of 9,500 mg/kg in the cooling fan and speaker of laptop computers.

Building Insulation Although HBCDs and PBDEs are the major BFRs used in housing insulation and construction materials, information on this application of FRs in products is extremely limited and concentration data are nonexistent. The types of products that use these FRs are indicated in Table 3.2. HBCDs and PBDEs have been the major flame retardants used in EPS and XPS, high density/flexible- and rigid PUF (Table 3.2). HBCD is the predominant BFR used in EPS and XPS placed in insulation foam boards, rigid PUF and PE (Janssen 2005). High density and flexible PUF can contain PentaBDE (Alaee et al. 2003) and rigid PUF can contain BDE-209 and other FRs (ICL-IP 2012). Many other building materials such as fuel hoses, weather seals, gaskets, laminates, drainage pipes, ducts and mechanical systems are flame-retarded with DecaBDE (Albemarle 2010).

3.3.3 Brominated Flame Retardant Concentrations in House Dust

House dust has been widely used as an important environmental matrix to monitor BFRs contamination in indoor environments. Dust from U.S. homes contains the highest measured amounts of total PBDEs worldwide (median range 3,600–42,000 nanograms per gram, ng/g dust) (Batterman et al. 2009; Sjödin et al. 2008a), followed by dust samples from the United Kingdom (UK) (median range 1,000–4,500 ng/g) (Ali et al. 2012; Harrad et al. 2008), and an e-waste recycling area in South China (geometric mean, GM, range 1,850–4,336 ng/g) (Wang et al. 2010).

BDE-209 is the major PBDE congener found in house dust, contributing ~66 % of total PBDEs. The highest BDE-209 concentrations were reported in dust from the UK (median 10,000 ng/g) (Ali et al. 2012). Lower BDE-209 concentrations were reported in house dust samples from the U.S. (highest median 3,257 ng/g) (Sjodin et al. 2008a) and other countries, reflecting the higher usage of DecaBDE in the UK. A recent study reported relatively high BDE-209 concentrations (median range 300 to 860 ng/g) in dust from an e-waste region in Vietnam (Tue et al. 2013).

Limited available data suggest that international differences in indoor contamination with HBCDs exist, but they are not substantial. The highest concentrations of HBCDs in dust were found in Japan (mean 6,570 ng/g) (Takigami et al. 2009) and the UK (median 1,300 ng/g) (Abdallah et al. 2008b), reflecting the higher usage of HBCD in those countries. Dust concentrations of HBCDs have been correlated with human body burdens (Roosens et al. 2009), suggesting that dust is a major determinant of human HBCD exposure.

Information on the presence of TBBPA in indoor dust is scarce. Generally, TBBPA concentrations are at the low end of those found for PBDEs and HBCDs, which is consistent with TBBPA's primary usage as a reactive FR. Dodson et al. (2012) reported median TBBPA concentrations of 260 ng/g and 200 ng/g in dust samples from California homes in 2006 and 2011, respectively. Dust concentrations were significantly related to the age of electronics present, suggesting that new electronics may contain less TBBPA than older products. Twofold higher TBBPA concentrations were reported in dust sampled from Japanese homes in 2006 (520 and 490 ng/g) (Takigami et al. 2009).

Higher TBBPA concentrations have been found in dust from swabbed surfaces of computers, microwaves, refrigerators, printers and other electronics (Di Napoli-Davis and Owens 2013). The highest concentrations were found in dust swabs from CRT computer monitors (mean 523 ng per milliliter, ng/ml), microwaves (206 ng/ml), and refrigerators (194 ng/ml). Lowest concentrations (below the LOQ) were found in dust from copiers and computer towers. Similarly, Kajiwara et al. (2011) reported high levels (up to 800 ng/g) in PC boards of electronics in Japan. High TBBPA concentrations (mean 2,400,000 ng/g) were also found in dust sampled inside CRT-TVs in Japan (Takigami et al. 2008).

Compared with the U.S. and Japan, TBBPA concentrations in dust samples from European homes are an order of magnitude lower. Median concentrations ranged from 48 ng/g in dust collected from German homes (Abb et al. 2011) to 62 ng/g in UK dust (Abdallah et al. 2008b) and 141 ng/g in Belgian dust (D'Hollander et al. 2010).

3.3.4 Brominated Flame Retardant Concentrations in Human Tissues

PBDEs PBDE concentrations in the North American general population are ten to100-fold higher than those in Europeans or Asians. In the U.S., average serum concentrations of total PBDEs range between 20 and 65 ng/g lipid weight (lw) for most of the population. In California, where flammability standards are most stringent, PBDE serum concentrations are higher (60–80 ng/g lw) than the rest of the country (Zota et al. 2008). The highest PBDE concentrations are found in U.S. children and toddlers compared to their parents as a result of exposure in breast milk and hand-to-mouth contact with indoor dust (Fischer et al. 2006; Toms et al. 2009). Populations at risk for occupational exposure to PBDEs such as firefighters (Shaw et al. 2013), foam recyclers and carpet installers (Stapleton et al. 2008a) have serum concentrations two to three times higher than the general population.

Studies conducted by the U.S. Centers for Disease Control and Prevention (CDC) suggest that serum PentaBDE concentrations may be declining in infants and young adults, but are still increasing in older adults. Among the participants of a National Health and Nutrition Examination Survey (NHANES) in 2003–2004, concentrations of BDE-47 in human serum samples significantly decreased in younger adults and increased in older adults ($\geq$60 yrs) (Sjödin et al. 2008a). Similarly, a statistically significant reduction in the concentrations of BDE-47, BDE-100, and $\sum$PBDEs were found in blood spot composites from 1,224 newborns in New York State) from 1997 to 2011 (Ma et al. 2013).

Average serum concentrations of total PBDEs in Europeans range from 1.0 to 5.0 ng/g lw in Greece, the Netherlands and Sweden (Kalantzi et al. 2011; Roze et al. 2009; Weiss et al. 2006, respectively) to 15–25 ng/g lw in Norway and France (Antignac et al. 2008; Thomsen et al. 2008). Breast milk concentrations in European women are an order of magnitude lower than those in U.S. women. In Asia, PBDE concentrations in milk and serum of Japanese and Philippino women are similar to those found in Europe (Haraguchi et al. 2009; Inoue et al. 2006; Malarvannan et al. 2013). However, in regions of China where many people dismantle and recycle e-waste in their homes, PBDE concentrations can be much higher than those found in the U.S. (Shi et al. 2013; Bi et al. 2007; Zhu et al. 2009). In Guiyu, China, a major e-waste recycling region, average serum PBDE concentrations were 580 ng/g lw (Bi et al. 2007).

Few data exist on PBDE concentrations in populations from Central and South America. PBDE concentrations reported in adipose samples from Brazil were relatively low (Kalantzi et al. 2009). In contrast, PBDE serum concentrations in children working and living at a waste disposal site in Nicaragua were among the highest ever reported (Athanasiadou et al. 2008). Serum BDE-47 levels in these children exceeded those in U.S. children (Fischer et al. 2006; Windham et al. 2010; Eskenazi et al. 2011), and were an order of magnitude higher than the median concentration for BDE-47 in the adult U.S. population (Sjödin et al. 2008b).

Women from Ghana and South Africa have breast milk PBDE concentrations similar to those found in women from Europe and Japan (Darnerud et al. 2011; Asante et al. 2011). In Australia, adult serum (9–20 ng/g lw) and milk concentrations (11 ng/g lw) are higher than those reported in Europe but lower than those reported in U.S. women (Toms et al. 2007, 2009).

HBCDs In contrast to PBDEs, HBCD concentrations in human tissues are higher in Europe and Asia than in North America. Average serum concentrations ranged from 0.5 in Sweden to 9.6 ng/g lw in Norway (Weiss et al. 2006; Thomsen et al. 2008). Average breast milk concentrations reported in European and Japanese women were similar ($\sim$2 ng/g lw) (Colles et al. 2008; Thomsen et al. 2010; Polder et al. 2008; Kakimoto et al. 2008). Higher HBCD levels have been found in milk of Chinese (4.3 ng/g lw) (Shi et al. 2013) and Australian women (10 ng/g lw), reflecting higher usage of this FR in those countries (Toms et al. 2012).

A recent study (Carignan et al. 2012) reported concentrations of HBCDs in milk samples from Boston, Massachusetts ranging from 0.36 to 8.1 ng/g lw (GM 1020 pg/g lw). HBCD body burdens in the mothers were positively associated with the number of stereo and video electronics in the home. A Norwegian study also found a weak positive trend between levels of HBCDs in mother's milk and the number of televisions in the home (Thomsen et al. 2010). In the Boston study, the association was stronger for stereo and video electronics other than televisions (e.g., CD players, DVD players, stereos), suggesting that electronics may be an important source of HBCD exposure in the indoor environment.

TBBPA There are very few reports of TBBPA concentrations in humans. TBBPA is a phenol that can be rapidly conjugated and subsequently excreted (Birnbaum and Staskal 2004), thus detection frequencies for this compound are typically low. Given that the biological half-life of TBBPA in human serum is estimated to be two to six days, continuous exposure may be required to maintain detectable levels in human tissue (Covaci et al. 2009).

Low concentrations of TBBPA, range not detected (ND) to 0.55 ng/g lw, were reported in milk samples collected in 2004–2005 from women in Boston, Massachusetts (Carignan et al. 2012). TBBPA was found in fewer (32 %) samples and at much lower concentrations than HBCD in the same samples. A weak positive relationship was found between TBBPA and HBCDs, suggesting possible related sources or routes of exposure. Johnson-Restrepo et al. (2008) found similar low concentrations of TBBPA in human adipose tissue from New York City residents (ND–0.46 ng/g lw). Concentrations of TBBPA were tenfold lower than HBCD concentrations and three to four orders of magnitude lower than PBDEs measured in the same samples.

Similar low TBBPA concentrations were reported in breast milk from European women (Abdallah and Harrad 2011; Cariou et al. 2008). Higher TBBPA concentrations ranging from ND to 12.46 ng/g lw (mean 0.41 ng/g lw) were found in milk samples from Beijing, China (Shi et al. 2013), reflecting the expanded production and use of TBBPA in Asia.

3.4 Health Effects of Brominated Flame Retardants

Exposure to PBDEs has been associated with and/or causally related to a wide range of health effects including endocrine disruption, reproductive/developmental effects, and neurotoxicity in animals and humans (Birnbaum and Staskal 2004; Costa and Giordano 2007; Shaw and Kannan 2009). HBCDs and TBBPA are also endocrine disruptors that are associated with similar adverse effects in laboratory animals (Covaci et al. 2006, 2009). Little is known about the human health effects of HBCDs and TBBPA.

3.4.1 Endocrine Disruption

BFRs have the potential to disrupt the endocrine system at multiple target sites, resulting in effects on thyroid, ovarian, and androgen function (Birnbaum and Staskal 2004). One of the primary toxic effects of BFRs is thought to be disruption of thyroid hormone homeostasis. Thyroid hormone alterations during development are of special concern because small changes in maternal and fetal thyroid levels can cause neurologic impairments and birth defects (Birnbaum and Staskal 2004; Costa and Giordano 2007).

Generally, human studies have reported positive associations between PBDE concentrations and serum thyroid hormone levels (Dallaire et al. 2009; Yuan et al. 2008). However, a recent study (Chevrier et al. 2010) reported an inverse relationship between serum PBDE concentrations and thyroid stimulating hormone (TSH) levels in pregnant women, suggesting that PBDE exposure throughout pregnancy may reduce thyroid hormone levels in fetuses and newborn babies. Kim et al. (2013) reported that PBDEs and, to a lesser extent, HBCDs were transferred from mothers to infants (both normal infants and infants with congenital hypothyroidism). ß-HBCD concentrations were correlated with reduced triiodothyronine (T3) levels in serum of mothers of infants with congenital hypothyroidism, but not in mothers of normal infants, suggesting that HBCDs may have influenced thyroid hormone function in the hypothyroid infants.

Evidence of TBBPA-induced thyroid alterations is limited to experimental data. TBBPA is structurally related to thyroxine (T4) and can act as a thyroid hormone agonist or antagonist in developing laboratory animals (Birnbaum and Staskal 2004). *Rana rugosa* tadpoles co-exposed to TBBPA and T3 exhibited suppression of T3-induced tail shortening, indicating a thyroid hormone antagonist effect (Kitamura et al. 2005) A reproductive developmental feeding study in rats reported an increase in T3 in TBBPA-exposed female rat offspring and a reduction in circulating total T4 in both genders (van der Ven et al. 2008). The developmental effects also included increased testis and pituitary weight.

PBDE and HBCD concentrations in house dust have been associated with altered hormone levels in men (Meeker et al. 2009; Johnson et al. 2013). PBDE levels in dust were related to reduced serum concentrations of the free androgen index, luteinizing hormone (LH), and follicle-stimulating hormone (FSH) and increased sex-hormone binding globulin (SHBG) and free T4. HBCD levels in dust were associated with decreased serum levels of SHBG and increased free androgen index in the men.

In rodents, HBCD exposure induces testicular lesions, reduced sperm count and sperm abnormalities, reduced ovarian follicles, and increased mortality of rat pups (van der Ven et al. 2009; Ema et al. 2007). Recent data (Fa et al. 2013) indicate that HBCD may exert sustained effects on cultured Leydig cells (Leydig cells are interstitial cells that produce testosterone).

As disruptors of adipocyte metabolism, PBDEs can predispose animals to obesity and diabetes (Hoppe and Carey 2007). Zhang et al. (2013) reported that low doses of BDE-209 impaired glucose homeostasis, which may contribute to Type 1 diabetes mellitus. In U.S. adults, background serum concentrations of the hexaBDE congener 153 were related to metabolic obesity syndrome and diabetes prevalence, and the authors suggested this provides some evidence that PBDEs may contribute to diabetes in the general population (Lim et al. 2008). Additional studies are needed to determine how different PBDE congeners may affect the risk of developing Type I and Type II diabetes in humans.

3.4.2 Reproductive and Developmental Effects

In a study of mother-son pairs from Denmark and Finland, elevated PBDE levels in breast milk were positively associated with cryptorchidism (undescended testicles) in boys (Main et al. 2007). The PBDE levels associated with cryptorchidism were also positively correlated with serum LH concentrations in the infants. Akutsu et al. (2008) reported that elevated blood levels of BDE-153 correlated with decreased sperm count and decreased testes size.

PBDE exposure was associated with reduced fertility in women from a predominantly Mexican-immigrant community in California (Harley et al. 2010). Increasing serum levels of PentaBDE congeners were significantly associated with longer time to pregnancy. Prenatal PBDE exposure of the infants of these women was associated with low birth weight, altered cognitive behavior, and reduced plasma levels of TSH (Harley et al. 2009). Elevated levels of PBDEs in breast milk of pregnant Taiwanese women were significantly associated with reduced weight, length, and chest circumference of their infants (Chao et al. 2007). In both studies, the effects were observed at levels lower than the average PBDE levels in the adult U.S. population. Elevated PBDE concentrations in umbilical cord blood of pregnant Chinese women involved in e-waste recycling were associated with premature delivery, low birth weight, and stillbirth among the infants (Wu et al. 2010). BDE-209 dominated the PBDE congener profiles in the Chinese women, suggesting continuous exposure to DecaBDE in e-waste operations.

3.4.3 Developmental Neurotoxicity

Experimental data indicate that PBDEs are able to cause neurotoxic effects, including aberrations in spontaneous behavior, habituation capability, learning, memory and changes in the cholinergic system in developing animals (Eriksson et al. 2006; Viberg et al. 2007; Johansson et al. 2008). Less is known about the developmental neurotoxicity of HBCDs and TBBPA (Covaci et al. 2006, 2009).

Increasing evidence suggests that prenatal PBDE exposure may cause neurodevelopmental deficits in children. Herbstman et al. (2010) reported that elevated PBDE concentrations in umbilical cord blood were associated with lower IQ performance scores (ranging from five to eight points lower) in children one through 6 years of age. In Dutch children, prenatal exposure to pentaBDEs and HBCDs was associated with adverse effects on motor, cognitive, and behavioral outcomes (Roze et al. 2009). Schreiber et al. (2010) demonstrated that the disruption of cellular thyroid hormone signaling by pentaBDE congeners profoundly affects the development of fetal human neural progenitor cells. Recently, Eskenazi et al. (2013) reported that both prenatal and childhood exposures to PBDEs were associated with reduced attention, fine motor coordination, and cognition including Verbal and Full-Scale IQ deficits in children at 5 and 7 years of age. This study, the largest to date, adds to growing evidence that PBDEs adversely affect child neurobehavioral development.

HBCDs and TBBPA are known to be neurotoxic to rat brain cells, causing inhibition of intracellular calcium regulation and neurotransmitter release (Reistad et al. 2006, 2007). A recent study found that low doses of HBCDs were highly cytotoxic to human neuroblastoma cells, inducing cell death and disregulation of Ca^{2+} signaling and mitochondrial function, along with the release of ß-amyloid peptide, similar to the neuronal degeneration observed in Alzheimer's disease (Al-Mouza and Michelangelli 2012).

3.4.4 Immunotoxicity

TBBPA is a cytotoxicant that can interfere with cellular signaling pathways, reducing cell viability and proliferation (Strack et al. 2007). TBBPA is also highly immunotoxic in vitro (Pullen et al. 2003) and inhibits T-cell activation by blocking the expression of CD25 proteins that are essential for the proliferation of activated T cells. TBBPA also affects human natural killer cells (Kibakaya et al. 2009) and induces reactive oxygen species-mediated toxicity (Choi et al. 2011). Thus, TBBPA may have a profound effect on an organism's innate and cell-mediated immune responses to bacteria, viruses, and possibly cancer.

3.4.5 Carcinogenicity and Genotoxicity

The carcinogenic potential of BFRs has not been adequately addressed in animal models or in human epidemiology studies. In 2-year studies conducted by the National Toxicology Program (NTP), exposure to high levels of DecaBDE caused liver and thyroid tumors in rodents (NTP 1986). The NTP is conducting studies to evaluate the carcinogenicity of DE-71, a commercial PentaBDE mixture (NTP 2013a). An earlier study reported an association between BDE-47 concentrations and an increased risk for non-Hodgkin's lymphoma (NHL) patients (Hardell et al. 2001). In the highest risk/highest exposure group, BDE-47 treatment was also significantly associated with elevated titers to Epstein Barr IgG, a human herpes virus that has been associated with certain subgroups of NHL. The incidence of thyroid cancer has been increasing in the U.S., especially among women and newborn babies, and part of the observed increase is hypothesized by some scientists to be related to exposure to PBDEs and other thyroid hormone disrupting compounds (Zhang et al. 2008; Johnson-Restrepo et al. 2007), but additional studies are needed to confirm this suspected relationship.

Cell line studies indicate that HBCD exposure can induce malignant transformation in mammalian cells by a non-mutagenic mechanism (Helleday et al. 1999). A recent NTP carcinogenicity study has reported a significant increase in the incidence of tumors in the liver, stomach, and other organs of TBBPA-exposed male mice and lung tumors in females. In male rats, TBBPA exposure resulted in a significant incidence in tumors of the pituitary gland, skin, testes, and other organs. In exposed female rats, tumor incidence was significantly increased in the lung, adrenal gland, mammary glands, pituitary, thyroid, uterus, and other organs (NTP 2013b).

3.5 Replacement Flame Retardants

With the 2005 phase-out of PBDEs, many other brominated, chlorinated, and organophosphate flame retardants (OPFRs) have been introduced as replacement FRs (USEPA 2005). Little is known about the composition, uses, human exposure levels and health effects of these chemicals now in everyday use. Of special concern are the replacement chemicals for PentaBDE, particularly tris (1,3-dichloro-2-propyl) phosphate (TDCPP or chlorinated Tris) and Firemaster 550® (FM 550) which contains two brominated components and some additional organophosphate OPFRs (Stapleton et al. 2011, 2012). TDCPP and FM 550 components were the most common FRs detected in PUF foam in U.S. couches purchased after 2005 and baby products such as changing pads and nursing pillows, suggesting that TDCPP and FM 550 are among the highest-volume FRs in households today.

TDCPP, and its brominated analogue tris (2,3-dibromopropyl) phosphate (TDBPP or brominated Tris) were previously removed from children's sleepwear after being found to be mutagenic (Blum and Ames 1977; Blum et al. 1978; Gold et al. 1978). Both compounds are also probable human carcinogens (Babich et al. 2006; NTP 2011). A recent study found that TDCPP concentrations in house dust were correlated with reduced free T4 and increased prolactin levels in men (Meeker and Stapleton 2010). TDCPP has been detected at concentrations similar to PentaBDEs in U.S. house dust, suggesting chronic exposure to the population (Stapleton et al. 2009).

Replacements for the OctaBDE and DecaBDE mixtures include 1,2-bis (2,4,6-tribromophenoxy) ethane (BTBPE) and decabromodiphenylethane (DBDPE). Both BTBPE and DBDPE have been found in U.S. house dust (Stapleton et al. 2008b), and the data indicate that DBDPE may be more persistent and bioaccumulative than DecaBDE (Gauthier et al. 2009).

3.6 Summary and Conclusions

Many halogenated FRs are added to household products to meet existing flammability standards and have a high potential for adverse health effects while lacking demonstrated fire safety benefits. Flammability standards for manufactured household products need to be updated so that they can be met without toxic chemicals.

PentaBDE and OctaBDE have not been produced in the U.S. since 2004, yet millions of pounds of these and other untested replacement flame retardants mixed with foam and plastic are present in household products and dust. The continued use of FRs with established health concerns and introduction of replacement FRs with limited toxicity data highlights the need to modernize U.S. chemical policies to require disclosure and safety testing of household product chemicals prior to sale.

According to the furniture industry, the average lifetime for foam-containing household furniture is $\sim$30 years, suggesting that only a fraction of the PBDEs used in furniture has migrated from indoor environments to the outdoor environment. The identification and responsible disposal of treated products is essential to prevent future dispersal into the environment and the food supply. Currently, the USEPA does not provide clear guidelines for the proper disposal of these products, however studies are in place to improve the process of recycling e-waste (USEPA 2013).

3.6.1 Data Gaps

For many FRs in use today, basic information on chemical identity, specific product applications, and volumes used, are typically not available, significantly limiting accurate assessment of their potential health effects.

For most household furniture, electronics, food packaging, and housing insulation foam, actual concentrations of BFRs in the polymers or finished products are unknown.

Human health effects of HBCDs and TBBPA are largely unknown, and experimental data for many health endpoints is limited. Similarly, the carcinogenic potential of BFRs has not been adequately addressed in animals or humans.

Toxicity data for numerous replacement FRs are lacking. Producers should be required to provide health and toxicity data on replacement FRs before the chemicals are marketed.

3.6.2 Recommended Future Research

Research is needed to determine actual concentrations of BFRs and replacement FRs in various polymers and in finished household products that are in widespread use.

Further research is warranted on pathways of human exposure to BFRs and the effectiveness of dust reduction strategies [wet mopping, hand washing, vacuuming with a high efficiency particulate air filter (also called a HEPA filter)] to reduce individual exposure to FRs in house dust. Studies are needed to investigate infant exposure to FRs in household products and house dust, since infants are in intimate contact with these materials and are more vulnerable to adverse effects than an older child or adult.

Experimental studies are warranted to generate toxicity data on replacement FRs before they are marketed. Epidemiological studies are needed to evaluate health effects of human exposure, particularly children's exposure to BFRs and replacement FRs in everyday use.

Because there is currently no strategy in place for the identification or safe disposal of treated furniture, in-use and discarded furniture and electronics will continue to be a source of indoor and outdoor exposure. More research is urgently needed to develop and refine sustainable end-of-life solutions, especially for e-waste.

References

Abb M, Stahl B, Lorenz W (2011) Analysis of brominated flame retardants in house dust. Chemosphere 85:1657–1663

Abdallah MA-E, Harrad S (2011) Tetrabromobisphenol-A, hexabromocyclododecane and its degradation products in UK human milk: relationship to external exposure. Environ Int 37(2):443–448

Abdallah MA-E, Harrad SJ, Covaci A (2008a) Hexabromocyclododecanes and tetrabromobisphenol-A in indoor air and dust in Birmingham, UK: implications for human exposure. Environ Sci Technol 42:6855–6861

Abdallah MA-E, Harrad SJ, Ibarra C et al (2008b) Hexabromocyclododecanes in indoor dust from Canada, the United Kingdom, and the United States. Environ Sci Technol 42:459–464
Akutsu K, Takatori S, Nozawa S et al (2008) Polybrominated diphenyl ethers in human serum and sperm quality. B Environ Contam Tox 80:345–350
Alaee M, Arias P, Sjodin A et al (2003) An overview of commercially used brominated flame retardants, their applications, their use patterns in different countries/regions and possible modes of release. Environ Int 29:683–689
Albemarle (2010–2011) Fire safety solutions: 2010–2011 product selector guide, http://alemarle.com. Accessed 24 Jan 2013
Ali N, Dirtu AC, Eede NVD et al (2012) Occurrence of alternative flame retardants in indoor dust from New Zealand: indoor sources and human exposure assessment. Chemosphere 88(11):1276–1282
Allen JG, McClean MD, Stapleton HM et al (2008) Linking PBDEs in house dust to consumer products using X-ray fluorescence. Environ Sci Technol 42:4222–4228
Al-Mousa F, Michelangelli F (2012) Some commonly used brominated flame retardants cause Ca^{2+} -ATPase inhibition, beta-amyloid peptide release and apoptosis in SH-SY5Y neuronal cells. PLoS ONE 7:1–8
Andra SS, Makris KC, Shine JP et al (2012) Co-leaching of brominated compounds and antimony from bottled water. Environ Int 38:45–53
Antignac J-P, Cariou R, Maume D et al (2008) Exposure assessment of fetus and newborn to brominated flame retardants in France: preliminary data. Mol Nutr Food Res 52:258–265
Asante KA, Adu-Kumi S, Nakahiro K et al (2011) Human exposure to PCBs, PBDEs and HBCDs in Ghana: Temporal variation, sources of exposure and estimation of daily intakes by infants. Environ Int 37(5):921–928
Athanasiadou M, Cuadra SN, Marsh G et al (2008) Polybrominated diphenyl ethers (PBDEs) and bioaccumulative hydroxylated PBDE metabolites in young humans from Managua, Nicaragua. Environ Health Perspect 116:400–408
Babich MA, Thomas TA, Hatlelid KM (2006) CPSC staff preliminary risk assessment of flame retardant (FR) chemicals in upholstered furniture foam. Consumer Product Safety Commission. https://www.cpsc.gov/library/foia/foia06/brief/uhff1.pdf. Accessed 08 May 2013
Batterman SA, Chernyak S, Jia C et al (2009) Concentrations and emissions of polybrominated diphenyl ethers from U.S. houses and garages. Environ Sci Technol 43:2693–2700
Bi X, Thomas G, Jones KC et al (2007) Exposure of electronics dismantling workers to polybrominated diphenyl ethers, polychlorinated biphenyls, and organochlorine pesticides in South China. Environ Sci Technol 41:5647–5653
Birnbaum LS, Staskal DF (2004) Brominated flame retardants: cause for concern? Environ Health Perspect 112:9–17
Blum A, Ames BN (1977) Flame-retardant additives as possible cancer hazards. Science 195:17–23
Blum A, Gold MD, Ames BN et al (1978) Children absorb tris-BP flame retardant from sleepwear: urine contains the mutagenic metabolite, 2,3-dibromopropanol. Science 201:120–123
Bromine Science Environmental Forum (BSEF) (2009) Factsheet hexabromocyclododecane. http//www.bsef.com. Accessed 10 Oct 2010
Bromine Science Environmental Forum (BSEF) (2012a) Factsheet Deca-BDE. http//www.bsef.com. Accessed 07 May 2013
Bromine Science Environmental Forum (BSEF) (2012b) Factsheet TBBPA. http//www.bsef.com. Accessed 07 May 2013
Carignan CC, Abdallah MA-E, Wu N et al (2012) Predictors of tetrabromobisphenol-A (TBBP-A) and hexabromocyclododecanes (HBCD) in milk from Boston mothers. Environ Sci Technol 46(21):12146–12153
Carignan CC, Heiger-Bernays W, McClean MD et al (2013) Flame retardant exposure among collegiate United States gymnasts. Environ Sci Technol 47(23):13848–13856

Cariou R, Antignac J-P, Zalko D et al (2008) Exposure assessment of French women and their newborns to tetrabromobisphenol-A: occurrence measurements in maternal adipose tissue, serum, breast milk and cord serum. Chemosphere 73:1036–1041

Chao H-R, Wang S-L, Lee W-J et al (2007) Levels of polybrominated diphenyl ethers (PBDEs) in breast milk from central Taiwan and their relation to infant birth outcome and maternal menstruation effects. Environ Int 33:239–245

Chen S-J, Ma Y-J, Wang J et al (2009) Brominated flame retardants in children's toys: concentration, composition, and children's exposure and risk assessment. Environ Sci Technol 43:4200–4206

Chevrier J, Harley KG, Bradman A et al (2010) Polybrominated diphenyl ether (PBDE) flame retardants and thyroid hormone during pregnancy. Environ Health Perspect 118:1444–1449

Choi JS, Lee YJ, Kim TH et al (2011) Molecular mechanism of tetrabromobisphenol A (TBBPA)-induced target organ toxicity in Sprague-Dawley male rats. Toxicol Res 27:61–70

Choi KI, Lee SH, Osako M (2009) Leaching of brominated flame retardants from TV housing plastics in the presence of dissolved humic matter. Chemosphere 74(3):460–466

Colles A, Koppen G, Hanot V et al (2008) Fourth WHO-coordinated survey of human milk for persistent organic pollutants (POPs): Belgian results. Chemosphere 73:907–914

Costa LG, Giordano G (2007) Developmental neurotoxicity of polybrominated diphenyl ether (PBDE) flame retardants. NeuroToxicol 28:1047–1067

Covaci A, Gerecke AC, Law RJ et al (2006) Hexabromo-cyclododecanes (HBCDs) in the environment and humans: a review. Environ Sci Technol 40:3679–3688

Covaci A, Voorspoels S, Abdallah MA-E et al (2009) Analytical and environmental aspects of the flame retardant tetrabromobisphenol-A and its derivatives. J Chromatography A 1216:346–363

D'Hollander W, Roosens L, Covaci A et al (2010) Brominated flame retardants and perfluorinated compounds in indoor dust from homes and offices in Flanders, Belgium. Chemosphere 81:478–487

Dallaire R, Dewailly E, Pereg D et al (2009) Thyroid function and plasma concentrations of polyhalogenated compounds in Inuit adults. Environ Health Perspect 117:1380–1386

Darnerud PO, Aune M, Larsson L et al (2011) Levels of brominated flame retardants and other persistent organic pollutants in breast milk samples from Limpopo province, South Africa. Sci Total Environ 409:4048–4053

Di Napoli-Davis G, Owens JE (2013) Quantitation of tetrabromobisphenol-A from dust sampled on consumer electronics by dispersed liquid-liquid microextraction. Env Pollut 180:274–280

Dodson RE, Perovich LJ, Covaci A et al (2012) After the PBDE phase-out: a broad suite of flame retardants in repeat house dust samples from California. Environ Sci Technol 46(24):13056–13066

Ema M, Fujii S, Hirata-Koizumi M et al (2007) Two-generation reproductive toxicity study of the flame retardant hexabromocyclododecane in rats. Reprod Toxicol 25:335–351

Eriksson P, Fischer C, Fredriksson A (2006) Poly-brominated diphenyl ethers, a group of brominated flame retardants, can interact with polychlorinated biphenyls in enhancing developmental neurobehavioral defects. Toxicol Sci 94:302–309

Eskenazi B, Chevrier J, Rauch SA et al (2013) In utero and childhood polybrominated diphenyl ether (PBDE) exposures and neurodevelopment in the CHAMACOS study. Environ Health Perspect 121:257–261

Eskenazi B, Fenster L, Castorina R et al (2011) A comparison of PBDE serum concentrations in Mexican and Mexican-American children living in California. Environ Health Perspect 119:1442–1448

European Union (2003a) European Union risk assessment report diphenyl ether, octabromo derivative. CAS No: 32536-52-0, EINECS No: 251-087-9. http://echa.europa.eu/documents/10162/5b10aa46-9a88-4aed-b338-1e06105b924c. Accessed 15 Dec 2012

European Union (2003b) European Commission 2003 WEEE 2002/96/EC (2003) Directive 2002/96/EC of the European Parliament and of the Council of January 27, 2003 on waste electrical

and electronic equipment (WEEE), Off J Eur Union 037 24. http://eur-lex.europa.eu/LexUriServ/LexUriServ.do?uri=OJ:L:2003:037:0024:0038:en:PDF. Accessed 15 June 2013
Fa S, Pogrimic-Majkic K, Dakic V et al (2013) Acute effects of hexabromocyclododecane on Leydig cell cyclic nucleotide signaling and steroidogenesis in vitro. Tox Letters 218(1):81–90
Fischer D, Hooper K, Athanasiadou M et al (2006) Children show highest levels of polybrominated diphenyl ethers in a California family of four: A case study. Environ Health Perspect 114:1581–1584
Gauthier LT, Potter D, Hebert CE et al (2009) Temporal trends and spatial distribution of non-polybrominated diphenyl ether flame retardants in the eggs of colonial populations of Great Lakes herring gulls. Environ Sci Technol 43:312–317
Gold MD, Blum A, Ames BN (1978) Another flame retardant, tris-(1,3-Dichloro-2-Propyl)-phosphate, and its expected metabolites are mutagens. Science 200:785–787
Goosey E, Abdallah M, Harrad S (2008) Dust from primary school and nursery classrooms in the UK: Its significance as a pathway of exposure for young children to PFOS, PFOA, HBCDs and TBBP-A. Organohalogen Compound 70:855–858
Hale RC, Alaee M, Manchester-Neesvig JB et al (2003) Polybrominated diphenyl ether flame retardants in the North American environment. Environ Int 29:771–779
Haraguchi K, Koizumi A, Inoue K et al (2009) Levels and regional trends of persistent organochlorines and polybrominated diphenyl ethers in Asian breast milk demonstrate POPs signatures unique to individual countries. Environ Int 35:1072–1079
Hardell L, Eriksson M, Lindstrom G et al (2001) Case-control study on concentrations of organohalogen compounds and titers of antibodies to Epstein-Barr virus antigens in the etiology of non-Hodgkin lymphoma. Leuk Lymph 42:619–629
Harley KG, Chevrier J, Bradman A et al (2009) Associations between maternal PBDE serum concentrations and birth weight and duration of gestation. Organohalogen Compound 71:002236(p1-1)
Harley KG, Marks AR, Chevrier J et al (2010) PBDE concentrations in women's serum and fecundability. Environ Health Perspect 118:699–704
Harrad S, de Wit CA, Abdallah MA-E et al (2010) Indoor contamination with hexabromocyclododecanes, polybrominated diphenyls ethers, and perfluoroalkyl compounds: An important exposure pathway for people? Environ Sci Technol 44:3221–3231
Harrad SJ, Ibarra C, Diamond ML et al (2008) Polybrominated diphenyl ethers in domestic indoor dust from Canada, New Zealand, United Kingdom and United States. Environ Int 34:232–238
Helleday T, Tuominen KL, Bergman Å et al (1999) Brominated flame retardants induce intragenic recombination in mammalian cells. Mutat Res 439:137–147
Herbstman JB, Sjodin A, Kurzon M et al (2010) Prenatal exposure to PBDEs and neurodevelopment. Environ Health Perspect 118:712–719
Hites RA (2004) Polybrominated diphenyl ethers in the environment and in people: a meta-analysis of concentrations. Environ Sci Technol 38(4):945–956
Hoppe AA, Carey GB (2007) Polybrominated diphenyl ethers as endocrine disruptors of adipocyte metabolism. Obesity 15:2942–2950
Innes J, Innes A (2003) Plastic flame retardants: technology and current developments. In: Expert overviews covering the science and technology of rubbers and plastics. Rapra Technology Limited, Shropshire, UK p 3–13
Inoue K, Harada K, Takenaka K et al (2006) Levels and concentration ratios of polychlorinated biphenyls and polybrominated diphenyl ethers in serum and breast milk in Japanese mothers. Environ Health Perspect 114(8):1179–1185
International Plastics (2013) Shrink wrap film. http://www.interplas.com/shrink-wrap. Accessed 24 May 2013
Israel Chemicals Limited Industrial Products (ICL-IP) (2012) Flame retardant brochure. http://icl-ip.com. Accessed 7 Feb 2013
Janssen S (2005) Brominated flame retardants: rising levels of concern. Health care without harm, Arlington, VA

Johansson N, Viberg H, Fredriksson A et al (2008) Neonatal exposure to deca-brominated diphenyl ether (PBDE 209) causes dose-response changes in spontaneous behaviour and cholinergic susceptibility in adult mice. Neuro Toxicology 29:911–919

Johnson PI, Stapleton HM, Mukherjee B et al (2013) Associations between brominated flame retardants in house dust and hormone levels in men. Sci Total Environ 445–446:177–184

Johnson PI, Stapleton HM, Sjodin A et al (2010) Relationships between polybrominated diphenyl ether concentrations in house dust and serum. Environ Sci Technol 44:5627–5632

Johnson-Restrepo B, Adams DH, Kannan K (2008) Tetrabromobisphenol A (TBBPA) and hexabromocyclododecanes (HBCDs) in tissues of humans, dolphins, and sharks from the United States. Chemosphere 70:1935–1944

Johnson-Restrepo B, Addink R, Wong C et al (2007) Polybrominated diphenyl ethers and organochlorine pesticides in breast milk from Massachusetts, USA. J Environ Monit 9:1205–1212

Kajiwara N, Noma Y, Takigami H (2008) Photolysis studies of technical decabromodiphenyl ether (DecaBDE) and ethane (DeBDethane) in plastics under natural sunlight. Environ Sci Technol 42:4404–4409

Kajiwara N, Noma Y, Takigami H (2011) Brominated and organophosphate flame retardants in selected consumer products on the Japanese market in 2008. J Hazard Mater 192:1250–1259

Kajiwara N, Takigami H (2010) Behavior of additive brominated flame retardants in textile products. Proceedings of the 5th international symposium on brominated flame retardants. 7-9 April 2010, Kyoto, Japan. http://www.Bfr2010.com/. Accessed 13 April 2010

Kajiwara N, Sueoka M, Ohiwa T et al (2009) Determination of flame-retardant hexabromocyclododecane diastereomers in textiles. Chemosphere 74:1485–1489

Kakimoto K, Akutsu K, Konishi Y et al (2008) Time trend of hexabromocyclododecane in the breast milk of Japanese women. Chemosphere 71:1110–1114

Kalantzi OI, Brown FR, Caleffi M et al (2009) Polybrominated diphenyl ethers and polychlorinated biphenyls in human breast adipose samples from Brazil. Environ Int 35:113–117

Kalantzi OI, Geens T, Covaci A et al (2011) Distribution of polybrominated diphenyl ethers (PBDEs) and other persistent organic pollutants in human serum from Greece. Environ Int 37(2):349–353

Kibakaya EC, Stephen K, Whalen MM (2009) Tetrabromobisphenal A has immunosuppressive effects on human natural killer cells. J Immunotoxicol 6:285–292

Kim U-J, Hong Y-H, Lee D-H et al (2013) PBDEs, HBCDs, TBBPA in infant-mother paired serum: focusing on investigating impact on thyroid hormone, relative proportion and relationship with environmental factors. Proceedings of the 6th international symposium on brominated flame retardants. 7-10 April 2013, San Francisco, USA. http://www.Bfr20103.com/. Accessed 13 April 2013

Kitamura S, Kato T, Iida M et al (2005) Anti-thyroid hormonal activity of tetrabromobisphenol A, a flame retardant, and related compounds: affinity to the mammalian thyroid hormone receptor, and effect on tadpole metamorphosis. Life Sci 18:1589–1601

Kuhn E, Arnet R, Känzig A et al (2004) Bromierte flammschutzmittel in kunststoffprodukten des Schweizer marktes; Swiss agency for the environment, forests and landscape. Bern, Switzerland. http://www.bafu.admin.ch/publikationen/publikation/00289/index.html?lang=de . Accessed 01 May 2013

Lim J-S, Lee D-H, Jacobs DRJ (2008) Association of brominated flame retardants with diabetes and metabolic syndrome in the U.S. population, 2003-2004. Diabetes Care 31:1802–1807

Lorber M (2008) Exposure of Americans to polybrominated diphenyl ethers. J Expo Sci Env Epid 18:2–19

Lowell Center for Sustainable Production (LCSP) (2005) Decabromodiphenylether: an investigation of non-halogen substitutes in electronic enclosure and textile applications. University of Massachusetts Lowell, MA. http://sustainableproduction.org/downloads/DecaBDE SubstitutesFinal4-15-05.pdf. Accessed 01 May 2013

Ma J, Addink R, Yun SH et al (2009) Polybrominated dibenzo-*p*-dioxins/dibenzofurans and polybrominated diphenyl ethers in soil, vegetation, workshop-floor dust, and electronic shredder residue from an electronic waste recycling facility and in soils from a chemical industrial complex in eastern China. Environ Sci Technol 43:7350–7356

Ma WL, Yun S, Bell EM et al (2013) Temporal trends of polybrominated diphenyl ethers (PBDEs) in the blood of newborns from New York State during 1997 through 2011: analysis of dried blood spots from the newborn screening program. Environ Sci Technol 47:8015–8021

Main KM, Kiviranta H, Virtanen HE et al (2007) Flame retardants in placenta and breast milk and cryptorchidism in newborn boys. Environ Health Perspect 115:1519–1526

Malarvannan G, Isobe T, Covaci A et al (2013) Accumulation of brominated flame retardants and polychlorinated biphenyls in human breast milk and scalp hair from the Philippines: Levels, distribution and profiles. Sci Total Environ 442:366–379

Mark F, Dresch EH, Bergfeld B et al (2006) Large scale demonstration of the treatment of electrical and electronic shredder residue, by co-incineration in the Würzburg municipal solid waste incinerator. Plastics Europe (40 pp). http://www.isopa.org/isopa/uploads/Documents/documents/TECHNICAL%20REPORT.pdf. Accessed 01 May 2013

Meeker JD, Johnson PI, Camann D et al (2009) Polybrominated diphenyl ether (PBDE) concentrations in house dust are related to hormone levels in men. Sci Total Environ 407:3425–3429

Meeker JD, Stapleton HM (2010) House dust concentrations of organophosphate flame retardants in relation to hormone levels and semen quality parameters. Environ Health Perspect 118(3):318–323

Mingwu S, Chao W, Yongjuan J et al (2010) Determination of selected polybrominated diphenylethers and polybrominated biphenyl in polymers by ultrasonic-assisted extraction and high-performance liquid chromatography-inductively coupled plasma mass spectrometry. Anal Chem 82:5154–5159

Morf LS, Tremp J, Gloor R et al (2005) Brominated flame retardants in waste electrical and electronic equipment: substance flows in a recycling plant. Environ Sci Technol 39:8691–8699

National Toxicology Program (NTP) (1986) Toxicology and carcinogenesis studies of decabromodiphenyl oxide (CAS No. 1163-19-5) in F344/N rats and B6C3F1 mice (feed studies). TR-309. Research Triangle Park, NC

National Toxicology Program (NTP) (2011) Substance profiles: tris(2,3-dibromopropyl) phosphate CAS No. 126-72-7. Report on carcinogens, 12th edition. http://ntp.niehs.nih.gov/ntp/roc/twelfth/profiles/Tris(dibromopropyl)Phosphate.pdf#search=Tris(2,3-dibromopropyl)-phosphate. Accessed 17 July 2013

National Toxicology Program (NTP) (2013a) Testing status of agents at NTP, pentabromodiphenyl oxide (technical) (DE 71). http://ntp.niehs.nih.gov/?objectid=BD73BA18-123F-7908-7BD4AEF6AB4318BF. Accessed 17 July 2013

National Toxicology Program (NTP) (2013b) Pathology, body weight, and survival tables for peer review, TR-587 tetrabromobisphenol A (TBBPA). http://ntp.niehs.nih.gov/?objectid=7DAAF343-BDB5-82F8-FDB55769687AF8EF. Accessed 01 July 2013

Polder A, Gabrielsen GW, Odland JØ et al (2008) Spatial and temporal changes of chlorinated pesticides, PCBs, dioxins (PCDDs/PCDFs) and brominated flame retardants in human breast milk from Northern Russia. Sci Total Environ 391:41–54

Posner S (2004) Survey and technical assessment of alternatives to decabromodiphenyl ether (decaBDE) in textile applications. Prepared by IFP Research for the Swedish Chemicals Inspectorate. http://www.kemi.se/upload/Trycksaker/Pdf/PM/PM5_04.pdf. Accessed 15 May 2013

Pullen S, Boecker R, Tiegs G (2003) The flame retardants tetrabromobisphenol A and tetrabromobisphenol A-bisallylether suppress the induction of interleukin-2 receptor alpha chain (CD25) in murine splenocytes. Toxicol 184(1):11–22

Reiss M, Ernst T, Popp R et al (2000) Analysis of flame retarded polymers and recycling materials. Chemosphere 40:937–941
Reistad T, Fonnum F, Mariussen E (2006) Neurotoxicity of the pentabrominated diphenyl ether mixture, DE-71, and hexabromocyclododecane (HBCD) in rat cerebellar granule cells in vitro. Arch Toxicol 80:785–796
Reistad T, Mariussen E, Fonnum F (2007) In vitro toxicity of tetrabromobisphenol-A on cerebellar granule cells: cell death, free radical formation, calcium influx and extracellular glutamate. Toxicol Sci 96:268–278
Richter H, Lorenz W, Bahadir M (1997) Examination of organic and inorganic xenobiotics in equipped printed circuits. Chemosphere 35:169–179
Roosens L, Abdallah MA-E, Harrad S et al (2009) Exposure to hexabromocyclododecanes (HBCDs) via dust ingestion, but not diet, correlates with concentrations in human serum: preliminary results. Environ Health Perspect 117:1707–1712
Roze E, Meijer L, Bakker A et al (2009) Prenatal exposure to organohalogens, including brominated flame retardants, influences motor, cognitive, and behavioral performance at school age. Environ Health Perspect 117:1953–1958
Saigal Polymers Private Limited (Pvt Ltd) (2013) Rubber foams, polyolefin foam. http://www.indiamart.com/tnrubber/rubber-foams.html#polyolefin-foam. Accessed 5 Feb 2013
Schecter A, Smith S, Colacino J et al (2011) Contamination of U.S. butter with PBDEs from wrapping paper. Environ Health Perspect 119(2):151–154
Schlummer M, Gruber L, Mäurer A et al (2007) Characterization of polymer fractions from waste electrical and electronic equipment (WEEE) and implications for waste management. Chemosphere 67(9):1866–1876
Schreiber T, Gassmann K, Götz C et al (2010) Polybrominated diphenyl ethers induce developmental neurotoxicity in a human in vitro model: evidence for endocrine disruption. Environ Health Perspect 118:572–578
Shaw SD, Berger ML, Harris JH et al (2013) Persistent organic pollutants including polychlorinated and polybrominated dibenzo-*p*-dioxins and dibenzofurans in firefighters from Northern California. Chemosphere 91(10):1386–1394
Shaw SD, Blum A, Weber R et al (2010) Halogenated flame retardants: do the fire safety benefits justify the risks? Rev Environ Health 25(4):261–305
Shaw SD, Kannan K (2009) Polybrominated diphenyl ethers in marine ecosystems of the American continents: foresight from current knowledge. Rev Environ Health 24:157–229
Shi Z, Jiao Y, Hu Y et al (2013) Levels of tetrabromobisphenol A, hexabromocyclododecanes and polybrominated diphenyl ethers in human milk from the general population in Beijing, China. Sci Total Environ 452–453:10–18
Sjödin A, Päpke O, McGahee EE et al (2008a) Concentration of polybrominated diphenyl ethers (PBDEs) in household dust from various countries. Chemosphere 73:S131–S136
Sjödin A, Wong L-Y, Jones RS et al (2008b) Serum concentrations of polybrominated diphenyl ethers (PBDEs) and polybrominated biphenyl (PBB) in the United States population: 2003-2004. Environ Sci Technol 42:1377–1384
Skinner N (2013) AB 127 fact sheet-Skinner safer building insulation. Assembly member Nancy Skinner: safer building insulation Feb 4 2013(p 1–1)
Stapleton HM, Allen JG, Kelly SM et al (2008a) Alternate and new brominated flame retardants detected in U.S. house dust. Environ Sci Technol 42:6910–6916
Stapleton HM, Dodder NG, Kucklick JR et al (2006) Determination of HBCD, PBDEs and MeO-BDEs in California sea lions (*Zalophus californianus*) stranded between 1993 and 2003. Mar Pollut Bull 52:522–531
Stapleton HM, Klosterhaus S, Eagle S et al (2009) Detection of organophosphate flame retardants in furniture foam and U.S. house dust. Environ Sci Technol 43:7490–7495
Stapleton HM, Klosterhaus S, Keller A et al (2011) Identification of flame retardants in polyurethane foam collected from baby products. Environ Sci Technol 45:5323–5331
Stapleton HM, Sharma S, Getzinger G et al (2012) Novel and high volume use flame retardants in US couches reflective of the 2005 PentaBDE phase out. Environ Sci Technol 46:13432–13439

Stapleton HM, Sjodin A, Jones R et al (2008b) Serum levels of polybrominated diphenyl ethers (PBDEs) in foam recyclers and carpet installers working in the United States. Environ Sci Technol 42:3453–3458

Stockholm Convention on Persistent Organic Pollutants (2009) http://chm.pops.int/. Accessed 05 May 2010

Strack S, Detzel T, Wahl M et al (2007) Cytotoxicity of TBBPA and effects on proliferation, cell cycle and MAPK pathways in mammalian cells. Chemosphere 67(9):S405–S411

Subcommittee on Flame-Retardant Chemicals (SFRC) (2000) Committee on toxicology, board on environmental studies and toxicology, National Research Council, toxicological risks of selected flame-retardant chemicals. National Academic Press, National Academy of Sciences. http://www.nap.edu/openbook.php?record_id=9841&page=R1. Accessed 15 Dec 2012

Takigami H, Sozuki G, Hirai Y et al (2008) Transfer of brominated flame retardants from components into dust inside television cabinets. Chemosphere 73:161–169

Takigami H, Suziki G, Hirai Y et al (2009) Brominated flame retardants and other polyhalogenated compounds in indoor air and dust from two houses in Japan. Chemosphere 76:270–277

Talsness CE, Andrade AJ, Kuriyama SN et al (2009) Components of plastic: experimental studies in animals and relevance for human health. Philos T Roy Soc B 364:2079–2096

Tanabe S, Ramu K, Isobe T et al (2008) Brominated flame retardants in the environment of Asia-Pacific: an overview of spatial and temporal trends. J Environ Monitor 10:188–197

Thomsen C, Knutsen HK, Liane VH et al (2008) Consumption of fish from a contaminated lake strongly affects the concentrations of polybrominated diphenyl ethers and hexabromocyclododecane in serum. Mol Nutr Food Res 52:228–237

Thomsen C, Stigum H, Frøshaug M et al (2010) Determinants of brominated flame retardants in breast milk from a large scale Norwegian study. Environ Int 36:68–74

Thuresson K, Bergman Å, Jakobsson K (2005) Occupational exposure to commercial decabromodiphenyl ether in workers manufacturing or handling flame-retarded rubber. Environ Sci Technol 35:1980–1986

Toms L-M, Guerra P, Eljarrat E et al (2012) Brominated flame retardants in the Australian population: 1993-2009. Chemosphere 89:398–403

Toms L-ML, Harden FA, Symons RK et al (2007) Polybrominated diphenyl ethers (PBDEs) in human milk from Australia. Chemosphere 68:797–803

Toms L-ML, Sjödin A, Harden FA et al (2009) Serum polybrominated diphenyl ether (PBDE) levels are higher in children (2–5 years of age) than in infants and adults. Environ Health Perspect 117:1461–1465

Tue NM, Takahashi S, Suzuki G et al (2013) Contamination of indoor dust and air by polychlorinated biphenyls and brominated flame retardants and relevance of non-dietary exposure in Vietnamese informal e-waste recycling sites. Environ Int 51:160–167

United Nations Environment Programme (UNEP) (2013) Recommendation by the persistent organic pollutants review committee to list hexabromocyclododecane in annex A to the Stockholm convention and draft text of the proposed amendment. Conference of the parties to the Stockholm convention on persistent organic pollutants sixth meeting, Geneva, Switzerland (28 April–10 May 2013)

United States Environmental Protection Agency (USEPA) (2005) Design for the Environment (DfE) U.S EPA partnership: furniture flame retardancy partnership. Environmental profiles of chemical flame-retardant alternatives for low-density polyurethane foam; Vol 1. http://www.epa.gov/dfe/pubs/flameret/altrep-v1/altrepv1-f1c.pdf. Accessed 15 Dec 2012

United States Environmental Protection Agency (USEPA) (2009) Announcements. http://www.epa.gov/oppt/existingchemicals/pubs/actionplans/deccadbe.html. Accessed 15 Dec 2012

United States Environmental Protection Agency (USEPA) (2013) Final report: automated removal of brominated flame retardant material from a mixed e-waste plastics recycling stream. http://cfpub.epa.gov/ncer_abstracts/index.cfm/fuseaction/display.abstractDetail/abstract/8974/report/F. Accessed 19 July 2013 van der Ven LT, van de Kuil T, Leonards PEG et al (2009) Endocrine effects of hexabromocyclododecane (HBCD) in a one-generation reproduction study in Wistar rats. Toxicol Lett 185(1):51–62

van der Ven LTM, van de Kuil T, Verhoef A et al (2008) Endocrine effects of tetrabromobisphenol-A (TBBPA) in Wistar rats as tested in a one-generation reproduction study and a subacute toxicity study. Toxicology 245:76–89

Vehlow J, Bergfeldt B, Jay K et al (2000) Thermal treatment of electrical and electronic waste plastics. Waste Manag Res 18:131–140

Viberg H, Fredriksson A, Eriksson P (2007) Changes in spontaneous behaviour and altered response to nicotine in the adult rat, after neonatal exposure to the brominated flame retardant, deca-brominated diphenyl ether (PBDE 209). Neuro Toxicol 28:136–142

Wäger P, Schluep M, Müller E (2010) RoHS Substances in mixed plastics from waste electrical and electronic equipment. Empa: St. Gallen. http://ewasteguide.info/Waeger_2010_Empa-WEEEForum. Accessed 01 May 2013

Wäger PA, Schluep M, Muller E et al (2012) RoHS regulated substances in mixed plastics from waste electrical and electronic equipment. Environ Sci Technol 46:628–635

Wang J, Ma Y-J, Chen S-J et al (2010) Brominated flame retardants in house dust from e-waste recycling and urban areas in South China: implications on human exposure. Environ Int 36:535–541

Watkins DJ, McClean MD, Fraser AJ et al (2011) Exposure to PBDEs in the office environment: Evaluating the relationships between dust, handwipes, and serum. Environ Health Perspect 119(9):1247–1252

Webster TF, McClean MD, Stapleton HM (2013) Exposure to polybrominated diphenyl ethers in the indoor environment. Fire Technol. doi:10.1007/s10694-013-0334-9

Weiss J, Wallin E, Axmon A et al (2006) Hydroxy-PCBs, PBDEs, and HBCDDs in serum from an elderly population of Swedish fishermen's wives and associations with bone density. Environ Sci Technol 40:6282–6289

Windham GC, Pinney SM, Sjodin A et al (2010) Body burdens of brominated flame retardants and other persistent organohalogenated compounds and their descriptors in US girls. Environ Res 110:251–257

Wong MH, Wu SC, Deng WJ et al (2007) Export of toxic chemicals-a review of the case of uncontrolled electronic-waste recycling. Environ Pollut 149:131–140

World Health Organization (WHO) (1994) Brominated diphenyl ethers. Environmental health criteria 162, World Health Organization, Geneva. http://www.inchem.org/documents/ehc/ehc/ehc162.htm. Accessed 15 Dec 2012

Wu K, Xu X, Liu J et al (2010) Polybrominated diphenyl ethers in umbilical cord blood and relevant factors in neonates from Guiyu, China. Environ Sci Technol 44:813–819

Wu N, Herrmann T, Paepke O et al (2007) Human exposure to PBDEs: associations of PBDE body burdens with food consumption and house dust concentrations. Environ Sci Technol 41:1584–1589

Yuan J, Chen L, Chen D et al (2008) Elevated serum polybrominated diphenyl ethers and thyroid-stimulating hormone associated with lymphocytic micronuclei in Chinese workers from an e-waste dismantling site. Environ Sci Technol 42:2195–2200

Zhang Y, Guo GL, Han X et al (2008) Do polybrominated diphenyl ethers (PBDE) increase the risk of thyroid cancer? Biosci Hyp 1:195–199

Zhang Z, Sun Z-Z, Xiao X et al (2013) Mechanism of BDE209- induced impaired glucose homeostasis based on gene microarray analysis of adult rat liver. Arch Toxicol. doi:10.1007/s00204-013-1059-8

Zhu L, Ma B, Hites RA (2009) Brominated flame retardants in serum from the general population in Northern China. Environ Sci Technol 43:6963–6968

Zota AR, Rudel RA, Morello-Frosch RA et al (2008) Elevated house dust and serum concentrations of PBDEs in California: unintended consequences of furniture flammability standards? Environ Sci Technol 42:8158–8164

Chapter 4
Nanoparticles in Polymer Nanocomposite Food Contact Materials: Uses, Potential Release, and Emerging Toxicological Concerns

Karthik V. Pillai, Piper R. Hunt and Timothy V. Duncan

Abstract Several types of nanotechnology-enabled plastics intended for the storage and transport of foods are close to commercialization. For food contact applications, nanocomposite plastics offer many advantages over traditional polymers. However, while the unique properties of engineered nanomaterials (ENMs) may be harnessed for many positive ends, there are concerns about whether ENMs pose risks to human health. The primary areas of interest for assessing safety of nanocomposite food contact materials (FCM) are the potential for migration of ENMs into food and the potential toxicity of such released ENMs. This chapter offers a review of theoretical and experimental methods to assess the likelihood of ENM release from nanotechnology-enabled materials into liquid media, as well as a brief overview of the potential toxicological considerations of ENMs likely to be used in FCMs. Because the use of nanotechnology in food contact applications is a developing field, this chapter also provides background information on some of the food-related applications of nanocomposites currently in development, and a discussion of current methods being used to assess the release of non-nanoscale food packaging additives or contaminants. The goal of this work is to provide readers with an appreciation for current activity in this field as well as an understanding of data gaps that may need to be addressed in order to ensure the safety of this emerging technology.

K. V. Pillai · T. V. Duncan (✉)
Division of Food Processing Science and Technology, Office of Food Safety,
Center for Food Safety and Applied Nutrition, U.S. Food and Drug Administration,
Bedford Park, IL, USA
e-mail: timothy.duncan@fda.hhs.gov

K. V. Pillai
e-mail: Karthik.Pillai@fda.hhs.gov

P. R. Hunt
Division of Toxicology, Office of Applied Research and Safety Assessment,
Center for Food Safety and Applied Nutrition, U.S. Food and Drug Administration,
Laurel, MD, USA
e-mail: Piper.Hunt@fda.hhs.gov

S. M. Snedeker (ed.), *Toxicants in Food Packaging and Household Plastics*,
Molecular and Integrative Toxicology, DOI: 10.1007/978-1-4471-6500-2_4,

Keywords Engineered nanomaterials · Food packaging · Polymer nanocomposites · Migration · Nanotoxicity · Nanoparticle toxicity · Safety of nanotechnology

4.1 Key Take Home Points

- Nanoscale materials exhibit unique properties not present in their corresponding bulk forms.
- Polymer nanocomposites (PNCs) are polymer materials (including plastics that may be used in food packaging) with engineered nanomaterials (ENMs) coated on the surface or dispersed internally.
- Compared to polymers with no ENM element, PNCs have enhanced physical properties including strength, flame retardancy and permeability.
- More research is needed to fully understand the potential of ENMs to migrate from food packaging PNCs to food and beverages.
- More research is needed to fully understand the potential toxicity of ENMs, particularly by oral routes of exposure.

4.2 Introduction

Nanotechnology is a relatively young branch of science dedicated to studying the attributes and interactions of matter on the nanoscale, usually defined as 1–100 nanometers (nm), where 1 nm is 1×10^{-9} m (one billionth of a meter). Matter with particles on the nanoscale has unique chemical and physical properties owing to quantum confinement of electrons and the high degree of relative surface area compared to macro-scale versions of the same basic material. Hence, ENMs are envisioned to have applications in many industries, including construction, biomedicine, and electronics, as well as use in consumer products like cosmetics, sports equipment, nutritional supplements, and food packaging.

Along with the anticipated benefits of these applications, there are concerns of whether there may be health and environmental risks associated with the widespread use of nanotechnology in the future. Use of any new technology in food production or packaging carries with it certain risks that need to be evaluated, but evaluating the toxicological properties and risk of exposure to ENMs involves unique methodological challenges atypical of other chemical substances. Since chemical and physical properties of ENMs are still an active area of research, there is uncertainty over the ways that ENMs may interact with biological systems.

Once regulated for use, ENMs may be added directly to food or incorporated into FCMs (including but not limited to packaging) for various functions. Foods naturally contain numerous structures on the nanoscale (e.g., proteins, micelles), which is why we refer here to those that are specifically engineered to have unique size-dependent (nanoscale) properties. Since ENM-based food packaging materials may be closer to commercialization than ENMs as direct food additives (Duncan 2011b), studies related to the safety of FCMs that contain ENMs have been identified as high priority.

This chapter will review ENM release from nanotechnology-enabled materials into liquid media, as well as provide a brief overview of the potential toxicological effects of ENMs if they migrate from ENM-packaging into foods. Because the use of nanotechnology in food contact applications is a developing field, this chapter also will provide background information on some of the potential applications of ENMs in FCM, and discuss current methodologies being used to assess the release of non-nanoscale food packaging additives or contaminants. The goal of this work is to provide readers with an appreciation for current activity in this field as well as an understanding of some data gaps in addressing the safety of this emerging technology.

4.3 Introduction to Uses and Functions of Food Packaging Materials

Food packaging must serve a diverse array of complex functions. Food packaging must protect the food from a host of contaminants including dirt, dust, chemicals and microbes. It also must be inert, inexpensive, lightweight, easy to dispose of or recycle, resistant to physical abuse, impervious to extreme conditions during processing, and able to withstand temperature variations during storage or transport. The packaging of foods usually must be impermeable to the diffusion of gases like water vapor or oxygen, and volatile organics like odors and flavors. It must serve a variety of other cosmetic or informational functions, such as offering convenience to the consumer, communicating nutritional information, and providing a canvas for marketing. Above all, it must be *safe*, which means the packaging should not introduce harmful substances into food under intended conditions of use or into the environment after disposal. Recent directions in food packaging development have extended these requirements even further, including developing antimicrobial functionality, enhanced biodegradability and/or the ability to release active components.

Food and beverage packaging today still makes use of traditional materials like metal (aluminum beverage cans and foils), paper (milk cartons, cereal boxes and most secondary packaging), and ceramic (glass jars and bottles). Wood continues to be used for pallets and other types of secondary containment. But it was the advent of plastics that completely revolutionized the storage and transport of packaged foods. Plastics are progressively dominating the modern packaging industry (Fig. 4.1), including the food-packaging sector. Packages that continue to be fabricated from traditional materials usually incorporate synthetic polymers to enhance their function. For example metal cans employ liners made of synthetic polymer epoxy to prevent reactions between foods and the metal surface, milk

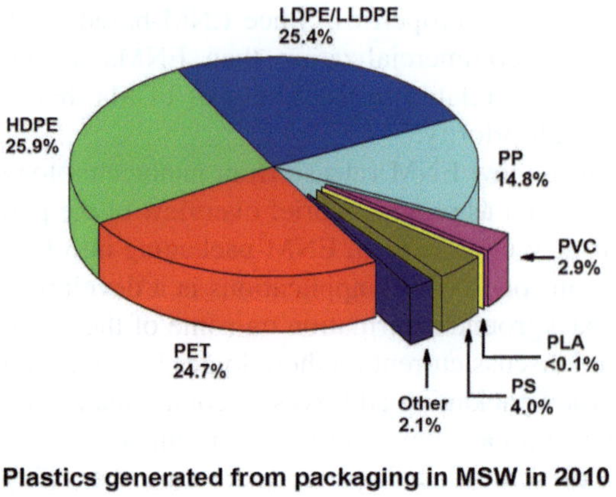

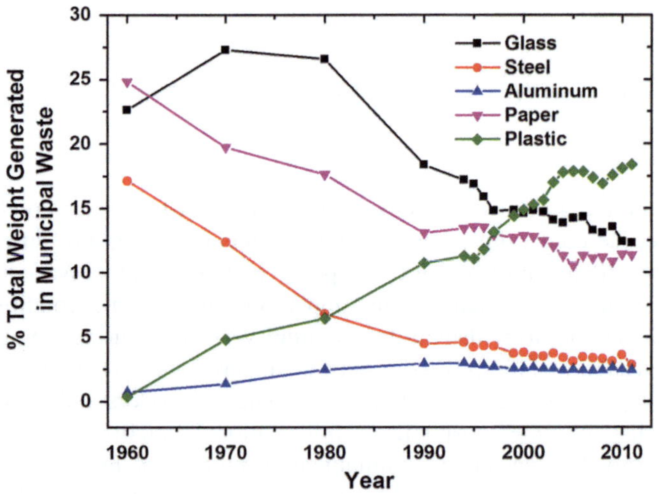

Fig. 4.1 Plastic packaging use in the U.S., represented as the amount generated in municipal solid waste (MSW). Data generated from the USEPA yearly MSW Characterization Reports accessible at http://www.epa.gov/epawaste/nonhaz/municipal/ (USEPA 2013). The *left panel* shows the types of polymers generated from plastic packaging materials in MSW in 2010. The *right panel* plots the percentage by weight of various classes of packaging materials generated in MSW by decade from 1960 to 2010. Wood, corrugated paperboard and some other materials typically comprising secondary packaging (e.g., pallets, boxes) are not represented in the figure

cartons are coated with polyethylene to prevent liquids from seeping through the packaging and causing spoilage, and lids and caps of glass jars and bottles usually include a polymer coating to improve the seal quality. These concepts are not new—even the first paper cartons and metal cans utilized organic coatings made of combinations of natural resins, gums, oils and waxes. The history of food packaging has recently been reviewed (Risch 2009).

Food packages manufactured completely from plastic polymers offer certain functional and economic advantages over those made from metal, glass or paper. Polymers are light-weight and low cost. They can be processed at low temperatures and are easy to form into a wide variety of shapes. Containers made from plastic are flexible and can tolerate forces (like dropping) that could permanently deform cans or shatter glass containers. Plastics can be engineered to have convenient functions for consumers (e.g., squeezable bottles, re-sealable caps, lids and zipping enclosures). Synthetic polymers also have a diverse array of physical properties that can be finely tuned, for example to allow specific levels of moisture or oxygen exchange. In addition, polymers can be blended or coextruded to give mixtures with intermediate properties, or additives can be incorporated into the polymer matrix to impart other properties difficult to achieve through polymer processing alone. This can include additives as simple as pigments or blockers of damaging ultra violet (UV) light, to more sophisticated materials like antimicrobials or molecules which can respond to the presence of pathogens, chemical contaminants or changes in environmental conditions.

Numerous polymers are used in food packaging (Fig. 4.1). The first historical examples were natural resins and waxes, but they were susceptible to decomposition and other chemical reactions, so they have been largely supplanted by synthetic polymers. *Polyolefins*, characterized by uninterrupted chains of carbon–carbon single bond linkages, were among the first such polymers to be used, and continue to be used widely today. *Polyethylene* (PE) is the simplest polymer in this class and is used in a variety of applications, from plastic sandwich bags and carton liners, to milk jugs and ice cream containers. There are many versions of PE produced (distinguished primarily by their relative densities), all highly non-polar materials with low moisture permeability. Other physical properties such as melting point, crystallinity and toughness can vary greatly depending on density and other molecular and processing parameters. Substituted polyolefins such as *polypropylene* (PP) and *polystyrene* (PS) are frequently used in food packaging for specific applications that demand higher heat resistance or where the various grades of PE are not hard or strong enough (e.g., microwavable PP containers and trays), or where specific types of processing are required (e.g., PS foams for egg cartons). *Polyvinyl chloride* (PVC) is another substituted polyolefin used frequently in food packaging (e.g., blister packs) because of its excellent thermoforming properties, although its use is currently in decline. Packaging materials that incorporate PE copolymers such as *ethylene vinyl alcohol* (EVOH) have emerged as popular alternatives to glass bottles and metal containers due to their high crystallinity, with the EVOH providing high barriers to oxygen diffusion as well as migration of other nonpolar volatile organics such as flavor compounds. *Non-polyolefinic* polymers, which include polyesters and polyamides, are frequently used as well. In particular, *polyethylene terephthalate* (PET) enjoys popularity as a material for carbonated beverage containers and single use water bottles due to its high degree of optical transparency, good barrier to gas migration, high strength and resistance to shattering. Bio-derived and bio-inspired polymers such as *thermoplastic starch*, *polylactic acid* (PLA) and others are currently under development as replacements for petroleum-based polymers because of their biodegradability. Several reviews are available on the properties of polymers used in food contact applications (Marsh and Bugusu 2007; Robertson 2006).

While synthetic polymers have revolutionized the food industry, they also possess a number of disadvantages that can limit their usefulness or in some cases raise concerns regarding their safety. Chemical stability and difficulty/expense of disposal or recycling rank high among these disadvantages, but the primary functional problem with plastics is their permeability, which is typically far greater than in glass or metal. The higher permeability gives polymers the ability to absorb flavors, colors and odors from foods. It also reduces the ability of polymer films to protect stored food from water vapor and oxygen, which is a major concern because these are leading agents of food spoilage. Beyond food quality, polymer permeability allows chemicals and other substances added to the polymer during synthesis to migrate into the food matrix, or the environment, during the product's lifecycle. Such substances are added to polymers during fabrication or processing, and may include intentional additives such as plasticizers to lend flexibility, antistatic agents, pigments, foaming or slip agents, and antioxidants or antiaging substances to prevent polymer degradation in the presence of light, oxygen and free radicals. Migrants from the interior of polymers may also include remnants or impurities left over from the formation of the polymer itself (Begley et al. 1995). While intelligent engineering and processing of polymers can minimize the permeability of these materials, there is a concern that any new polymer additive or food contact substance could pose a migration hazard. These concerns apply as well to nanoscale additives, which are being developed for a wide range of targeted food contact applications, and which have a number of uncertainties related to their potential to migrate from plastics to the food being eaten by consumers.

4.4 Nanocomposites: State of the Science

A polymer composite is a hybrid material that incorporates a non-polymeric component into the bulk polymeric material in order to improve or alter its function. Fiberglass is an example that integrates glass fibers into the interior of a polymeric matrix (polyester) to increase the polymer's strength. **A *polymer nanocomposite* (PNC) is a polymer composite in which the non-polymeric component has at least one dimension in the nanoscale.** PNCs can be divided into two major categories: those that incorporate the nano-elements as fillers (in the interior of the matrix) and those that incorporate the nano-elements as coatings (on the exterior surface of the matrix) (Fig. 4.2).

Nanomaterials may be dispersed into a polymer matrix to impart specific functions or to modify the properties of the polymer. One example would be the addition of nanoscale fillers with unique optical properties to the polymer matrix; colored, fluorescent or UV-absorbing nanoscale particles can be distributed within a polymer matrix to make colored, fluorescent or UV-absorbing plastics. Also, the addition of nanofillers to polymers increases the strength (Podsiadlo et al. 2007), gas

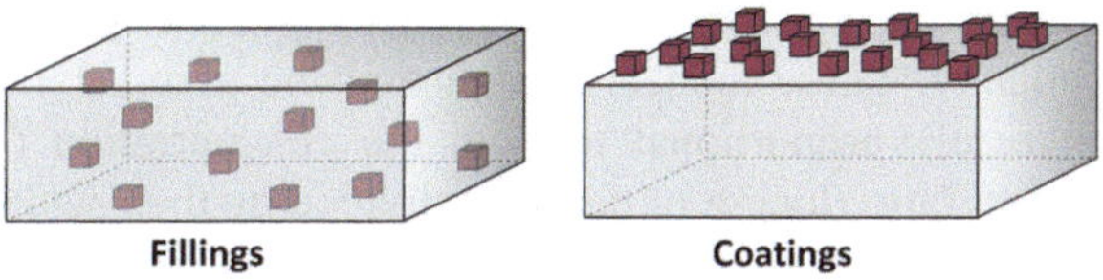

Fig. 4.2 Polymer nanocomposites, depicting filled (*left*) and coated materials (*right*)

permeability (de Moura et al. 2011), and biodegradability (Ray et al. 2002) of the bulk materials compared to analogous materials that have no fillers added. While similar changes to bulk polymer properties can be achieved with fillers possessing dimensions larger than the nanoscale (fiberglass example), the extremely large relative surface area of nanoparticles can magnify the desired property enhancements. Also, because nanoparticles have spatial dimensions of the same or smaller magnitude as visible light wavelengths, nanocomposite plastics are often visibly transparent, enhancing the product's appearance. Owing to the diversity of physical properties exhibited by PNCs, these materials are anticipated to find use in a number of industries and consumer product sectors. The following sections provide examples of specific uses and functions of PNC plastics used in FCMs, including packaging.

4.4.1 High Barrier Polymer Nanocomposite Plastics

Oxygen or water vapor present in the uncontrolled outside environment are primary causes of food spoilage. A chief function of a food packaging material is to inhibit large-scale movement of gases from the outside through to the package's interior. The type of material(s) used in the packaging plays a large role in determining how well the package meets this requirement. The permeability of metal and glass to gases is practically zero because the structures of these materials consist of extensive three-dimensional networks of strongly bonded atoms. Most organic polymers are comparatively porous, sometimes even to large molecules, and so permeability of polymers in food contact applications presents a significant engineering challenge.

In a conventional thin polymer film, the overall rate at which a gas can cross from one side to the other depends on the molecular structure of the polymer itself and a range of processing parameters. In a thin film of uniform thickness that separates a region of high gas concentration from one of low concentration, the gas will migrate through the film from the high to the low concentration region. There is typically a simple inverse relationship between the transmission time and the film thickness. Thicker films provide better diffusion barriers.

The introduction of fillers such as nanoparticles into the polymer matrix alters the previous example by distributing the impermeable nanoparticles within the diffusion space while maintaining, or enhancing, other physical properties important in packaging materials. While gas molecules take the shortest path possible through a filler-free polymer, in a PNC gas molecules must take detours around the impermeable fillers. This *tortuous path effect* has been modeled theoretically. These models predict that the permeability of tortuous gas diffusion is inversely proportional to both the concentration (volume fill fraction) and the aspect ratio (length of particles divided by their width) of the filler elements. Other factors found to affect the degree of tortuosity include the filler shape, orientation within the matrix, uniformity of size and aggregation state (Choudalakis and Gotsis 2009). Additionally, distributed filler elements can affect gas transmission rates by attenuating the permeability of the polymer itself in *interfacial regions*, where polymer chains are bound to or otherwise affected by particle surfaces.

When the filler element has at least one dimension on the nanoscale, the available surface area (and hence the interfacial volume) is maximized and there is the potential for higher dimensional aspect ratios. Because of the importance of the particle aspect ratio, researchers have identified platelet-shaped particles as the ideal fillers for high barrier PNCs. The most popular candidates are clays such as montmorillonite (MMT), kaolinite, and saponite. These materials are natural aluminosilicates that consist of hundreds of layers, with each layer up to several microns wide but only a single nanometer thick, oriented in stacked *tactoids*. Breaking up the clay tactoids and dispersing individual clay platelets allows for a uniform distribution of the clay platelets in the polymer matrix. This complex composite structure reduces the permeability of the packaging to gases, including oxygen and water vapor. The degree to which the gas permeability is affected is a complex function of the processing methodology, clay type (natural clays have different aspect ratios), use of organic additives, and degree of clay dispersal (Duncan 2011a).

Clay-based PNCs (CPNCs) targeted for food contact applications make use of materials that are also commonly utilized food contact polymers, and there are numerous reports in the literature in which the developed CPNC is clearly intended for food packaging use. Most of these reports have focused on improving oxygen or water moisture barriers in various clay-polymer systems rather than testing whether they can affect the quality or safety of actual foods. One study that did assess the efficacy of CPNCs with a real food substance was a report in 2007: apple slices stored in plastic containers composed of a 3 weight percent (wt %) nano-$CaCO_3$/PP showed almost no more oxygen-mediated oxidation after 10 days than freshly sliced specimens, as measured by (+)-catechin content (Avella et al. 2007). Additional research is needed to confirm that the improved gas barriers of CPNCs improve the quality and safety of food items.

4.4.2 Antimicrobial PNC Plastics

Food packaging reduces the prevalence of pathogenic organisms by restricting microbial access to the food matrix and/or by maintaining an environment that is not conducive to proliferation of microbes post-packaging. There has been recent interest in developing packaging materials that actively limit the growth of microbes by inclusion of antimicrobial substances on the food contact surface. This approach is especially desirable in the case of packages that are intended to contain foods that cannot be heat-sterilized before sealing and for packages that rely on semi-permeable polymer films as the barrier material (e.g., packages intended for storage of fresh produce and meat products). Silver and metal oxide nanoparticles have been identified as promising materials that may be combined with polymers to form antimicrobial PNCs for use in food packaging (Duncan 2011a; de Azeredo 2013).

Elemental and ionic silver have long been known to hold antimicrobial properties, and within the last decade it has been discovered that nanoparticulate silver likewise can kill or inactivate a wide range of bacteria, viruses, fungi, and other microscopic organisms. It continues to be a matter of debate whether *silver nanoparticles* (AgNPs) differ from free silver ions in their mechanism of action or antimicrobial efficacy (Seil and Webster 2012; Sweet and Singleton 2011; Duncan 2011a).

AgNPs can be easily incorporated into polymer matrices to yield plastic materials that inhibit microbial growth on the material's surface. For example, AgNP-based PNCs can limit the growth of methicillin-resistant *Staphylococcus aureus* in woven hospital dressings (Strohal et al. 2005) and urinary catheters (Paladini et al. 2012). In food packaging and processing materials, AgNPs have been incorporated into *low-density polyethylene* (LDPE) multilayer films with efficacy against the fungus *Aspergillus niger* (Sanchez-Valdes et al. 2009), cellulose pads intended to be utilized for fresh food and meat storage as protection against yeast and bacteria (Fernandez et al. 2010), in LDPE-based plastic bags to reduce fruit decay (Li et al. 2009), and in edible alginate coatings for fresh produce (Fayaz et al. 2009). In the latter article, the authors reported that the AgNP-PNC coated produce had higher consumer acceptance (based on taste and appearance) than controls (uncoated vegetables), effectively showing that the use of nanosilver-incorporated packaging materials can potentially improve the quality as well as safety of packaged foods.

Several classes of *metal oxide nanoparticles* (MONPs) have also been shown to have antimicrobial properties and likewise have been incorporated as active filler elements in PNC materials targeted for food packaging. MONPs shown to have antimicrobial properties include those composed of titanium dioxide (TiO_2), zinc oxide, magnesium oxide, copper oxide, and others. TiO_2 particles, in particular, have been identified as promising antimicrobial materials because of their unique photoactivated mechanism, in which absorption of UV-light stimulates reactive oxygen species production that in turn tears apart cellular membranes in the immediate vicinity of the nanoparticle. It may be possible to engineer PNC food packaging materials in which the antimicrobial effect could simply be turned on and off literally with the flip of a light switch. Cerrada et al. dispersed TiO_2 nanoparticles in EVOH films and demonstrated photoactive biocidal properties of these films against numerous microorganisms of relevance to food safety (Cerrada et al. 2008), and a food packaging study by Chawengkijwanich and Hayata showed that TiO_2/polypropylene PNCs inhibit *E. coli* growth on fresh cut lettuce under UV illumination (Chawengkijwanich and Hayata 2008).

Current research needs in the area of antimicrobial PNCs for food packaging applications include achieving a better understanding of the factors (intrinsic to particles or related to the composite material, the stored food, or external conditions) that affect the antimicrobial efficacy, discovery, and optimization of alternative types of ENMs with antimicrobial properties, and assessing the performance of antimicrobial PNCs in real food systems. The first goal is particularly important because a majority of scientific literature reports the use of PNC materials fabricated from particles and host polymers with uncontrolled or unknown properties. Not only does this approach make it difficult to design more efficacious antimicrobial PNCs, but it also hinders efforts to assess the safety of currently existing materials or predict the safety of materials that may be produced in the future. Such efforts are the focus of the remaining sections, starting with an overview of how safety of traditional FCMs are evaluated before looking at how these methods may apply to nanomaterials.

4.5 Overview of FDA Migration Guidelines for Food Packaging Materials

The United States (U.S.) Food and Drug Administration's (FDA) Center for Food Safety and Applied Nutrition (CFSAN) has regulatory oversight over the majority of consumer food and beverage products manufactured in or imported into the U.S. This regulatory oversight extends to:

> any substance the intended use of which or may reasonably be expected to result, directly or indirectly, in its becoming a component or otherwise affecting the characteristics of any food (including any substance intended for use in producing, manufacturing, packing, processing, preparing, treating, packaging, transporting, or holding food; and including any source of radiation intended for any such use), if such substance is not generally recognized, among experts qualified by scientific training and experience to evaluate its safety, as having been adequately shown through scientific procedures (or, in the case of a substance used in food prior to January 1, 1958, through either scientific procedures or experience based on common use in food) to be safe under the conditions of its intended use (United States Code 2010).

This tract, encoded in U.S. Law in Section 321 of the *Federal Food, Drug and Cosmetic Act* (FFDC), is the legal definition of a *food additive* in the U.S.[1] This definition specifies that any substance that directly contacts food and may, under conditions of intended use, become a component of the food is required by law to be safe.

The standard of safety mandated by the U.S. Congress is a "reasonable certainty that no harm will result from the proposed use of an additive", which has been interpreted by the FDA to mean "that there is a reasonable certainty in the minds of competent scientists that the substance is not harmful under the intended conditions of use" (Code of Federal Regulations 2013). The burden of ensuring that a prospective FCM meets this safety standard is the responsibility of manufacturers, who must prove to the FDA that all materials coming in contact with food are safe under

[1] There are additional exemptions (e.g., pesticides and color additives), which have their own separate regulatory definitions.

their intended conditions of use before they are permitted to be marketed in the U.S. The manufacturers, then, provide the FDA with sufficient exposure and toxicity data to make this case. Readers who are interested in a more detailed explanation of FDA's food additive regulatory process should consult the resources provided on FDA's webpage (http://www.fda.gov/Food/IngredientsPackagingLabeling/default.htm) or see a review available in the literature (Rulis and Levitt 2009).

The FDA has published a guidance document for manufacturers titled *Guidance for Industry: Preparation of Premarket submissions for food contact substances: chemistry recommendations* that describes what information should be submitted by the manufacturer in support of a safety claim when soliciting FDA's approval of a new FCM (FDA 2007). A key component of this guidance document is FDA's recommendation that manufacturers provide sufficient information to permit estimation of daily consumer exposure of potential migrants from a FCM. Exposure is an important component in a safety evaluation because toxicity of any food additive substance is based on the likelihood that a consumer will actually ingest it. When the FCM additive in food is indirect and unintended, exposure is a function of both the expected frequency with which a consumer is likely to come into contact with the substance (*frequency of use*) and the expected degree to which the potential additive is likely to be released into the food matrix from the contact substance under the intended conditions of use (*migration*). The Premarket Submission guidance document contains detailed information on the recommended way to estimate or measure migration of small organic molecules (See section II(D) *Migration Testing and Analytical Methods*).

The recommended approach for assessing migration includes preparing a known quantity of the FCM (typically a section or sections of polymer film or other material) and exposing it to a precise volume of a liquid, called a *food simulant*, that mimics the chemical or physical properties of a food class. The exposure times and temperatures used are based on a careful consideration of how the food contact substance is likely to be used before and/or after consumers have purchased the food product. For instance, if the food contact substance is intended to be used as a packaging material for storage of a fatty food (e.g., peanut butter) at room temperature, the guidance document recommends that migration be assessed by storing the material in a known quantity of fractionated coconut oil (the *fatty food simulant*) at a temperature of 40 °C for 10 days, conditions which have been established to be equivalent to storage at 6–12 months at 20 °C. After the extraction experiment is completed, the food simulant is then assayed for the presence of the migrant using an appropriate analytical chemical test. The FDA Premarket Submission guidance document provides specific information about what FDA feels is the best way to handle and report analytical data, validate analytical methods, and tailor the migration experiment to the circumstances related to the material under consideration. Migration experiments are usually designed to reflect a "worst case scenario", so that the maximum likely exposure can be estimated.

Experimental determinations of migration are time consuming, expensive, and prone to error due to the complexity of the experimental design and the low levels of migrant typically being measured. As an alternative to migration experiments, the FDA has specified that the use of mathematical models to estimate diffusion rates can be acceptable, which can be used to predict migration. These models are based on the physicochemical properties of the host material and basic diffusion physics. Modern FCMs are usually made of porous organic polymers, which may be approximated as very viscous fluids. Classic physics, such as Fick's Laws of Diffusion, predicts that the diffusion of small molecules through this fluid matrix is dependent on the substance's concentration gradient, its solubility in the polymer, and temperature. However, many other factors can affect diffusion rates, including the shape, weight, and polarity of the migrant, as well as other additives and plasticizers that may be present in the polymer matrix. A number of semi-empirical models have been developed to estimate migration of FCS from a polymer matrix to food (Brandsch et al. 2002; Begley et al. 2005), and refining semi-empirical models is an active area of research (Welle 2013; Fang et al. 2013). However, many of these models were not developed to explicitly handle nano-particulate migrants, and so the extent to which they can be used to estimate migration rates of nano-sized filler elements from nanocomposites needs to be further studied.

4.6 Assessment of Nanoparticle Release from Polymer Nanocomposites

Experimental literature that focuses on the release of nanoparticles from nano-composites into liquid media is limited, but several studies have evaluated the release of nanosilver from nanocomposites intended for food contact applications. Song et al. (2011) studied the migration of silver into food simulants 3 % acetic acid and 95 % ethanol from AgNP/polyethylene composite films using inductively coupled plasma mass spectrometry (ICP-MS). The experiment in 3 % acetic acid resulted in a *migration ratio* of 5.6 % at 70 °C after 9 hours (h) of exposure.[2] Use of 95 % ethanol as the simulant resulted in a migration ratio of 0.22 %. The authors offered that the higher migration in acetic acid was due to the higher solubility of silver in the acidic medium. Huang et al. (2011) also performed migration experiments with food simulants, using a AgNP/polyethylene plastic bag commercially available in China. They claimed to observe release of whole AgNP nanoparticles of sizes up to 300 nm into food simulants (4 % acetic acid, 95 % ethanol or hexane) during 15-day experiments, as observed by scanning electron microscopy (SEM) and confirmed by energy-dispersive X-ray spectroscopy (EDX). Given the size of these particles, their release via a diffusion mechanism is contrary to most theoretical

[2] *Migration ratio* was defined as the ratio of amount of silver in the food simulant at the end of the migration experiment to the amount of silver in the film prior to the migration experiment.

estimations of nanoparticle diffusion rates, so it is possible these particles were already located on the film surface and simply desorbed into the external matrix.

Silver nanoparticle migration has also been studied in actual foodstuffs. Cushen et al. (2013) assessed the migration of AgNP from AgNP/PVC composites to chicken meat. AgNP/PVC composites with two sizes of nanoparticles (mean diameters of 10 and 50 nm) were prepared via solvent casting. Lacking a means to distinguish between whole nanoparticles and silver ions, they used a worst-case scenario approach wherein the entire detected silver was assumed to be from AgNPs. Silver content in chicken as a function of different migration conditions such as nanoparticle size, concentration of AgNP in the film (0.5 and 5.0 % w/w), time of exposure (1.1, 2.0, 3.1 and 4.0 days), and temperature (6.59–4.13 °C) was found to range between 0.03–8.4 milligrams per kilogram (mg/kg). A significant positive correlation was found between increasing AgNP concentration ($R^2 = 0.77$, $p < 0.01$) and time ($R^2 = 0.18$, $p < 0.01$) of exposure to migration. However, a negative correlation was found between migration and increasing temperature ($R^2 = -0.13$, $p < 0.05$). No correlation was seen between particle size and migration. The negative correlation was speculated to be due to crosslinking of AgNP to PVC polymer chains. Sufficient explanation was not provided for the lack of correlation between size of the AgNP and migration, as the authors rightfully pointed out that there is a lack of information with respect to distinguishing between whole AgNP versus dissolved silver ions. Emamifar et al. (2011) analyzed silver release into orange juice from LDPE films containing either 5 % P105 (AgNP on TiO_2 powder) or 1 % ZnO nanoparticles, produced in house by melt mixing in a twin-screw extruder. They detected release of silver and zinc at very low levels (after 112 days storage at 4 °C, 0.13 ± 0.005 and 0.54 ± 0.005 micrograms per liter (μg/L), respectively) by ICP-MS. As before, their methodology prohibited the distinction between nanoparticles and dissolved ions.

Conducting release studies on clay/polymer nanocomposites has presented technical challenges. Unlike AgNPs, which are nanoscale in three dimensions, clays are platelet shaped, with a single nanometer in one direction and as much as several microns in the other two dimensions. Entire clay platelets are probably too large and irregularly shaped to move about through the rigid polymer environment and become released as whole particles. However, the very same high mechanical sheer forces that are typically used to effect good dispersion of clays into polymers during manufacture could break apart clay platelets into residual fragments with a very broad size distribution. The constituent elements that make up most aluminosilicate clays (silicon and aluminum) resist many chemical digestion procedures and have high natural abundance, which makes them challenging materials to analyze by ICP-based elemental analysis techniques.

These challenges haven't prevented some researchers from attempting to assess release of nanoclays in migration experiments. One of the earlier attempts (Avella et al. 2005) examined vegetables in contact with clay/starch nanocomposites and found that they exhibited elevated levels of silicon compared to controls, although no trends in aluminum or iron levels were observed. A more thorough migration study of MMT/wheat-gluten composites (Mauricio-Iglesias et al. 2010) found elevated levels of aluminum (as measured by ICP) when these materials were exposed to 3 % acetic acid for 10 days at 40 °C. Analogous experiments with silicon showed slightly different results, with migration evident in 3 % acetic acid, and in water and olive oil as well. The authors found an enhanced release of silicon when the composite materials were processed under high-pressure treatment, an emerging food processing technology used to inactivate pathogenic microorganisms. A follow-up study by the same authors (Mauricio-Iglesias et al. 2011) used Fourier-transform infrared spectroscopy (FTIR) to show that high-pressure treatment actually changes the molecular structure of embedded clays, suggesting that migration of nanoscale additives into foods may be explicitly linked to the type of processing used. More recent work by (Farhoodi et al. 2013) prepared MMT/PET composites by melt-mixing, followed by blow molding into bottles. Three percent acetic acid was stored in these bottles at 24 and 45 °C respectively for time durations ranging from 7 to 90 days, an apparent simulation of the use of MMT/PET materials in carbonated beverage (soda) bottle packaging. Inductively coupled plasma atomic emission spectroscopy (ICP-AES analysis) of the stored simulant revealed aluminum content after 90 days to be 0.18 and 0.34 mg/kg at 25 and 45 °C, respectively; observed silicon content was significantly higher, in agreement with the fact that silicon is typically a higher proportion (by weight) of MMT clays and the fact that the silicon fraction is closer to the MMT clay and polymer interfaces (Duncan 2011a).

A limitation of many of these studies is that they rely on ICP-MS (or ICP-AES), which only measures the pure elemental content of the metal. Because any whole nanoparticles that migrate into the simulant from the polymer would be destroyed either by the preliminary acid digestion or during passage into the 6,000+ °C plasma, any information about the migrated nanoparticles (if indeed they are particles at all) is lost. There have been few solutions to this critical issue. One workaround included covalently attaching fluorescent dyes to organically modified MMT nanoclay platelets, incorporating them into PP melts and then transforming them into films by compression molding (Diaz et al. 2013). The intention was to be able to track the clay particles during the course of the migration tests. Any fluorescence detected in the simulant after the test was completed was taken by the authors as evidence of whole particle (or fractional particle) migration. The authors performed migration tests in 100 % ethanol at 80 °C for 4 h and examined the simulant residue (after evaporation) by confocal laser scanning microscopy (CLSM). The CLSM images showed characteristic fluorescence signals of the attached dyes from well-defined particles which the authors claim was due to the migration of clay particles to the food simulant.

Other studies evaluating nanoparticle release from polymers that are relevant to FCM are provided in Table 4.1. The work summarized in this table shows that valuable research has been done in this area, but little of it has been done systematically. The release literature at this time implies that nanoparticles, or at least nanoparticle residuals, can be released into external liquid media under conditions relevant to food storage and processing, although whether this release is true migration or just release of residual ENMs adhered to the surface of FCSs remains unclear. Either way, this body of work provides little predictive information on what conditions and materials are likely to yield more migration than others.

Table 4.1 Summary of selected studies related to the release of ENMs from nanocomposite polymer films into liquid media or foodstuffs

ENM	Size (diameter unless specified)	Matrix/ Substrate	Conditions[b]	External matrix	ENM loading[b]	Maximum level of migration[b]	Reference
Theoretical nanoparticle	5 nm	LDPE, HDPE, PP, PET, PS	25 °C	n/a[a]	1 kg/m^3	1.3 mg/m^2 per year	(Simon et al. 2008)
Ag	20–80 nm	Poly(dl-lactide-co-glycolide)	37 °C, 100 days	Water	7 wt%	~1.2 ppm[c]	(Fortunati et al. 2011)
Ag	100–300 nm	PE	50 °C, 15 days	Water, 4 % acetic acid 95 % ethanol, hexanes	100 µg/g	~4.0 µg/dm^2 [c]	(Huang et al. 2011)
Ag	7 nm	PE	70 °C, 9 h	3 % acetic acid 95 % ethanol	234 mg/kg	5.6 %	(Song et al. 2011)
Ag	Dia. = 4–8 nm Thick. = 1.3–8.3 nm	PTFE	3 days	Distilled water	Surface coating	92.71 %	(Alissawi et al. 2012)
Ag	30, 70 nm	Polysulfone	37 °C, 24 h	0.9 % NaCl	2 wt%	0.2 ppm	(Mollahosseini et al. 2012)
Ag	10, 50 nm	PVC	20 °C, 4 days	Chicken	5 wt%	3.94 mg/kg	(Cushen et al. 2013)
Ag, Cu	1–20 nm	Silicone	37 °C, 85 days	Water	0.1 wt%	Cu = 1,300 ng/cm^2 Ag = 250 ng/cm^2	(Hahn et al. 2011)
Cu	50 nm	LDPE	37 °C, 120 days	Dilute nitric acid	25 wt%	19.1 µg/day	(Xia et al. 2011)
Cu	36 ± 9 nm	PLA	24 h	Saline solution	1.5 wt%	~2,000 ppb[c]	(Longano et al. 2012)

(continued)

Table 4.1 (continued)

ENM	Size (diameter unless specified)	Matrix/ Substrate	Conditions[b]	External matrix	ENM loading[b]	Maximum level of migration[b]	Reference
Mg/Al layered double hydroxide	Dia. ~ 200 nm Thickness < 20 nm	PLA	40 °C, 10 days	95 % ethanol	5.5 wt%	2.2 mg/dm^2	(Schmidt et al. 2011)
Ag based nanoclay	Ag Dia. ~ 20 nm	PLA	Orbital shaking, 150 rpm, 8 days	2.0×10^{-3} M nitric acid	10 wt%	~6 mg/kg[c]	(Busolo et al. 2010)
Clay	n/a	Starch	40 °C, 10 days	Vegetables	4 %	1.9 mg/100 g	(Avella et al. 2005)
Clay	Aspect ratio ~ 320	PLA	40 °C, 10 days	95 % ethanol	5 wt%	6.7 mg/dm^2 [d]	(Schmidt et al. 2009)
Clay	n/a	Wheat gluten	High pressure treatment plus 10 days at 40 °C	Water, 3 % acetic acid, 15 % ethanol, olive oil	5 wt%	Al ~ 1 mg/kg food simulant[c] Si ~ 4.5 mg/kg food simulant[c]	(Mauricio-Iglesias et al. 2010)
Clay	Width = 25–27 nm Length = 130–200 nm	PET	45 °C, 90 days	3 % acetic acid	3 wt%	Al = 0.34 mg/kg Si = 9.5 mg/kg	(Farhoodi et al. 2013)

[a] n/a indicates that the information was not available. [b] *ENM loading*, the *Maximum Migration Level* refers to the highest observed degree of migration observed in the study; the *Conditions* column refers to the conditions under which the maximum migration was observed. [c] This value was estimated by the authors from data depicted in a figure. [d] In this study the authors observe 6.7 mg/dm^2 of particles by MALS, but were unable to confirm that these particles were clay by ICP-MS

Research studies are needed using well-characterized test materials, where particle and polymer characteristics such as particle size and shape, polymer molecular weight and dispersion morphology are well defined. Such studies are important because they will begin to reveal key structure function-relationships that can shed light on what is the best way to predict migration rates, and what mechanisms are likely to be responsible for migration that is observed. This will help manufacturers target materials that are least likely to give rise to nanoparticle release, and it will help regulators better understand data provided by manufacturers when applying for permission to bring nanocomposites to market. For more information on the topic of ENM release from FCMs, particularly related to detection methods, the reader is directed to a recent publication by (Noonan et al. 2014).

4.7 Toxicological Considerations of Released Nanoparticles

The remainder of the chapter will address the question: *if ENMs do become released into foods, what is their potential to cause harm*?

Although the number of papers published in the area of nanotoxicology is still far exceeded by the number of general nanotechnology papers, the proportion of papers dealing with toxicity has increased rapidly in the last decade (Fig. 4.3), underscoring efforts to better understand the health implications of nanotechnology proliferation. Evaluating ENM toxicity does involve some unique challenges, including the lack of well-established techniques to measure their properties and a lack of general agreement pertaining to the kind of information that is needed to evaluate toxicological risks to humans. Validated detection and characterization methods are in turn required to help us better understand the toxicological properties of ENMs. Some of the questions that robust methods could help answer include:

- *What is the minimum characterization data needed to assess toxicity?*
- *Can toxicological data for one type of ENM be applied to the same type of ENM with different structural or surface characteristics?*
- *Can impurities in ENMs pose a toxicological concern and what is the best way to test for the presence of such impurities?*
- *Do ENMs have nano-specific toxicity mechanisms?*
- *Under what circumstances might toxicological data for a macroscale analog of an ENM be relevant?*
- *How do ENMs interact with complex biological systems, and how do their properties change as a function of gut microenvironment, diet, and an individual's state of health?*
- *Are specific population subgroups (e.g., children, immune compromised, the aged) more susceptible to effects of ENMs?*
- *What animal models or other model systems are most appropriate to study ENM toxicity?*

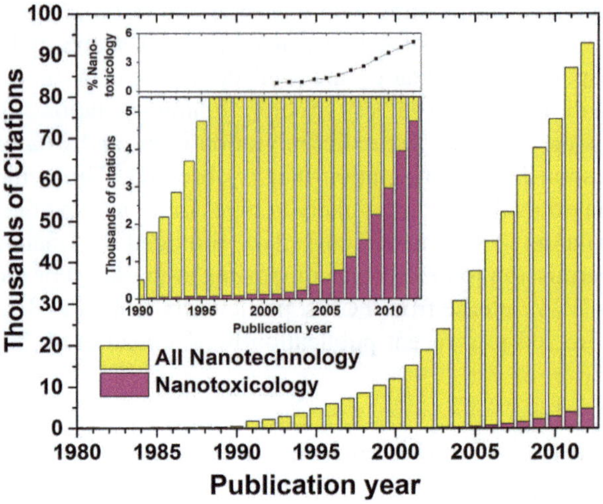

Fig. 4.3 Nanoparticle and nanotoxicology citations from 1980 to 2013. The number of scientific articles published each year on toxicity of nanoparticles (*pink*) and all nanotechnology topics (*yellow*), as indexed by ISI Web of Knowledge. The *bottom inset* shows an expanded view. The *top inset* shows the proportion (in %) of nanotoxicology articles to total nanotechnology articles per year, and indicates that this proportion has been growing significantly over the last decade. Values in *yellow* were determined by a topic search string of "nano* OR ultrafine particles" and values in *pink* were determined by a search string of "(nano* OR ultrafine particles) AND (toxic* OR cytotoxic* OR genotoxic*)"

- *Can the presence of ENMs in foods or in the gut attenuate the toxicological properties of other (non-nano) substances?*
- *Do ENMs have detrimental effects on the gut microbiome?*
- *Can chronically ingested ENMs induce effects that are not detected in acute and subchronic studies?*
- *What is the best way to ensure that toxicological results are reported in a consistent and meaningful way?*

To underscore the importance of developing good protocols for evaluating the toxicity of ENMs, Card and Magnuson (2010) recently developed a scoring system to rate the quality of nano-toxicological studies relevant to food-based routes of exposure based on a combination of factors, including adequacy of experimental design, completeness of documentation, and thoroughness of nanoparticle characterization. In a subsequent work (Card et al. 2011), they concluded that a majority of existing nanotoxicity studies were either deficient in some critical component of study design, inadequately characterized the nanomaterials being evaluated or failed to discuss the reliability of traditional oral toxicological assays in the wake of the unique chemical properties of nanomaterials. While additional oral toxicity studies have been published since then, the issues raised by Card and colleagues are still relevant; a primary need in this area remains a better methodological toolset, including a better understanding of what is the most appropriate way to use the tools we have to ensure the safety of nanomaterials in FCMs. Given the lack of agreement over the best way to experimentally assess the toxicity of ENMs, it is difficult at this time to provide a critical analysis of the nanotoxicological literature and arrive at *general* conclusions or recommendations with respect to which types and characteristics of ENMs are most likely to pose risks to humans, particularly by oral routes of exposure. This is perhaps best demonstrated by considering the body of literature related to the in vitro and in vivo toxicity of AgNPs.

AgNPs have been found to exhibit cytotoxic and/or genotoxic effects in human cell lines in vitro, including isolated testicular cells (Asare et al. 2012), mesenchymal stem cells (Hackenberg et al. 2011), liver cells (Piao et al. 2011), hepatoma cells (Kim et al. 2009), and lymphocytes (Eom and Choi 2010). Additional studies have shown that changes in AgNP characteristics such as decreased particle size (Gaiser et al. 2012; Liu et al. 2010), alterations in surface coating (Lin et al. 2012), and advanced age (Kittler et al. 2010) can increase toxicity of AgNPs to human cells, but more attention needs to be focused on the effect of other factors such as aggregation state, shape, and surface charge on toxicity in human or other mammalian cells lines.

The mechanism of AgNP toxicity at the cellular or microbial level is still a matter of considerable controversy, with some researchers believing that AgNPs are simply vehicles for silver cation desorption and others arguing that AgNPs have nanoparticle-specific toxicity mechanisms. Xiu et al. reported that AgNPs showed no significant cytotoxicity (to *E. coli*) under strict anaerobic conditions, which the authors claim suppressed Ag^+ desorption from AgNP surfaces (Xiu et al. 2012), while Kim et al. provided evidence that Ag^+ ion desorption from AgNPs is negligible during their toxicity assays and concluded that the mechanism of AgNPs was independent of Ag^+ toxicity (Kim et al. 2009). Neither of these points of view is necessarily wrong. Given the limited number of experimental studies available and the complexity of ENM behavior in biological systems, it is certainly possible that AgNPs manifest differing toxicity profiles depending on the cell lines used, the physical characteristics of the particles, or the experimental test conditions.

Regardless of the cellular toxicity mechanism, most studies have found that AgNPs display some degree of cytoxicity or genotoxicity to mammalian cells in vitro. This is in contrast with in vivo acute single-dose and sub-chronic studies, which indicate a more complicated picture for oral toxicity of AgNPs. In oral toxicity assessments in vivo, systematic toxicological effects are usually moderate or not observed at all, even in acute, high-dose toxicity studies, despite evidence of AgNP accumulation in various organ systems. Examples of these acute and sub-chronic toxicity studies are discussed below. In addition, summaries of selected 28 to 90 days subchronic rodent toxicity studies are provided in Table 4.2.

In acute, single-dose toxicity studies, Kim et al. observed no abnormal gross findings or signs of mortality in rats fed a single dose of 2,000 mg/kg of 10 nm AgNPs (Kim et al. 2013), and Maneewattanapinyo et al. reported no abnormal gross findings or histopathological signs in mice exposed to a single dose of 5,000 mg/kg of 10–20 nm AgNPs (Maneewattanapinyo et al. 2011). In three 28-day sub-chronic rat toxicity studies using small (10–17 nm mean diameter)

Table 4.2 Selected in vivo sub-chronic oral toxicity studies related to ENMs most likely to be incorporated into food contact materials

ENM	NP[a] size coating	Species, age	Route of exposure	Dose	Study duration (days)	Pathology summary	ENM tissue distribution[b]	Reference
Ag[c]	14 nm PVP[d]	Rat 5 weeks	Oral gavage	12.6 mg/kg b.w.[e] per day	28	n/a	Intestine, kidney, liver, lung, brain	(Loeschner et al. 2011)
Ag	14 nm PVP	Rat 4 weeks	Oral gavage	2.25, 4.5, 9.0 mg/kg b.w. per day	28	Normal b.w., relative organ weights, food and water intake, urine and blood parameters (including ALP[f] and WBC[g])	n/a	(Hadrup et al. 2012)
Ag	10 nm PVP 17 nm uncoated	Rat 5 weeks	Oral gavage	90 mg/kg b.w. per day	28	Normal b.w., behavior, serum ALP and AST[h], inflammatory indicators	Intestine, liver, spleen, testis, kidney, brain, lung	(van der Zande et al. 2012)
Ag	42 nm uncoated	Mouse 7 weeks	Oral	0.25, 0.5, 1.0 mg/kg b.w. per day	28	Elevated inflammatory indicators and elevated blood AST and ALP at high dose Elevated blood ALT[i] at high dose in females only Dose dependent elevated serum TGF-β Liver histopathology: none detected	n/a	(Park et al. 2010)
Ag	56 nm CMC[j]	Rat 5 weeks	Oral gavage	30, 125, 500 mg/kg b.w. per day	90	Normal food and water intake, organ weights, WBC Very slightly reduced b.w. at high dose in males only Elevated blood cholesterol and ALP at high dose Liver histopathology: increased levels of minimal bile duct hyperplasia at all treatment levels	Dose dependent: kidney, liver, lung, testes, brain, blood Kidney Ag deposition more than twice as high in females than males	(Kim et al. 2010)

(continued)

Table 4.2 (continued)

ENM	NP[a] size coating	Species, age	Route of exposure	Dose	Study duration (days)	Pathology summary	ENM tissue distribution[b]	Reference
Ag	60 nm CMC	Rat 6 weeks	Oral gavage	30, 300, 1,000 mg/kg b.w. per day	28	Normal b.w., food intake, organ weights Elevated blood cholesterol and ALP at mid- and high dose Liver histopathology: increased incidence of bile duct hyperplasia and inflammatory cell infiltration (dose and quantification not provided)	Dose dependent: stomach, liver, kidney, lung, testes, brain, blood Kidney Ag deposition more than twice as high in females than males	(Kim et al. 2008)
Al_2O_3[k]	35 nm	Mouse 7 weeks	Oral	15, 30, 60 mg/kg b.w. per day	28	Increased feed and water intake at all dosing levels Decreased b.w. and decreased WBC at high-dose Dose dependent elevated serum TGF-β Histopathology: none detected in brain, kidney, liver or lung	Thymus, lung, brain only at high dose No Al detected in kidney, heart, liver, spleen, or testis	(Park et al. 2011a)
Au[l]	14 nm GG	Rat (140–180 g b.w.)	Oral	75, 150, 300 parts per million (ppm)	28	Normal b.w., organ weights, food intake, and blood and urine parameters	n/a	(Dhar et al. 2011)
Se[m]	20–60 nm	Rat 5 weeks	In feed	2–5 ppm in feed	90	Growth inhibition with normal food and water intake at 4 ppm and above Elevated ALT at highest dose	n/a	(Jia et al. 2005)
TiO_2[n]	5 nm	Mouse (22 ± 2 g b.w.)	Oral gavage	62.5, 125, 250 mg/kg b.w. every second day	30	Normal food and water intake, behavior Decreased weight gain at mid- and high dose with increased relative weight of liver, kidney, spleen and thymus Dose-dependent elevation of blood ALT, ALP, and AST Dose-dependent decrease in blood WBC	n/a	(Duan et al. 2010)

(continued)

Table 4.2 (continued)

ENM	NP[a] size coating	Species, age	Route of exposure	Dose	Study duration (days)	Pathology summary	ENM tissue distribution[b]	Reference
TiO_2	75 nm	Rat 3 and 8 weeks	Oral gavage	10, 50, 200 mg/kg b.w. per day	30	Normal b.w., food and water intake, and activity for all treatment groups Liver edema in young but not adult rats at mid- and high dose	Stomach and intestine at high dose No Ti detected in blood, liver, kidney or spleen	(Wang et al. 2013)
ZnO[o]	40 nm	Rat 8 weeks	Oral	67, 134, 268, 537 mg/kg per day	90	Normal food and water intake and organ weights Mild pancreatitis at high dose Elevated ALP and reduced b.w. gain at high-dose in males only	n/a	(Seok et al. 2013)

[a] mean nanoparticle diameter
[b] listed in order of reported relative concentration
[c] Ag, silver
[d] PVP, polyvinylpyrrolidone
[e] b.w., body weight
[f] ALP, alkaline phosphatase
[g] WBC, white blood cell count
[h] AST, aspartate transaminase
[i] ALT, alanine transaminase
[j] CMC, carboxymethylcellulose
[k] Al_2O_3, aluminum oxide
[l] Au, gold
[m] Se, selenium
[n] TiO_2, titanium oxide
[o] ZnO, zinc oxide

AgNPs, no toxicological signs were observed at concentrations of 9.0 or 90 mg orally administered AgNP per kg body weight (b.w.) per day, despite evidence of silver accumulation in multiple organs of the same or similarly treated animals (Hadrup et al. 2012; Loeschner et al. 2011; van der Zande et al. 2012). In contrast, in 28- and 90-day rat oral toxicity studies using 60 and 56 nm carboxymethylcellulose (CMC) coated AgNPs, respectively, reported evidence for mild liver damage included elevated cholesterol and alkaline phosphatase levels and bile duct hyperplasia (Kim et al. 2008, 2010). These two studies also found dose-dependent distribution patterns of AgNP accumulation in various organs, and a gender difference with female kidneys showing more than twice the level of accumulated silver than male kidneys. The AgNPs used in studies that found evidence for mild liver damage differed in size and coating from the AgNPs used in studies that did not find liver damage, corroborating in vitro findings that nanoparticle physicochemical characteristics can alter bioavailability and toxicity, but also making it difficult to compare data from one in vivo study to the next.

The relationship of AgNP size to oral toxicity was addressed in a 14-day study in mice using 1 mg/kg b.w. per day of 22, 42, 71, and 323 nm AgNPs. No silver was detected in brain, lung, liver, kidney, or testis after exposure to the 323 nm AgNPs, while exposure to the smaller AgNPs resulted in detection of silver in all of these tissues (Park et al. 2010). Interestingly, for brain, lung, and testis, decreasing AgNP size correlated to increasing deposition, suggesting that smaller AgNPs are more bioavailable than larger ones. These authors also found that the smaller AgNPs induced significant increases in serum TGF-β, a marker of tissue damage and immune response (Park et al. 2010). In a separate 28-day mouse oral toxicity study reported in the same article, 42 nm AgNP at 1 mg/kg b.w. per day resulted in elevated serum levels of inflammatory cytokines, TGF-β, alkaline phosphatase (ALP, an indicator of bile duct obstruction), and aspartate transaminase (AST, a general indicator of inflammation or trauma); serum alanine transaminase (ALT, an indicator of liver inflammation) was also elevated, but only in females. This would seem to support the rat studies which reported that AgNP exposure resulted in mild liver damage, but no histopathology of the liver was detected (Park et al. 2010).

Only one oral AgNP treatment study investigating in vivo genotoxicity has been published, and no change was reported in the number of cells with micronuclei in bone marrow (Kim et al. 2008). It should be noted that this study did not include a positive control for genotoxicity, and that the treatment duration of 4 weeks may not be sufficient for manifestation of genotoxic effects, stressing the importance of validated, consistent methodology as well as the need for longer-term investigations.

Other studies have examined the bioavailability, excretion patterns, and tissue distribution of various types of ENMs, with only some comparing results to data for respective nanomaterial components in bulk or ionic form. While most AgNP studies that examined silver deposition detected it in multiple organs after oral AgNP exposure, the bioavailability of AgNPs is lower than that of ionic silver. Three rat studies have reported that the majority of orally administered AgNPs are excreted in feces, with corresponding higher tissue deposition and lower fecal concentrations after ionic silver treatment (Park et al. 2011b; Loeschner et al. 2011; van der Zande et al. 2012). For aluminum oxide nanoparticles (Al_2O_3NP), bioavailability is higher than for bulk Al_2O_3, with high levels of aluminum detected in rat liver, kidney, brain, and other tissues and low levels in feces after 14 days of oral exposure to 30 or 40 nm Al_2O_3NP, while exposure to powdered bulk Al_2O_3 resulted in far lower tissue deposition and higher fecal concentrations (Balasubramanyam et al. 2009). In contrast, there appears to be little absorption of titanium dioxide (TiO_2) nanoparticles after oral exposure, with titanium detected in rat fecal samples and stomach and intestinal linings, but not other tissues (Cho et al. 2013; Wang et al. 2013). Given that nanoparticle size, treatment duration, and dosing differed in each of these studies, they cannot be used together for the purpose of comparing bioavailability among nanomaterials of differing composition.

While toxicity studies using many different model systems are available for ENMs composed of elemental silver and copper, aluminosilicate clays, and metal/metalloid oxides such as aluminum, silicon, copper, titanium and zinc, there is need for a better understanding of how toxicity data from these studies might be applied to predicting safe exposure levels for humans. This could be facilitated by more systematic, long-term studies on the oral toxicity of nanoparticles, and subsequent meta-analysis comparing data for each ENM among various in vitro and in vivo toxicological model systems. Currently, published chronic ENM studies in mammals are lacking. Table 4.2 summarizes sub-chronic mammalian oral toxicology studies of ENMs with potential for incorporation into FCMs. Unfortunately, standardized procedures to evaluate exposure effects, particularly with respect to the types of model organisms, the characteristics of tested particles, and the conditions under which toxicity is assessed (dosage, etc.), have not been established. There is an urgent need for more systematic examinations of the effect of particle characteristics like size and shape, aggregation/agglomeration state, surface characteristics (including bound ligands and charge), age, and crystal structure on oral toxicity. Though many in vivo studies exist which probe these various factors, the majority utilize exposure routes with little relevance to food consumption, or model organisms with unproven predictive utility for human risk assessment.

Given these data gaps, it is very difficult at this time to make broad conclusions about the toxicity of ENMs likely to be incorporated into FCMs, which presents a challenge to both manufacturers of these products and to regulatory scientists. The aforementioned lack of robust methods to address these data gaps only increases the level of this challenge. Readers seeking more information on this subject are referred to several in-depth reviews (Bouwmeester et al. 2009; Frohlich and Roblegg 2012; Albanese et al. 2012; Luque-Garcia et al. 2013; Alger et al. 2014). The European Food Safety Authority (EFSA) has published an opinion article which describes some outstanding toxicological issues on ENMs (EFSA Scientific Committee 2009, 2011). The FDA has also published a recent draft guidance document on how to assess whether manufacturing changes—including changes in particle size—are likely to impact safety and regulatory status of food contact substances (FDA 2011).

4.8 Conclusions and Knowledge Gaps

This chapter has reviewed some of the health and environmental safety concerns related to the use of ENMs in food contact applications. Polymer nanocomposites are an emerging technology with applications in many commercial sectors, and already they are appearing in certain markets. Therefore it is important to be proactive about assessing the safety of these materials, particularly as they apply to foods, which are directly consumed by the public and therefore have a higher potential for direct exposure than products in other commercial sectors. The material covered herein has mostly been concerned with the risk of ENM release from FCMs, but the topic of toxicology of nanomaterials also has been addressed. While important progress has been made in both of these areas with respect to our understanding of the potential risks polymer nanocomposite FCMs pose to the public, there are still significant informational gaps that need to be filled. With respect to migration and release, better models for ENM diffusion through plastics need to be developed and better methods to actually measure the release of particles (versus dissolved ions) need to be identified. The need for improved methodology also applies to the area of toxicology, where identifying, quantifying and characterizing low levels of nanoscale matter in complex food systems remains a steep challenge. Potential effects of long-term, low-level oral exposure to ENMs require further investigation. Good quality reference materials and standardized, validated detection methods are key to meeting these challenges so as to ensure the continued health and well-being of the public.

References

Alger H, Momcilovic D, Carlander D et al (2014) Methods to evaluate uptake of engineered nanomaterials by the alimentary tract. Compr Rev Food Sci Food Saf. doi:10.1111/1541-4337.12077

Albanese A, Tang PS, Chan WCW (2012) The effect of nanoparticle size, shape, and surface chemistry on biological systems. Annu Rev Biomed Eng 14:1–16

Alissawi N, Zaporojtchenko V, Strunskus T et al (2012) Tuning of the ion release properties of silver nanoparticles buried under a hydrophobic polymer barrier. J Nanopart Res 14(7):1–12

Asare N, Instanes C, Sandberg WJ et al (2012) Cytotoxic and genotoxic effects of silver nanoparticles in testicular cells. Toxicology 291(1–3):65–72

Avella M, De Vlieger JJ, Errico ME et al (2005) Biodegradable starch/clay nanocomposite films for food packaging applications. Food Chem 93(3):467–474

Avella M, Bruno G, Errico ME et al (2007) Innovative packaging for minimally processed fruits. Packag Technol Sci 20(5):325–335

Balasubramanyam A, Sailaja N, Mahboob M et al (2009) Evaluation of genotoxic effects of oral exposure to aluminum oxide nanomaterials in rat bone marrow. Mutat Res 676(1–2):41–47

Begley TH, Gay ML, Hollifield HC (1995) Determination of migrants in and migration from nylon food-packaging. Food Addit Contam 12(5):671–676

Begley T, Castle L, Feigenbaum A et al (2005) Evaluation of migration models that might be used in support of regulations for food-contact plastics. Food Addit Contam 22(1):73–90

Bouwmeester H, Dekkers S, Noordam MY et al (2009) Review of health safety aspects of nanotechnologies in food production. Regul Toxicol Pharmacol 53(1):52–62
Brandsch J, Mercea P, Ruter M et al (2002) Migration modelling as a tool for quality assurance of food packaging. Food Addit Contam 19:29–41
Busolo MA, Fernandez P, Ocio MJ et al (2010) Novel silver-based nanoclay as an antimicrobial in polylactic acid food packaging coatings. Food Addit Contam 27A(11):1617–1626
Card JW, Magnuson BA (2010) A method to assess the quality of studies that examine the toxicity of engineered nanomaterials. Int J Toxicol 29(4):402–410
Card JW, Jonaitis TS, Tafazoli S et al (2011) An appraisal of the published literature on the safety and toxicity of food-related nanomaterials. Crit Rev Toxicol 41(1):20–49
Cerrada ML, Serrano C, Sánchez-Chaves M et al (2008) Self-sterilized EVOH-TiO_2 nanocomposites: interface effects on biocidal properties. Adv Funct Mater 18(13):1949–1960
Chawengkijwanich C, Hayata Y (2008) Development of TiO_2 powder-coated food packaging film and its ability to inactivate *Escherichia coli* in vitro and in actual tests. Int J Food Microbiol 123(3):288–292
Cho W-S, Kang B-C, Lee JK et al (2013) Comparative absorption, distribution, and excretion of titanium dioxide and zinc oxide nanoparticles after repeated oral administration. Part Fibre Toxicol 10:9
Choudalakis G, Gotsis AD (2009) Permeability of polymer/clay nanocomposites: a review. Eur Polym J 45(4):967–984
Code of Federal Regulations (CRC) (2013) 21CFC170.3, Title 21, Vol. 3, Part 170, Food Additives; Section 170.3(i) Definitions; definition of safety. http://www.accessdata.fda.gov/scripts/cdrh/cfdocs/cfCFR/CFRSearch.cfm?fr=170.3. Accessed 20 Oct 2013
Cushen M, Kerry J, Morris M et al (2013) Migration and exposure assessment of silver from a PVC nanocomposite. Food Chem 139(1–4):389–397
de Azeredo HMC (2013) Antimicrobial nanostructures in food packaging. Trends Food Sci Technol 30(1):56–69
de Moura MR, Lorevice MV, Mattoso LHC et al (2011) Highly stable, edible cellulose films incorporating chitosan nanoparticles. J Food Sci 76(2):S25–S29
Dhar S, Mali V, Bodhankar S et al (2011) Biocompatible gellan gum-reduced gold nanoparticles: cellular uptake and subacute oral toxicity studies. J Appl Toxicol 31(5):411–420
Diaz CA, Xia Y, Rubino M et al (2013) Fluorescent labeling and tracking of nanoclay. Nanoscale 5(1):164–168
Duan Y, Liu J, Ma L et al (2010) Toxicological characteristics of nanoparticulate anatase titanium dioxide in mice. Biomaterials 31(5):894–899
Duncan TV (2011a) Applications of nanotechnology in food packaging and food safety: barrier materials, antimicrobials and sensors. J Colloid Interface Sci 363(1):1–24
Duncan TV (2011b) The communication challenges presented by nanofoods. Nat Nanotechnol 6(11):683–688
Emamifar A, Kadivar M, Shahedi M et al (2011) Effect of nanocomposite packaging containing Ag and ZnO on inactivation of *Lactobacillus plantarum* in orange juice. Food Control 22(3–4):408–413
Eom HJ, Choi J (2010) p38 MAPK activation, DNA damage, cell cycle arrest and apoptosis as mechanisms of toxicity of silver nanoparticles in Jurkat T cells. Environ Sci Technol 44(21):8337–8342
European Food Safety Authority (EFSA) Scientific Committee (2009) The potential risks arising from nanoscience and nanotechnologies on food and feed safety. EFSA J 958:1–39. doi:10.2903/j.efsa.2009.958, http://www.efsa.europa.eu/en/efsajournal/pub/958.htm. Accessed 24 Oct 2013
European Food Safety Authority (EFSA) Scientific Committee (2011) Guidance on the risk assessment of the application of nanoscience and nanotechnologies in the food and feed chain. EFSA J 9(5):2140. doi:10.2903/j.efsa.2011.2140, http://www.efsa.europa.eu/en/efsajournal/pub/2140.htm. Accessed 24 Oct 2013

Fang XY, Domenek S, Ducruet V et al (2013) Diffusion of aromatic solutes in aliphatic polymers above glass transition temperature. Macromolecules 46(3):874–888

Farhoodi M, Mousavi SM, Sotudeh-Gharebagh R et al (2013) Migration of aluminum and silicon from PET/clay nanocomposite bottles into acidic food simulant. Packag Technol Sci. doi:10.1002/pts.2017

Fayaz AM, Balaji K, Girilal M et al (2009) Mycobased synthesis of silver nanoparticles and their incorporation into sodium alginate films for vegetable and fruit preservation. J Agric Food Chem 57(14):6246–6252

Fernández A, Picouet P, Lloret E (2010) Reduction of the spoilage-related microflora in absorbent pads by silver nanotechnology during modified atmosphere packaging of beef meat. J Food Prot 73(12):2263–2269

Food and Drug Administration (FDA) (2007) Guidance for industry: preparation of premarket submissions for food contact substances: chemistry recommendations. http://www.fda.gov/Food/GuidanceRegulation/GuidanceDocumentsRegulatoryInformation/IngredientsAdditivesGRASPackaging/ucm081818.htm. Accessed 20 Oct 2013

Food and Drug Administration (FDA) (2011) Considering whether an FDA-regulated product involves the application of nanotechnology. http://www.fda.gov/regulatoryinformation/guidances/ucm257698.htm. Accessed 20 Oct 2013

Fortunati E, Latterini L, Rinaldi S et al (2011) PLGA/Ag nanocomposites: in vitro degradation study and silver ion release. J Mater Sci Mater Med 22(12):2735–2744

Fröhlich E, Roblegg E (2012) Models for oral uptake of nanoparticles in consumer products. Toxicology 291(1–3):10–17

Gaiser BK, Fernandes TF, Jepson MA et al (2012) Interspecies comparisons on the uptake and toxicity of silver and cerium dioxide nanoparticles. Environ Toxicol Chem 31(1):144–154

Hackenberg S, Scherzed A, Technau A et al (2011) Cytotoxic, genotoxic and pro-inflammatory effects of zinc oxide nanoparticles in human nasal mucosa cells in vitro. Toxicol In Vitro 25(3):657–663

Hadrup N, Loeschner K, Bergström A et al (2012) Subacute oral toxicity investigation of nanoparticulate and ionic silver in rats. Arch Toxicol 86(4):543–551

Hahn A, Brandes G, Wagener P et al (2011) Metal ion release kinetics from nanoparticle silicone composites. J Control Release 154(2):164–170

Huang YM, Chen S, Bing X et al (2011) Nanosilver migrated into food-simulating solutions from commercially available food fresh containers. Packag Technol Sci 24(5):291–297

Jia X, Li N, Chen J (2005) A subchronic toxicity study of elemental Nano-Se in Sprague-Dawley rats. Life Sci 76(17):1989–2003

Kim YS, Kim JS, Cho HS et al (2008) Twenty-eight-day oral toxicity, genotoxicity, and gender-related tissue distribution of silver nanoparticles in Sprague-Dawley rats. Inhalation Toxicol 20(6):575–583

Kim S, Choi JE, Choi J et al (2009) Oxidative stress-dependent toxicity of silver nanoparticles in human hepatoma cells. Toxicol In Vitro 23(6):1076–1084

Kim YS, Song MY, Park JD et al (2010) Subchronic oral toxicity of silver nanoparticles. Part Fibre Toxicol 7:20

Kim JS, Song SS, Sung JH et al (2013) Genotoxicity, acute oral and dermal toxicity, eye and dermal irritation and corrosion and skin sensitisation evaluation of silver nanoparticles. Nanotoxicology 7(5):953–960

Kittler S, Greulich C, Diendorf J et al (2010) Toxicity of silver nanoparticles increases during storage because of slow dissolution under release of silver ions. Chem Mater 22(16):4548–4554

Li H, Li F, Wang L et al (2009) Effect of nano-packing on preservation quality of Chinese jujube *(Ziziphus jujuba Mill. var. inermis (Bunge) Rehd)*. Food Chem 114(2):547–552

Lin J-J, Lin W-C, Dong R-X et al (2012) The cellular responses and antibacterial activities of silver nanoparticles stabilized by different polymers. Nanotechnology 23(6):065102

Liu W, Wu Y, Wang C et al (2010) Impact of silver nanoparticles on human cells: effect of particle size. Nanotoxicology 4(3):319–330

Loeschner K, Hadrup N, Qvortrup K et al (2011) Distribution of silver in rats following 28 days of repeated oral exposure to silver nanoparticles or silver acetate. Part Fibre Toxicol 8:18

Longano D, Ditaranto N, Cioffi N et al (2012) Analytical characterization of laser-generated copper nanoparticles for antibacterial composite food packaging. Anal Bioanal Chem 403(4):1179–1186

Luque-Garcia JL, Sanchez-Díaz R, Lopez-Heras I et al (2013) Bioanalytical strategies for in vitro and in vivo evaluation of the toxicity induced by metallic nanoparticles. TrAC Trends Anal Chem 43:254–268

Maneewattanapinyo P, Banlunara W, Thammacharoen C et al (2011) An evaluation of acute toxicity of colloidal silver nanoparticles. J Vet Med Sci 73(11):1417–1423

Marsh K, Bugusu B (2007) Food packaging—roles, materials, and environmental issues. J Food Sci 72(3):R39–R55

Mauricio-Iglesias M, Peyron S, Guillard V et al (2010) Wheat gluten nanocomposite films as food-contact materials: migration tests and impact of a novel food stabilization technology (high pressure). J Appl Polym Sci 116(5):2526–2535

Mauricio-Iglesias M, Gontard N, Gastaldi E (2011) Impact of high pressure treatment on the structure of montmorillonite. Appl Clay Sci 51(1–2):174–176

Mollahosseini A, Rahimpour A, Jahamshahi M et al (2012) The effect of silver nanoparticle size on performance and antibacteriality of polysulfone ultrafiltration membrane. Desalination 306:41–50

Noonan GO, Whelton AJ, Carlander D et al (2014) Measurement methods to evaluate engineered nanomaterial release from food contact materials. Comp Rev Food Sci Food Saf. doi:10.1111/1541-4337.12079

Paladini F, Pollini M, Talá A et al (2012) Efficacy of silver treated catheters for haemodialysis in preventing bacterial adhesion. J Mater Sci Mater Med 23(8):1983–1990

Park EJ, Bae E, Yi J et al (2010) Repeated-dose toxicity and inflammatory responses in mice by oral administration of silver nanoparticles. Environ Toxicol Pharmacol 30(2):162–168

Park E-J, Kim H, Kim Y et al (2011a) Repeated-dose toxicity attributed to aluminum nanoparticles following 28-day oral administration, particularly on gene expression in mouse brain. Toxicol Environ Chem 93(1):120–133

Park K, Park E-J, Chun IK et al (2011b) Bioavailability and toxicokinetics of citrate-coated silver nanoparticles in rats. Arch Pharmacol Res 34(1):153–158

Piao MJ, Kang KA, Lee IK et al (2011) Silver nanoparticles induce oxidative cell damage in human liver cells through inhibition of reduced glutathione and induction of mitochondria-involved apoptosis. Toxicol Lett 201(1):92–100

Podsiadlo P, Kaushik AK, Arruda EM et al (2007) Ultrastrong and stiff layered polymer nanocomposites. Science 318(5847):80–83

Ray SS, Yamada K, Okamoto M et al (2002) Polylactide-layered silicate nanocomposite: a novel biodegradable material. Nano Lett 2(10):1093–1096

Risch SJ (2009) Food packaging history and innovations. J Agric Food Chem 57(18):8089–8092

Robertson GL (2006) Food packaging: principles and practice, 2nd edn. Taylor and Francis Group, LLC, Boca Raton

Rulis AM, Levitt JA (2009) FDA's food ingredient approval process, safety assurance based on scientific assessment. Regul Toxicol Pharmacol 53(1):20–31

Sánchez-Valdes S, Ortega-Ortiz H, Ramos-de Valle LF et al (2009) Mechanical and antimicrobial properties of multilayer films with a polyethylene/silver nanocomposite layer. J Appl Polym Sci 111(2):953–962

Schmidt B, Petersen JH, Koch CB et al (2009) Combining asymmetrical flow field-flow fractionation with light-scattering and inductively coupled plasma mass spectrometric detection for characterization of nanoclay used in biopolymer nanocomposites. Food Addit Contam 26A(12):1619–1627

Schmidt B, Katiyar V, Plackett D et al (2011) Migration of nanosized layered double hydroxide platelets from polylactide nanocomposite films. Food Addit Contam 28A(7):956–966

Seil JT, Webster TJ (2012) Antimicrobial applications of nanotechnology: methods and literature. Int J Nanomed 7:2767–2781

Seok SH, Cho W-S, Park JS et al (2013) Rat pancreatitis produced by 13-week administration of zinc oxide nanoparticles: biopersistence of nanoparticles and possible solutions. J Appl Toxicol 33(10):1089–1096

Šimon P, Chaudhry Q, Bakoš D (2008) Migration of engineered nanoparticles from polymer packaging to food—a physicochemical view. J Food Nutr Res 47(3):105–113

Song H, Li B, Lin Q-B et al (2011) Migration of silver from nanosilver-polyethylene composite packaging into food simulants. Food Addit Contam 28(12):1758–1762

Strohal R, Schelling M, Takacs M et al (2005) Nanocrystalline silver dressings as an efficient anti-MRSA barrier: a new solution to an increasing problem. J Hosp Infect 60(3):226–230

Sweet MJ, Singleton I (2011) Silver nanoparticles: a microbial perspective. Adv Appl Microbiol 77:115–133

United States Code (2010) 21 United States Code, 2010 Edition, Title 21—Food and Drugs, Chapter 9 – Federal Food, Drug, and Cosmetic Act, Subchapter II – Definitions; Section 321, Definition for food additive. http://www.gpo.gov/fdsys/pkg/U.S.CODE-2010-title21/html/U.S.CODE-2010-title21-chap9-subchapII.htm. Accessed 20 Oct 2013

United States Environmental Protection Agency (USEPA) (2013) Municipal solid waste. U.S. Environmental Protection Agency. http://www.epa.gov/epawaste/nonhaz/municipal/. Accessed 30 June 2014

van der Zande M, Vandebriel RJ, Van Doren E et al (2012) Distribution, elimination, and toxicity of silver nanoparticles and silver ions in rats after 28-day oral exposure. ACS Nano 6(8):7427–7442

Wang Y, Chen Z, Ba T et al (2013) Susceptibility of young and adult rats to the oral toxicity of titanium dioxide nanoparticles. Small 9(9–10):1742–1752

Welle F (2013) A new method for the prediction of diffusion coefficients in poly(ethylene terephthalate). J Appl Polym Sci 129:1845–1851

Xia X, Tang Y, Xie C et al (2011) An approach to give prospective life-span of the copper/low-density-polyethylene nanocomposite intrauterine device. J Mater Sci Mater Med 22(7):1773–1781

Xiu Z-M, Zhang Q-B, Puppala HL et al (2012) Negligible particle-specific antibacterial activity of silver nanoparticles. Nano Lett 12(8):4271–4275

Chapter 5
The Alkylphenols Nonylphenol and Octylphenol in Food Contact Materials and Household Items: Exposure and Health Risk Considerations

Suzanne M. Snedeker and Anthony G. Hay

Abstract The alkylphenols (APs) nonylphenol (NP) and octylphenol (OP) are not approved for use in food contact materials (FCM) in Europe or the United States, however, tris(nonylphenyl)phosphite (TNPP) is used as a stabilizer and antioxidant in a variety of food contact plastics, including polyvinyl chloride- and linear low-density polyethylene films, polystyrene, high-impact polystyrene, and rubber-based products. It is not known if the NP detected in foods is a trace contaminant of TNPP formulations, is from TNPP breakdown in FCMs, or is derived from food contact with NP-laden biosolids that are sometimes used as fertilizers. Both NP and OP have been identified in indoor air, and NP has been detected in household items and house dust. NP and OP, as well as TNPP, are weakly estrogenic. Adverse effects on the target organs (liver, kidney, and heart), immune and neurological function, the developing reproductive system, and metabolic pathways related to obesity and diabetes risk, have been observed in laboratory animals or cell culture systems exposed to NP or OP. While there is some evidence of NP acting as a tumor promoter in animal models (lung and mammary gland), induction of apoptosis has been observed in lung-, gastric-, and bone-tumor cell lines, and in Sertoli cells. Additional research is needed to determine sources of NP and OP in FCMs in the global food supply, identify factors that affect migration of these APs from food packaging to foodstuffs, and better define exposure to human populations.

S. M. Snedeker (✉)
Independent Consultant, Ithaca, NY, USA
e-mail: smsnedeker@gmail.com

A. G. Hay
Department of Microbiology, Cornell University, Ithaca, NY, USA
e-mail: agh5@cornell.edu

A. G. Hay
Graduate Field of Environmental Toxicology, Cornell University, Ithaca, NY, USA

S. M. Snedeker (ed.), *Toxicants in Food Packaging and Household Plastics*,
Molecular and Integrative Toxicology, DOI: 10.1007/978-1-4471-6500-2_5,

Keywords Alkylphenols · Nonylphenol · Octylphenol · Food contact materials · Food packaging · Polyvinyl chloride cling films · PVC · Tris(nonylphenyl) phosphite · TNPP · High-impact polystyrene · HIPS · High-density polyethylene · HDPE · Linear low-density polyethylene · LLDPE · Estrogenicity · Immune function · Neurological function · Obesity · Diabetes · Reactive oxygen species · Apoptosis · Vinyl shower curtains · Household dust · Household air · Migration

5.1 Key Take Home Points

- Nonylphenol and octylphenol are alkylphenols that have been detected in the food supply.
- One likely source of nonylphenol is the use of tris(nonylphenyl)phosphite (TNPP) as a stabilizer and antioxidant in food contact plastics.
- Nonylphenol, octylphenol, and TNPP are all weak estrogens.
- Research evidence indicates that nonylphenol or octylphenol can adversely affect the functioning of various organ systems, the developing reproductive system, immune function, neurological function and behavior, and obesity and diabetes risk.
- More research is need on the conditions affecting the migration of various alkylphenols from food packaging and processing materials to foods and beverages.
- More research is needed on the level of nonylphenol and octylphenol in the global food supply and the extent of human exposure to these chemicals.

5.2 Exposure to Alkylphenols

5.2.1 Overview of Selected Exposures

Alkylphenols (APs) are used widely in plastic manufacturing as surfactants and antioxidants. Both modified polystyrene (PS) labware and polyvinyl chloride (PVC) medical tubing were identified as sources of *para*-nonylphenol (4-NP) ($C_{15}H_{24}O$), and NP migrating out of the plastic was identified as having estrogenic properties in an estrogen-dependent MCF-7 human breast tumor cell line (Soto et al. 1991; Inoue et al. 2002; Hill et al. 2003). Other studies conducted in Germany (Guenther et al. 2002; Raecker et al. 2011) and Asia (Li et al. 2008) detected low levels of APs, including isomers of 4-NP and 4-*tert*-octylphenol (4-*tert*-OP) ($C_{14}H_{22}O$), in a variety of foods leading to speculation on the source of these chemicals (Fig. 5.1). While alkylphenol ethoxylates (APEs) are used in pesticides as stabilizers of suspensions and as non-ionic surfactants in detergents, and these

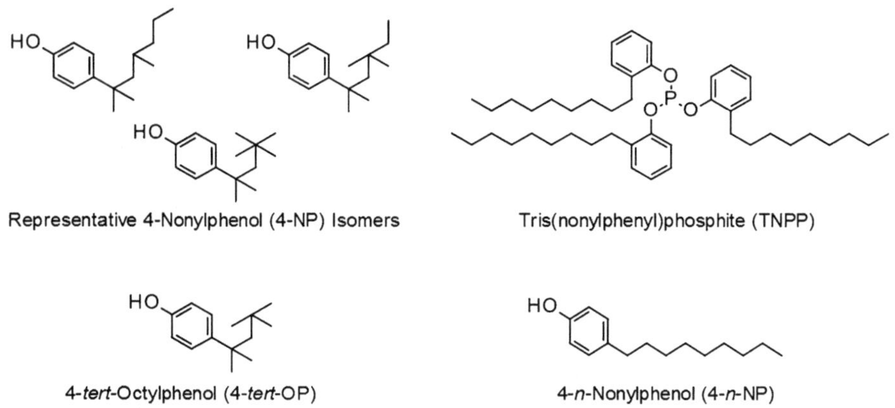

Fig. 5.1 Chemical structures of selected alkylphenols and tris(nonylphenyl)phosphite

uses may explain some of the AP residues detected in fresh and packaged produce (Li et al. 2008; Cacho et al. 2012; Guenther et al. 2002), research is now examining to what extent food contact materials (FMCs) are a source of APs in food. Evidence that APs are detected in FCMs, foodstuffs, indoor air, and household dust are summarized below.

5.2.2 Alkylphenols in Food Contact Materials

While NP is not approved for direct use in food contact plastics in either the United States (U.S.) or Europe, there is evidence that its presence in FCMs may be due to the use of APEs as dispersing and stabilizing agents in food packaging and the use of TNPP ($C_{45}H_{69}O_3P$, CAS No. 264223-78-4, Fig. 5.1) in FCMs (Bradley et al. 2010; Howe et al. 2001). TNPP is approved for use as a stabilizer/antioxidant in food plastic polymers both in Europe and the U.S. (Howe et al. 2001). It is commonly used as an antioxidant in PVC films (Howe et al. 2001), as a stabilizer in PS, high-impact polystyrene (HIPS), rubber materials used in food packaging (Fernandes et al. 2008), and in polyolefins, including ethylene-vinyl acetate copolymers (EVA), and linear low-density polyethylene (LLDPE) films and containers (Howe et al. 2001). TNPP is used in food-contact rubber (hoses and conveyor belts) that repeatedly come into contact with food. While TNPP is used as a stabilizer in HIPS precursors, TNPP is not added to food-contact HIPS directly, hence 4-NP is present in HIPS as a contaminant (Howe et al. 2001). It has been estimated that TNPP used in FCMs contains 1–4 % 4-NP as an impurity (Howe et al. 2001). The oxidation of TNPP in PVC films can release NP and serve as a possible source of the NP in food wrapped in PVC films or containers made of PVC (Casajuana and Lacorte 2004). TNPP also can release 4-NP under acidic conditions (Howe et al. 2001). Therefore, NP may be present as a *non-intentionally added substance* (NIAS) in FCMs as an impurity or from the degradation of permitted plastic additives (Bradley et al. 2010).

Several studies have examined whether different types of FCMs contain NP. Fernandes and colleagues (Fernandes et al. 2008) determined 4-NP concentrations in a wide variety of plastic FCMs, and found the highest concentrations of 4-NP in PVC-based cling film (287 micrograms per gram, μg/g), a HIPS food tray (265 μg/g), and a PS ice-tray (64.2 μg/g). Low levels (0.3–1.5 μg/g) of NP were detected in a vinyl acetate bread bag, a vinyl acetate wine stopper, a polycarbonate (PC) baby bottle, and various rubber-based products. Other types of FCMs, including LDPE cling film, a hard polyethylene terephthalate (PET) water container, a high-density polyethylene (HDPE) milk container, melamine, coated papers, a nylon roasting bag, a cellulose fudge wrapper, and silicone products contained very low (<0.07 μg/g) to undetectable levels of NP.

While some studies have found trace or undetectable levels of 4-NP and/or 4-*tert*-OP in water stored in PET, PC, and HDPE bottles (Guart et al. 2011), levels vary widely according to bottle composition and geographic location. Median levels of APs in bottled water samples packaged in PET plastic were 7.9 nanograms per liter (ng/L) for 4-NP, and <2 ng/L for 4-*tert*-OP in a study conducted in Greece (Amiridou and Voutsa 2011). A study conducted in China reported 4-NP in all of the 21 bottled water samples analyzed (range 108–298 ng/L), though the bottle composition was not specified (Li et al. 2010b). In a study on NP levels in bottled water from three types of plastic, NP was not detected in water from PET bottles, but NP levels in water sampled from PVC and HDPE containers averaged 300 and 180 ng/L, respectively (Loyo-Rosales et al. 2004). Since the NP in the bottled water samples could have resulted from AP detergents used to wash the bottles or been present in the water itself, additional migration studies were conducted. Containers were rinsed with deionized water and refilled with a food simulant (10 % ethanol), and samples were taken for NP analysis during a time-course storage at 20 or 40 °C for up to 360 hours (h). At 240 h at 40 °C, levels of NP were 2-fold higher in the simulant sampled from the HDPE bottles (approximately 240 ng/L) than the PVC bottles (approximately 120 ng/L).

There have been comparatively few studies that have examined NP levels in source waters used for bottled water. One study in Spain analyzed AP levels in source waters, and found a detectable level of 4-NP in one out of 131 samples at 0.058 μg/L, while 4-*tert*-OP was detected in four samples at a mean level of 0.0024 μg/L (Bono-Blay et al. 2012). Exposure to NP from tap water in China has been estimated (based on consumption of 2 L/day) to be as high as 1.4 μg/day (Li et al. 2010b).

Other studies have evaluated levels of APs in FCMs and their migration into foods. Some of the highest levels of NP in FCMs have been found in PVC film and HIPS. The Food and Environment Research Agency, York, UK determined the level of NP in a variety of PVC films and HIPS samples (Bradley et al. 2010). Of the ten PVC samples tested, two contained NP at 2,590 and 4,150 milligrams per kilogram (mg/kg), and three out of ten HIPS samples contained NP at 28, 925, and 1,760 mg/kg. The migration of NP from these FCMs was determined, and while NP migrated from PVC film that was in contact with cake (0.3–0.6 mg/kg) or cheese (0.2–0.8 mg/kg), there was very little migration of NP from the three positive HIPS samples into test foods, below the limits of detection (LOD) at 0.2 mg/kg, suggesting that NP migration potential is partly dependent on the type of food contact plastic.

Others researchers have found a range of 4-NP present in PVC films used commercially and domestically, and that migration into food simulants is influenced by test conditions. Of the ten samples of domestic PVC film, 4-NP was detected in two samples; of the 22 commercial PVC films, ten samples were found to have detectable levels of 4-NP, with 4-NP levels in positive samples ranging from 500 to 3,300 μg/g (Inoue et al. 2001). In the PVC films with positive detections of 4-NP, migration rates into food simulants were higher into *n*-heptane (62.5 % migrated) than into 4 % acetic acid or distilled water (0.23 % migrated). There was a wide range of 4-NP migration in cooked rice samples that were wrapped in PVC films and then microwaved for 1 min, from not detectable to 410 ng/g. Other studies have evaluated NP migration from other types of food packaging, but the type of plastic or potential source of the NP used in the packaging was not always specified. While NP was detected migrating from a tuna can to food simulant medium at a mean level of 3,912 ng/L, the source of the NP was not identified (Fasano et al. 2012). Similarly, OP was detected migrating from a plastic wine top in this study at a mean level of 26,629 ng/L, but the type of plastic was not specified. While NP has been detected in milk samples packed in Tetra Brik (ultra high temperature processing) (16.5–34.5 μg/kg) and HDPE packaging (23.6–27.7 μg/kg), it is not known if the source of NP was from the packaging, FCMs used in processing milk, or present in the milk via APs absorbed by the dairy cows from bio-solids applied to pastures as fertilizer amendments (Casajuana and Lacorte 2004).

These studies do suggest that FCMs, especially PS, HIPS, HDPE, and PVC films and containers, may be sources of APs, especially NP. Further research is needed to determine the conditions that favor and hinder migration of APs from FCMs, including temperature, length of storage, and the fat and moisture content of the food.

5.2.3 Levels of Alkylphenols in Households

Studies conducted in Japan and the U.S. have detected APs in indoor air, household dust, and household items. Both 4-NP and 4-*tert*-OP were detected in indoor air with median levels at 47.5 and 3.2 ng per cubic meter (m^3), respectively, in households of Tokyo, Japan (Saito et al. 2004). Additional monitoring identified PVC-based materials used in household floor and wall coverings as a possible source of APs in indoor air. Studies by Rudel and colleagues indicate that NP as well as certain nonylphenol-ethoxylates (NPEs) and octylphenol-ethoxylates

(OPEs) are indoor air contaminants (Rudel et al. 2003). Levels of APs in household dust and air were monitored in 120 homes located in Cape Cod, Massachusetts. Median levels of 4-NP were 110 ng/m^3 in indoor air and 2.58 μg/g in dust samples, while median levels of mono-NPE and mono-OPE were 17 and 8.6 ng/m^3, respectively, in air. Median AP and APE levels were below 6 μg/g in dust samples. In a California-based study conducted several years later, median household indoor air levels of 4-*tert*-NP were half (53.0 ng/m^3) (Rudel et al. 2010) of levels detected on Cape Cod (Rudel et al. 2003), reflecting either different usage patterns or changes in formulation of products with NP. Median indoor air levels of NP in the Rudel et al. (2010) California study were similar to those previously reported by Satio et al. (2004) for Japanese households.

In an attempt to identify sources of APs, levels of 4-*tert*-NP, 4-*tert*-OP and some APEs were determined in a variety of household cleaners, cosmetics, personal care products, and household products; a vinyl shower curtain had the highest level of 4-*tert*-NP (in the range of >100 to 1,000 μg/g) in the items surveyed (Dodson et al. 2012). To what extent FCMs could be a source contributing to household air and dust levels of NP or OP is not known.

5.3 Bioavailability, Pharmacokinetics, and Biomonitoring

5.3.1 Bioavailability and Pharmacokinetics

A study conducted by Green and colleagues that investigated the absorption, metabolism, and excretion of 4-NP in male and female CD rats, found that up to 75 % of an orally (gavage) or intravenously administered dose of ^{14}C radiolabled-NP appeared in the feces in the first 24 h after dosing (Green et al. 2003). About 65 % (in males) and 80 % (in females) of the 10 mg/kg oral dose was absorbed, with 50 % being absorbed in both genders at the higher 100 mg/kg dose. Absorbed NP was metabolized in the liver and appeared in the bile as the glucuronide conjugate form of NP. There were gender differences in peak blood levels after dosing, with males displaying levels several fold higher than female rats. While the authors state there was not evidence of tissue retention of NP, in repeated dose experiments low levels of radiolabeled orally administered NP was detected in the fat after 14 days of dosing.

In humans, radiolabled $^{13}C_6$-NP was administered to two male volunteers (5 mg as a oral dose to one volunteer, and 1 mg intravenously to the second volunteer) and blood, urine, and fecal samples were monitored over time (Muller et al. 1998). Because less than 1 % of the dose was excreted in the feces as NP, and the bioavailability of administered NP was calculated to be about 20 % (oral dose), the authors suggested that after absorption by the gastrointestinal tract, NP was metabolized in the gut wall as well as in the liver. Studies using microsomes prepared from rat intestinal tissue found that while NP was glucuronidated within the gut wall, this form remained for a long time in the intestinal wall, probably due to the long side-chain of NP which may impair transport (Daidoji et al. 2006). Studies conducted in male and female Wistar rats found that very little of the orally administered radiolabled 4-*n*-NP was detected in the feces, whereas most of the elimination was in the urine as metabolites from beta-oxidation of the nonyl side chain as well as conjugation to the sulfate or glucuronide forms (Zalko et al. 2003). The unbranched alkyl chain present in 4-*n*-NP, however, is not representative of the highly branch alkyl chain of technical grade NP (see Fig. 5.1), thus it is unlikely that beta-oxidation plays a role in the metabolism in technical grade NP.

Other metabolic studies in Sprague-Dawley (SD) rats using a mixture of branched-chain alkyl NPs have demonstrated evidence of NP glucuronidation (predominate form in the blood after oral dosing), however, NP aglycone also was detected in tissues of the rats (Doerge et al. 2002). In addition, when NP was administered to pregnant dams by gavage, there was evidence of placenta transfer of NP aglycone to the fetal serum and brain tissue. The authors suggest that while NP glucuronides predominate in the blood after metabolism, the aglycone form can accumulate in estrogen-responsive tissues in the rat, and they suggested tissue accumulation of endocrine active-forms of NP could occur in humans. This study is in contrast to a metabolic study conducted in Wistar rats that found very little of the 4-*n*-NP administered to pregnant dams from day 3 to 19 of gestation reached the fetuses (Zalko et al. 2003). Few studies have examined the ability of NP to cross the placenta in humans. In one study, human placentas were perfused with 4-NP, and there was some evidence of a slow rate of NP transfer across the placenta (Balakrishnan et al. 2011). Some of these differences may have been influenced by variability in the rates of glucuronidation between strains and species, but may also be affected by variations in gut microbiota and enterohepatic cycling (Snedeker and Hay 2012).

5.3.2 Biomonitoring of Alkylphenols

Compared to other endocrine disrupting compounds such as bisphenol-A and various phthalates, there is comparatively limited biomonitoring data on levels of APs in urine, blood, adipose tissue, and human breast milk. Because APs are metabolized in the liver relatively quickly to their sulfonated and glucuronide forms, and are rapidly cleared by the body, very low levels are detected in human blood (Asimakopoulos et al. 2012) and low levels in urine reflect recent exposure (CDC 2013a). This section will summarize available data on biomonitoring of 4-*tert*-OP and 4-NP in human populations.

Urine The Centers for Disease Control and Prevention (CDC) have included urinary levels of 4-*tert*-OP in their environmental biomonitoring program as a part of the 2006–2006, 2007–2008, and 2009–2010 National Health and Nutrition Examination Survey (NHANES) (CDC 2013b). Urinary levels of 4-*tert*-OP were below the LOD for the 50th, 75th, and most of the 90th percentile in NHANES results for all three surveys. At the 95th percentile (geometric means), urinary levels of 4-*tert*-OP were 0.2–0.3 μg/L for all three surveys; creatinine-adjusted means ranged from 0.56–64 μg/g creatinine. Females tended to excrete higher levels of 4-*tert*-OP than males, which may be reflective of more use of household or personal care products that contain 4-*tert*-OP by females. While the CDC does not routinely monitor urinary levels of NP as a part of their biomonitoring program, urinary levels of the linear NP, 4-*n*-NP, were determined in the NHANES III survey, with a frequency of detection of 51 %, with median concentrations less than 0.1 mg/L and levels higher in males at 0.17 mg/L (Calafat et al. 2005). Both 4-NP and OP have been detected in a small study of 60 Japanese subjects (Jing et al. 2011). OP was detected in 17/60 urine samples at 0.41–11.10 ng per milliliter (ml) while 4-NP was detected at a higher frequency (29/60) at 16.9–27.8 ng/ml. Similar levels of 4-NP were reported in a study of 287 children and young adults from Guangzhou, China (geometric mean 15.92 μg/g creatinine) (Li et al. 2013). In contrast, researchers were unable to detect NP in any of the 131 urine samples collected from a general population in Belgium (Pirard et al. 2012). Whether these results represent true geographic differences in the U.S., Asian, and European exposures to APs, or whether results differ because of differences in methodology is not known.

Blood Given the transient nature of APs after ingestion and their rapid metabolism by the liver, blood levels of APs are not considered to be reflective of human exposure (Asimakopoulos et al. 2012). Studies measuring blood levels of APs are limited to several method development studies with a small number of samples (Kawaguchi et al. 2004; Tan and Ali Mohd 2003). Research is needed, however, to determine AP levels in cord blood samples to evaluate whether APs may be crossing the placental barrier and exposing the fetus.

Breast Milk European and Taiwanese researchers have monitored NP and OP levels in human breast milk. In a study conducted in Italy, NP and OP levels were determined in ten samples of mature human milk, collected over 24 h from both breasts; mean levels for NP and OP were 32.0 and 12.0 ng/ml, respectively (Ademollo et al. 2008). In contrast, 4-NP levels in 59 breast milk samples obtained from Taiwanese women were considerably lower (geometric mean for NP 2.26 ng/g) with even lower reported levels for 4-OP (geometric mean for 4-OP 0.02 ng/g) (Chen et al. 2010). While these studies demonstrate that NP and OP are detectable in human breast milk, considerably more research needs to be done analyzing a larger number of milk samples from different geographic locations to characterize levels in this biological medium.

Adipose Tissue Levels of APs have been monitored in the adipose tissue from human subjects, and have varied according to the geographical location of the subjects. In a study conducted in Italy, levels of APs in adipose tissue obtained from 16 subjects undergoing bariatric surgery ranged from 10 to 266 ng/g for NP, with considerably lower levels reported for OP (6–80 ng/g) (Ferrara et al. 2011). Several NPEs were also detected in adipose tissue, with the frequency of detection decreasing as the ethoxylation level increased. In a study of adipose tissue samples obtained from surgery patients from southern Spain, the median levels in the 20 samples obtained were 57 ng/g for NP and 4.5 ng/g for OP, respectively (Lopez-Espinosa et al. 2009). Considerably lower levels of unbranched 4-*n*-NP in human adipose tissue was reported in a study from Belgium, with detection in eight out of 11 subjects, mean levels of 4-*n*-NP were 0.025 ng/g (Geens et al. 2012). Reports of AP adipose tissue levels in subjects from North America and Asia were not located. International standards for analytical methods to measure APs, or a study where one centralized laboratory analyzes samples from multiple geographic locations, could help clarify whether the wide differences in adipose levels of NP and OP are due to differences in exposure or collection and analytical methodology.

5.4 Health Effects of Alkylphenols

While there is some evidence that exposure to NP or OP may have adverse effects on specific organs, including the kidney, heart, and liver, the majority of the studies evaluating toxic effects of APs have focused on male and female reproduction and endocrine effects, immune function, and neurological and behavioral endpoints. This section will briefly highlight some of the studies that have assessed effects of AP exposure on these organs and functional systems, especially those from mammalian laboratory animal models and tissue culture systems.

5.4.1 Toxicological Effects on Organs

Liver There were no toxic effects on hepatic tissues in male or female SD rats fed up to 150 mg/kg body weight of 4-NP (as commercial grade NP, CAS No. 84852-15-3) in a 90-day subchronic study (Cunny et al. 1997), however, a 28-day study conducted in male and female Crj:CD(SD)IGS rats using the same commercial grade of NP did show evidence of centrilobular liver cell hypertrophy at the high 250 mg/kg dose (Woo et al. 2007). In a drinking water study conducted in SD rats, at the level of 25 parts per million (ppm) neither NP (as 4-NP, CAS No. 25154-52-3) or 4-OP (CAS No. 1806-26-4) showed evidence of hepatic cytotoxic effects, however, both NP and 4-OP induced an increase in apoptosis in the liver, and hepatocellular gluconeogenesis was observed in 4-OP-treated animals (Hernandez-Rodriguez et al. 2007). It has been suggested that the induction of apoptosis may involve certain metabolic pathways, including mitochondria-dependent and Fas-Fas-L pathways (Jubendradass et al. 2012b). Others have suggested that 4-NP-induced hepatic necrosis in Wistar rats may be the result of oxidative damage (Korkmaz et al. 2010).

Kidney Effects of NP on the kidney have been one of the most consistently reported toxicological effects in rodent feeding studies, though effects are dependent on dose and diet composition. For instance, in a 90-day feeding study with male and female SD rats, dose-related decrease in kidney weights and reduced haline levels in the high-dose 150 mg/kg 4-NP-group of male rats were the only NP treatment-related effects reported (Cunny et al. 1997). In contrast, 4-NP administered at a higher dose (250 mg/kg body weight) in a 28-day oral dosing (via gavage) study resulted in renal toxicity in male and female rats, including basophilic tubules (male and females) with haline casts (only males), and lymphocyte infiltration (females) as well as cortical tubular dilation (females) (Woo et al. 2007). Studies conducted in male Wistar rats suggest that NP-induced renal necrosis may be related to oxidative damage in the kidney (Korkmaz et al. 2011). Polycystic kidney disease (PCK) has been reported in male rats with development exposure to NP (2,000 ppm, gestation day seven to postnatal day 50) (Cooper et al. 2006; Latendresse et al. 2001), with diet playing a role in the severity of the PCK; soy-based diets protected against NP-induced PCK compared to casein-based and other soy-free diets (Cooper et al. 2006). Several other multi-generational reproductive toxicology studies have reported depressed kidney weights and histopathological lesions with developmental exposure to NP. A consistent adverse effect of 4-NP administration (as branched 4-NP, CAS No. 84852-15-3) in SD rats across three generations in a reproductive toxicity study was an increase in kidney weights in both genders, and various renal lesions at the mid- (650 ppm) and high-doses (2,000 ppm) of NP, including renal medullary tubular dilatation and cysts, focal mineralization, granular casts, and hydronephrosis (Chapin et al. 1999). These renal lesions were dose-related in the males, and prevalent in the high-dose 2,000 ppm females. Increase in kidney weights and similar renal lesions have been reported in other reproductive toxicology studies assessing the effect of 4-NP administration (Tyl et al. 2006; Han et al. 2004; Nagao et al. 2001).

Heart and Vasculature 4-NP, and to a greater extent 4-OP and 17β-estradiol, had the capacity to increase coronary perfusion pressure when administered at 100 nanomolar (nM) to perfused rat hearts (Ruehlmann et al. 1998). Investigators further demonstrated that 4-OP and estradiol can act on vasculature by inhibition of L-type Ca^{2+} channels in smooth muscle cells, suggesting an estrogen-related effect of APs on cardiac tissue.

5.4.2 Toxicological Effects on Systems

5.4.2.1 Endocrine Disruption

Thyroid Gland While a number of studies have examined effects of NP exposure on thyroid function in aquatic species, there are relatively few studies available in mammalian models. Dose, duration of treatment, and gender all appear to influence whether APs affect thyroid function. For instance, serum levels of T3 and T4, but not thyroid stimulating hormone (TSH), were elevated in ovariectomized female SD rats fed 4-NP at 20 or 80 mg/kg via a soy-free diet for up to 12 weeks (Schmutzler et al. 2004). In contrast, in an oral dosing (by gavage) 28-day subchronic study, serum TSH levels were significantly elevated in intact female SD rats in the high 250 mg/kg per day 4-NP-group, but not lower dose groups (10 or 50 mg/kg per day), compared to controls (Woo et al. 2007). There were no NP-treatment effects on TSH in males, and T3 and T4 serum levels were unaffected in both genders. While thyroid weights were significantly elevated in 50 and 250 mg/kg 4-NP-treated males (not females) in the Woo study, no effects on thyroid weight were reported in a 90-day subchronic diet study in rats fed up to 2,000 ppm 4-NP (Cunny et al. 1997). Treatment of the thyroid hormone-dependent rat pituitary cell line GH3 with NP or OP resulted in inhibition of T3-induced cell growth. While these studies suggest some potential for NP or OP to affect thyroid hormones and function, much more research is needed in both animal models and cell cultures to clarify divergent findings as well as to identifying mechanisms by which APs may affect thyroid function.

Estrogenicity The ability of 4-NP leaching from PS labware to elicit an estrogenic response in vitro was first demonstrated using MCF-7 cells, an estrogen-dependent human breast tumor line (Soto et al. 1991). This initial observation was confirmed in vivo using uterotrophic and uterine permeability assays (Shelby et al. 1996; Milligan et al. 1998), and in vitro using transcriptional assays with mammalian or yeast cell lines transfected with estrogen receptor-alpha (ER-α) (Shelby et al. 1996; Van den Belt et al. 2004), and competitive ER-binding assays (Shelby et al. 1996; Milligan et al. 1998). Later studies using the estrogen-dependent MCF-7 cell line demonstrated that both 4-*n*-NP (CAS No. 104-40-5) and 4-*n*-OP (CAS No. 1806-26-4) show similar estrogenic-responses while 4-*tert*-OP (CAS No. 140-66-9) elicited a weaker estrogenic response (Isidori et al. 2010). Both 4-NP and 4-*tert*-OP were found to be 100,000 less potent than 17β-estradiol in the mouse uterotrophic assay (Milligan et al. 1998). Potency may depend on the test system, since OP has been found to be 1,000 fold less potent in human estrogen-dependent cell lines (White et al. 1994), while 4-*tert*-OP was 5,000 fold less potent than 17β-estradiol in a yeast transfection assay (Miller et al. 2001). Potency and ER-binding appears to be dependent on AP-side chain length, branching, and test species (e.g. human- verses trout-ER) (Olsen et al. 2005). Structurally, estrogenic activity of APs has been associated most closely with phenols with R-groups in the *para*-position (4-position) (Miller et al. 2001). Other assays have used gene-responsiveness to assess AP estrogenicity. Using microarray analysis in human breast cancer cells, researchers have found both 4-NP and 4-*tert*-OP can up-regulate estrogen-responsive genes, especially genes involved in proliferation, transcription, and transport (Terasaka et al. 2006).

TNPP, which may oxidize to release 4-*n*-NP in food contact applications, has been shown to have estrogenic activity in a yeast-two hybrid assay (Ogawa et al. 2006). Additional studies are needed to confirm whether TNPP is estrogenic in the in vivo uterotrophic assay, and if TNPP has the capacity to bind to ERs and up-regulate estrogen-responsive genes.

5.4.2.2 Female Reproduction and Mammary Gland Development

Female Reproduction in Rodents The effect of oral AP administration on female reproductive toxicology endpoints has been evaluated in several multigenerational rodent studies (Nagao et al. 2001; Chapin et al. 1999; Willoughby et al. 2005). 4-NP administration to laboratory rats orally via gavage at 50 mg/kg (Nagao et al. 2001), and in the diet at 650 and 2,000 ppm (Chapin et al. 1999) resulted in accelerated vaginal opening in all generations of offspring. Other adverse effects on females in the 650 ppm and 2,000 ppm 4-NP dose groups in the Chapin et al. (1999) study included increased uterine weights (F_1 generation only) and decreased ovarian weights (F_2 generation only), and increased length of the estrous cycle (adult females in the F_1 and F_2 generations) in the high-dose 2,000 ppm group. While there were no treatment effects on fertility, or female anogenital distance in any generation, the effects on earlier vaginal opening and increased length of the are clear indications of an in vivo estrogenic effect of developmental 4-NP exposure (Chapin et al. 1999)

Others have reported the effects of post-natal 4-*tert*-OP administration (50 mg/kg, post-natal day 0 to10, via sc injection) on female reproductive function in the rat, including accelerated vaginal opening, decreased ovarian weight, and increased uterine weight (Willoughby et al. 2005). This study, however, did not see earlier vaginal opening with 5 mg/kg 4-*tert*-OP or with 4-NP at 5 or 50 mg/kg administered during this early post-natal period, suggesting that critical windows of exposure may exist for AP's effect on female reproductive development. Since very few animal modeling studies have evaluated whether OP exposure affects female reproduction, additional studies are needed to determine OP's effect on female reproductive endpoints using an oral route of exposure.

Menarche Onset in Humans Since most rodent developmental studies have reported accelerated vaginal opening with 4-NP administration, researchers have compared secondary sexual characteristics and menarche onset in Taiwanese girls to levels of 4-NP in collected urine in 786 female students (Chen et al. 2009). After controlling for age and body mass index, urinary 4-NP levels were significantly inversely related to age of menarche, suggesting that further study is needed to determine if the earlier onset of menarche is related to NP exposure.

Mammary Gland Development in Rats There is some evidence that NP exposure may affect mammary gland development. Moon and colleagues treated Long Evans pregnant rats with NP (branched alkyl chains, 85 % *para*-isomer) at 10 or 100 mg/kg (oral gavage) during gestational days 15–19, and observed advanced lobular development in the high-, but not low-dose NP group's mammary glands on post-natal day 22 (Moon et al. 2007). The long-term effects of NP administration on mammary gland development and function are not known.

5.4.2.3 Male Reproduction

Male Reproduction in Rodents Chapin and colleagues administered branched 4-NP (CAS No. 84852-15-2) in the diet (NIH-07 feed) to SD rats in a 3.5 generational reproductive toxicology study (Chapin et al. 1999). 4-NP was present in the diet at 0, 200 ppm (9–35 mg/kg per day), 650 ppm (30–100 mg/kg per day), and 2,000 ppm (100–350 mg/kg per day). There were no treatment-related changes in a number of male reproductive endpoints, including anogenital distance, preputial separation, testicular or epididymal weight, or in functional measures including pup number or viability in any generation. While there were reduced epididymal sperm densities at the mid- and high-NP doses, this effect was only observed in the F_2 generation, and not the F_0, F_1 or F_3 generation. A lack of effects from 4-NP administration (CAS No. 84852-15-2) on male reproductive endpoints, including no effect on epididymal sperm counts, has been reported in a three-generation study of SD rats, confirming Chapin and colleagues findings (Tyl et al. 2006). While a two-generation reproduction study that administered NP (CAS No. 25154-52-3, 99 % pure) by gavage at 2–50 mg/kg also reported a lack of effects on male reproductive endpoints or sperm indices in male Crg:CD(SD) IGS rats, 4-NP treatment did affect levels of hormones (Nagao et al. 2001). This included decreased levels of T3 and increased levels follicle stimulating hormone (FSH) in males at postnatal day 22 in the F_1 50 mg/kg males. In contrast, several other reproductive studies have reported changes in male reproductive tissues in 4-NP-treated male rats, especially at high-dose levels (250 mg/kg per day), including statistically significant decreases in testicular weight (de Jager et al. 1999a), epididymal weight (de Jager et al. 1999a, b; Hossaini et al. 2001; Han et al. 2004), and seminiferous tubule diameter (de Jager et al. 1999a).

There is some evidence that exposure to NP can effect antioxidant-related enzymes in exposed adult male rats. Adult male Wistar rats gavaged with NP at doses up to 300 mg/kg per day showed a decreased in the activity of superoxide dismutase (SOD), catalase (CAT), and glutathione peroxidase (GPx) in epididymal sperm (Aly et al. 2012).

Cell Culture Studies In cultured Sertoli cells from male SD rats exposed to 4-NP for 24 h, the cytotoxicity (cell death) observed may have been due to accumulation of reactive oxygen species (ROS) (Gong and Han 2006). Subsequent studies by the same research group found changes in the plasma membrane of 4-NP-treated Sertoli cells and in intracellular calcium homeostasis, suggesting that these changes may partly explain 4-NP-induced cytotoxicity (Gong et al. 2008). These authors also suggested that the changes in calcium signaling may be important in triggering apoptosis. Cell culture studies and in vivo animal models have implicated 4-NP-exposure in apoptosis induction in Sertoli cells (Li et al. 2010a; Wu et al. 2009) and germ cells (spermatogonia and spermatocytes) (McClusky et al. 2007). Models and proposed molecular mechanisms by which NP and other environmental estrogens may induce apoptosis in germ cells have been reviewed (Lagos-Cabre and Moreno 2012). Others have proposed that the generation of ROS in NP-treated Sertoli cells may be mediated by miR-135 via the Wnt/beta-catenin signaling pathway (Choi et al. 2011).

Male Sexual Development in Humans There have been few studies assessing whether NP-exposure in human populations affects reproductive or sexual development endpoints in males. A study of Korean students examined the association between urinary and plasma levels of *n*-NP and *tert*-OP in boys with and without hypospadias (urinary tract opening is at a place other than the tip of the penis) (Choi et al. 2012). They reported a highly significant positive association between urinary levels of *n*-NP in the boys with hypospadias. More research is needed to examine this possible association of APs and developmental effects on male reproductive organs in human populations.

5.4.2.4 Neurological Function and Behavior

Several studies have investigated whether APs can affect neurological development and behavior, as well as the synthesis of neurological hormones and neurotransmitters. While oral exposure to NP from gestational day 3 to postnatal day 20 in male rats did not affect spontaneous motor activity or performance in an elevated plus-maze test, NP-treated pups did display fewer avoidance responses, as well as showing evidence of adversely affecting monoaminergic neural pathways (Negishi et al. 2004). NP administered to 5-day old rat pups intracisternally can increase motor hyperactivity, though the mechanism for this effect was not identified (Masuo et al. 2004a, b). In cell culture studies, while treatment of neural stem cells with 4-NP resulted in DNA fragmentation due to apoptosis in one in vitro study (Kudo et al. 2004), there was no evidence of DNA fragmentation when rat pheochromocytoma cells were exposed to unbranched APs, 4-*n*-NP or 4-*n*-OP, in another study (Talorete et al. 2001). However, there was evidence that incubation of the pheochromocytoma cells for an extended period of time (24 h) with 4-*n*-NP and especially with 4-*n*-OP resulted in inhibition of acetylcholinesterase, even at levels as low as 0.8 μM (Talorete et al. 2001). These authors noted that human exposure to NP in drinking water is at a similar level. Using bovine adrenal medullary cells, 4-NP administration had a marked effect on the reduction of catecholamine synthesis at a NP concentration similar to that observed in U.S. rivers (Yanagihara et al. 2005). Short-term treatment of this cell line with 4-NP also caused elevations in various MAP-kinases, and these effects were estrogen-independent.

Since the hormone estrogen is known to play an important role in the sexual differentiation of central nervous system structures, the effects of 17β-estradiol and 4-NP as a weak estrogen on dendritic and synaptic development was investigated using cultures of fetal rat hypothalamic cells (Yokosuka et al. 2008). Synaptic densities increased in the mid-dose (4-NP at 10 nM; estradiol at 100 nM), but not high-dose level (1 μM) for either 4-NP or estradiol. Similar effects were observed on glial development. Since 4-NP may affect development of neuronal structures in the hypothalamus, further studies are needed to assess to what extent exposures to NP or OP may affect neurological development or behavior in humans.

5.4.2.5 Immune Function and Allergic Response

Many of the effects of NP and OP on immune cells and allergic response have been summarized by Suen and colleagues (Suen et al. 2012), including effects on various cytokines, enhancement of mast cell degranulation, increased expression of tumor necrosis factor-alpha (TNF-α), and increased Th2 differentiation. Some of these effects are partially estrogen-dependent, while others appear to be estrogen independent. The induction of TNF-α in myeloid dendritic cells by NP or 4-OP is partly reversible by an ER-antagonist (Hung et al. 2010). The decrease of ERK1/2 phosphorylation in differentiated THP-1 cells exposed to 4-*n*-NP also can be suppressed with an ER-antagonist (Bennasroune et al. 2012). There is also evidence that the increased mast cell degeneration induced by NP is mediated by ER-α (Narita et al. 2007). In contrast, the ability of 4-NP and 4-*tert*-OP to suppress the development of Th1 and enhance the development of Th2 T cell thymocytes appears to be ER-independent (Iwata et al. 2004).

In an in vivo two-generational oral feeding study conducted in SD rats, immunologic effects of NP (CAS No. 84852-15-3) on natural killer cells (NK) and leukocytes in the F_1 generation were gender specific; while NP induced a dose-dependent (at 500 and 2,000 ppm NP) increase in spleenic NK activity in the F_1 females, this effect was not observed in NP-treated F_1 males (Karrow et al. 2004). In vitro studies have demonstrated the ability of NP to suppress the lytic function of NK cells (Udoji et al. 2010), supporting evidence of NP's effect on suppressing immune function. Tissue culture studies using chorionic villous explants from human placenta have investigated whether treatment with 4-NP affects cytokine secretion (Bechi et al. 2010). Using environmentally relevant low doses of 4-NP (10^{-13}, 10^{-11}, and 10^{-9} M), 4-NP-exposed cultures had increased release of granulocyte-macrophage colony-stimulating factor, interferon-γ (IFN-γ), interleukin (IL)-1β, IL-4, and IL-10, while reducing the release of TNF-α and having no effect on IL-2 and IL-5, compared to vehicle control cultures. To what extent APs may affect implantation and pregnancy outcomes via effects on cytokines, or if APs may affect allergic diseases in the human population such as asthma, is not known.

5.4.2.6 Obesity and Diabetes Risk

There is emerging evidence that APs may affect risk of developing obesity and diabetes. Developmental perinatal exposure to 4-NP via administration to dams by oral gavage from gestational day 12 to lactation day 7 resulted in a larger fat mass, and higher blood cholesterol and glucose levels in mouse pups at 60 days of age (Hao et al. 2012). Perinatal exposure to 4-NP with obesity in the F_1 generation also has been demonstrated in male and female Wistar rats (Zhang et al. 2014). In addition, there was evidence of transgenerational obesogenic effects in the F_2 offspring not exposed to 4-NP. The authors suggest this obesity effect in the F_1 and F_2 generations, may be mediated by ER-α signaling pathways since there was a parallel downregulation of ER-α imprinted genes. These 4-NP-induced generational obesity transmission effects were enhanced in animals fed high-fat diets. Resistin gene expression was increased in 3T3-LI adipocytes exposed to OP (Lee et al. 2008). The hormone resistin decreases differentiation and increases insulin resistance in adipocytes. APs appear to have direct effects on insulin signaling pathways. NP stimulated insulin secretion from rat pancreatic cells in vitro, and this effect was blocked by an ER-inhibitor (Adachi et al. 2005). The insulin-induction effect may be dose-dependent. Song and colleagues reported a U-shaped response curve for both NP and OP in rat pancreatic cell islets (Song et al. 2012). Both APs caused mitochondrial swelling, disrupting the expression of genes that regulate mitochondrial functioning of beta cells. NP may affect insulin signaling in other tissues including the liver. When rats received NP by gavage (15 to 1,500 μg/g body weight per day for 45 days), NP-treatment induced lipid peroxidation in the liver and decreased activity of hepatic antioxidant enzymes, as well as increasing blood levels of insulin and glucose (Jubendradass et al. 2012a). The authors suggest that the NP-induced reduction in insulin signaling in the liver may be due to oxidative damage and the generation of ROS. Because of this laboratory-based evidence suggesting that NP and OP may affect pathways related to obesity risk or diabetes risk, analyses of NHANES human biomonitoring data are needed to see if measures of AP exposure, such as urinary excretion, are related to the risk of diabetes and obesity human populations.

5.4.2.7 Carcinogenicity and Genotoxicity

Neither NP nor OP has been rated as a human carcinogen or suspected carcinogen by the International Agency on Cancer or the National Toxicology Program. The following section summarizes studies that have examined NP's effect on cell proliferation, tumor promotion, and induction of apoptosis in vivo and in vitro studies in various organ systems.

Acevedo and colleagues hypothesized that NP could affect the incidence of mammary tumors in MMTVneu mice by elevating hepatic and serum levels of the potent estrogen, estriol (E3) (Acevedo et al. 2005). They administered technical grade 4-NP at 0, 35 and 45 mg/kg per day orally to female mice for 32 weeks. While the incidence of mammary tumors (carcinomas) was higher in the high-dose 4-NP animals compared to controls or mice treated with 17β-estradiol (10 μg/kg per day, level equivalent to NP's potency based on ER binding), levels of serum E3 were not elevated in the 4-NP-treated mice, suggesting the 4-NP-induction of mammary tumors was probably due to a different mechanism than elevation of circulating E3.

There is additional evidence that NP may have tumor promoting effects in the mammary gland. C-Ha-*ras* transgenic (TG) rats were treated with a single dose of the carcinogen 7,12-dimethylbenz[*a*]anthracene (DMBA), and then were fed non-soybean based diets containing 0, 10, 25, 100, or 250 ppm of NP for 18 weeks (Fukamachi et al. 2004). While the overall percent of animals with mammary tumors did not differ between controls and NP-treated animals at 18 weeks of age, tumor multiplicity was significantly higher for mammary adenocarcinomas in the 10 ppm NP TG rats compared to controls, but there was not a dose-response effect since number of adenocarcinomas per rat were similar in the 250 ppm NP-group compared to controls. In non-TG animals, tumor induction was monitored for up to 18 weeks post-DMBA induction, and while there was an effect of NP on decreased latency of mammary tumors in the 10 and 25 ppm NP groups compared to controls, there were no treatment effects on the percent of animals with mammary tumors nor on the multiplicity of tumors. The significance of this study is difficult to interpret since the time course was short and therefore did not allow for development of tumors at a more advanced age, potentially masking effects on late-stage mammary gland tumor induction.

NP has the capacity to promote rat lung tumors possibly by stimulating cell proliferation and inducing oxidative damage in DNA (Seike et al. 2003). Male F344/DuCrj rats pretreated with multiple carcinogens, then fed NP (mixture of branched side-chains) in the diet at 25 or 250 ppm for 28 weeks, had significantly higher incidences of lung adenomas and carcinomas at either NP dose compared to controls. This tumor promotion effect was accompanied by an increase in a marker for oxygen-mediated DNA-damage in the lung, suggesting that NP may have an effect both on lung cell proliferation and oxidative damage (Seike et al. 2003).

NP has wide range of effects in different cancers from promoting cell proliferation to inducing cytoxicity via apoptosis in various tumor cell lines. 4-NP increased cell proliferation of estrogen-dependent BG-1 ovarian cells in both cell culture experiments and in vivo in xenographed mice (Park and Choi 2012). In contrast, in a DMBA-induced model of ovarian cancer, female SD rats fed a diet with 25–250 ppm 4-NP ($\sim$97 % purity) for 50 weeks had an 86 % reduction in ovarian carcinoma compared to DMBA-only controls (Tanaka et al. 2002). A549 lung cancer cells exposed to 4-NP were unaffected at low concentrations (10 and 25 μM), but displayed inhibition of cell proliferation and cytotoxicity in the range of 50–100 μM 4-NP (Andreescu et al. 2005). This apoptosis effect was not observed for several other environmental estrogens tested, suggesting that 4-NP has unique structural characteristics that may affect its cytotoxic effects in cultured lung tumor cells. This effect of NP on inducing cytotoxic apoptosis has also been demonstrated in human osteosarcoma MG63 cells, and the apoptotic effect may be due to changes in calcium homeostasis, since changes in intracellular and extracellular Ca^{2+} are a trigger for apoptosis (Wang et al. 2005). Similarly, in a human gastric cancer cell line (SCM1), NP induced Ca^{2+}-dependent apoptosis involving the activation of p38 MAPK-associated caspase-3 (Kuo et al. 2010). When the SCM1 cells were pretreated with BAPTA/AM, a chelator of a Ca^{2+}, apoptosis did not occur when the cells were then treated with NP.

5.5 Regulatory Considerations

Most regulatory actions that have been taken or recommended to reduce NP exposure have focused on limiting use of NPEs used as nonionic surfactants, especially those used in industrial detergents, dust control, and deicing agents. The U.S. Environmental Protection Agency (USEPA) released an Action Plan on August 10, 2010 that supports the voluntary phase out of NPEs in detergents, and initiates rule making that would add NP to the Toxic Release Inventory that requires facilities to report releases of inventoried chemicals to the environment (USEPA 2010). In Europe, a directive developed in 2003 restricted the use and sale of NP and NPEs to levels less than 0.1 % by weight in a variety of applications, including use in industrial and domestic detergents, textile and leather processing, pesticides and biocides, cosmetics and personal care products (except spermicides), metal working, and in the manufacturing of paper and pulp (EC 2003). These same restrictions are listed in the European Union's *2006 Registration, Evaluation, Authorisation and Restriction of Chemical Regulation* (commonly called "REACH") legislation in Annex XVII, section number 46 (EC 2006).

Assessment of whether the level of NP migrating from FCMs should be of concern has not reached a consensus in regulatory agencies, partly because NP is predominantly present as an impurity or degradation product as a NIAS. The U.S. Food and Drug Administration (FDA) includes "nonylphenol, nonylphenol,p−, octylphenol, octylphenol,p−, and tris(nonylphenyl)phosphite" (see Table 5.1) in their *List of Indirect Additives Used in Food Contact Substances* (FDA 2014), but NP and OP are not approved by the FDA for use as direct food additives. Neither NP nor OP are listed in FDA's *Cumulative Estimated Daily Intake* database for food contact substances (FDA 2013).

In Europe, no *specific migration limit* has been assigned to NP by European regulatory agencies. However, the Danish Institute of Safety and Toxicology has proposed a *Tolerable Daily Intake* (TDI) of 5 μg/kg body weight for NP (DanishEPA 2000). Several studies have calculated NP-exposures from FCMs. A study conducted in Japan that determined migration of NP from PVC film to cooked rice samples estimated that exposure from 4-NP could reach 171.9 ng/g (Inoue et al. 2001). With a consumption of 200 g of rice per day by a 50 kg adult (projected for a Japanese population), NP ingestion was estimated to be 0.7 μg/kg body weight per day, which is below the proposed Danish TDI of 5 μg/kg body weight. One study has attempted to estimate potential exposure to 4-NP from TNPP used in food contact applications (Howe et al. 2001). Using FDA methods, these researchers determined the migration of 4-NP from PVC-film and LLDPE, both TNPP plastics, and other plastics into FDA-approved food simulants (10–95 % ethanol) under different storage conditions and times (2 h to 10 days). The overall dietary exposure to NP from LLDPE, HIPS, PVC-film and ethylene-vinyl acetate was calculated to be 25.5 parts per billion (as expressed per gram of diet). Based on a diet containing 3,000 g of food (solids and liquids), 4-NP

Table 5.1 Selected indirect additives used in food contact substances (FCS), Regnum listings under the U.S. Code of Federal Regulations (US CRF), Title 21, compiled from (FDA 2014; USCFR 2013)

Chemicals in FDA indirect additives in FCS database	Other chemical names	CAS No.	Doc. No.	Regnum US CFR
Nonylphenol	4-Nonylphenol	25154-52-3	6025	176.180 176.200 176.210 177.1200
Nonylphenol,p–	4-*n*-Nonylphenol	104-40-5	6026	175.300 178.2010
Octylphenol	4-*tert*-Octylphenol	27193-28-8	6062	175.105
Octylphenol,p–	4-*n*-Octylphenol	1806-26-4	6064	175.300 177.2410
Tris(nonylphenyl)phosphite		26523-78-4	6956	177.2600

exposure was estimated to be 0.0765 mg per person per day. Using a reference 60 kg individual, authors estimated that the exposure to 4-NP from TNPP-products would be less than 0.0013 mg/kg body weight per day. This is equivalent to 1.3 μg/kg body weight per day, which is less than the proposed Danish TDI of 5 μg/kg body weight. In contrast, researchers from the UK's Food Standards Agency have detected levels of 4-NP in PVC film as high as 4,150 mg/kg, with migration into several high fat foods in the range of 0.2–0.8 mg/kg. These researchers have suggested that NP migration from PVC-film to food approaches or may exceed the proposed Danish TDI for NP (Bradley et al. 2010).

5.6 Conclusions

While the use of APEs as surfactants in detergents, and subsequent breakdown to estrogenic APs during sewage treatment has received considerable attention and regulatory action, far less attention has been paid to APs originating in FCMs, especially food contact plastics. Isomers of 4-NP and 4-*tert*-OP have been detected in foods, beverages, and a variety of food contact plastics. One possible source of APs in FCMs is TNPP, which is used as a stabilizer and antioxidant in food contact plastics. Preliminary studies suggest that 4-NP can migrate from TNPP-containing PVC-plastic films to food. TNPP is also used in PS, HIPS, and LLDPE-based films, though less is known about its migration from these and other food plastics. While it is likely that the 4-NP released from TNPP would have a linear alkyl chain (4-*n*-nonylphenol), the isomeric composition of 4-NP (branched vs. linear alkyl chain) found in foods is not always reported. The nature of the 4-NP alkyl chain may affect both the pharmacokinetics and pharmacodynamics of NP. More research is needed to determine if APs are migrating from FCM or coming from other sources such detergent breakdown products, and under what conditions migration to foodstuffs is likely to occur. Neither NP nor OP are approved for use in food contact substances by European or U.S. regulators. Since 4-NP from either TNPP and/or other sources is being found in foods and FCMs, regulators should reevaluate whether migration limits and other regulatory action is warranted to prevent AP contamination of the food supply, especially in light of the wide range of adverse health effects that have been summarized here.

5.7 Research Needs and Recommendations

- Most studies reporting levels of 4-NP and 4-*tert*-OP in foods or beverages have originated in Europe or Asia; more studies are needed to determine levels of these APs in the global food supply.
- Research is needed to determine the chemical structures of the APs found in the food supply (e.g., branched versus linear alkyl groups) and how this may affect metabolism and health effects on human populations.
- Research is need to determine the conditions that affect the migration of APs, including the type of packaging material, storage conditions (time and temperature), and fat and moisture content of the foodstuff.
- Since TNPP appears to be a source of 4-NP food contamination, and 4-NP exposure has been associated with a wide range of health effects in laboratory animal, cell culture, and human studies, regulatory agencies should review whether there should be limits on the use of TNPP in FCMs.
- A specific migration limit should be established for 4-NP by food safety regulatory agencies.
- While 4-*tert*-OP has been detected in the food supply, current information is very limited, and more research is needed to document the extent of contamination as well as identification of the sources of this AP, including exposures from household items and FCMs.

References

Acevedo R, Parnell PG, Villanueva H et al (2005) The contribution of hepatic steroid metabolism to serum estradiol and estriol concentrations in nonylphenol treated MMTVneu mice and its potential effects on breast cancer incidence and latency. J Appl Toxicol 25(5):339–353

Adachi T, Yasuda K, Mori C et al (2005) Promoting insulin secretion in pancreatic islets by means of bisphenol A and nonylphenol via intracellular estrogen receptors. Food Chem Toxicol 43(5):713–719

Ademollo N, Ferrara F, Delise M et al (2008) Nonylphenol and octylphenol in human breast milk. Environ Int 34(7):984–987

Aly HA, Domenech O, Banjar ZM (2012) Effect of nonylphenol on male reproduction: analysis of rat epididymal biochemical markers and antioxidant defense enzymes. Toxicol Appl Pharmacol 261(2):134–141

Amiridou D, Voutsa D (2011) Alkylphenols and phthalates in bottled waters. J Hazard Mater 185(1):281–286

Andreescu S, Sadik OA, McGee DW (2005) Effect of natural and synthetic estrogens on A549 lung cancer cells: correlation of chemical structures with cytotoxic effects. Chem Res Toxicol 18(3):466–474

Asimakopoulos AG, Thomaidis NS, Koupparis MA (2012) Recent trends in biomonitoring of bisphenol A, 4-*t*-octylphenol, and 4-nonylphenol. Toxicol Lett 210(2):141–154

Balakrishnan B, Thorstensen E, Ponnampalam A et al (2011) Passage of 4-nonylphenol across the human placenta. Placenta 32(10):788–792

Bechi N, Ietta F, Romagnoli R et al (2010) Environmental levels of *para*-nonylphenol are able to affect cytokine secretion in human placenta. Environ Health Perspect 118(3):427–431

Bennasroune A, Rojas L, Foucaud L et al (2012) Effects of 4-nonylphenol and/or diisononylphthalate on THP-1 cells: impact of endocrine disruptors on human immune system parameters. Int J Immunopathol Pharmacol 25(2):365–376

Bono-Blay F, Guart A, de la Fuente B et al (2012) Survey of phthalates, alkylphenols, bisphenol A and herbicides in Spanish source waters intended for bottling. Environ Sci Pollut Res Int 19(8):3339–3349

Bradley EL, Leon I, Speck D et al (2010) Nonylphenol in food contact plastics and migration into foods, Report FD 09/05, FSA Project A03057. Food Standards Agency. http://www.foodbase.org.uk/results.php?f_report_id=510. Accessed 20 Feb 2014

Cacho JI, Campillo N, Vinas P et al (2012) Determination of alkylphenols and phthalate esters in vegetables and migration studies from their packages by means of stir bar sorptive extraction coupled to gas chromatography-mass spectrometry. J Chromatogr A 1241:21–27

Calafat AM, Kuklenyik Z, Reidy JA et al (2005) Urinary concentrations of bisphenol A and 4-nonylphenol in a human reference population. Environ Health Perspect 113(4):391–395

Casajuana N, Lacorte S (2004) New methodology for the determination of phthalate esters, bisphenol A, bisphenol A diglycidyl ether, and nonylphenol in commercial whole milk samples. J Agric Food Chem 52(12):3702–3707

Centers for Disease Control (CDC) (2013a) Biomonitoring summary, 4-*tert*-octylphenol, CAS No. 140-66-9. Centers for Disease Control and Prevention. http://www.cdc.gov/biomonitoring/Octylphenol_BiomonitoringSummary.html. Accessed 19 Dec 2013

Centers for Disease Control (CDC) (2013b) Fourth national report on human exposure to environmental chemicals, Updated tables, September 2013, Urinary 4-*tert* octylphenol. Centers for Disease Control and Prevention. http://www.cdc.gov/exposurereport/. Accessed 19 Dec 2013

Chapin RE, Delaney J, Wang Y et al (1999) The effects of 4-nonylphenol in rats: a multigeneration reproduction study. Toxicol Sci 52(1):80–91

Chen GW, Ding WH, Ku HY et al (2010) Alkylphenols in human milk and their relations to dietary habits in central Taiwan. Food Chem Toxicol 48(7):1939–1944

Chen ML, Lee HY, Chuang HY et al (2009) Association between nonylphenol exposure and development of secondary sexual characteristics. Chemosphere 76(7):927–931

Choi H, Kim J, Im Y et al (2012) The association between some endocrine disruptors and hypospadias in biological samples. J Environ Sci Health A Tox Hazard Subst Environ Eng 47(13):2173–2179

Choi JS, Oh JH, Park HJ et al (2011) miRNA regulation of cytotoxic effects in mouse Sertoli cells exposed to nonylphenol. Reprod Biol Endocrinol 9:126

Cooper S, Latendresse JR, Doerge DR et al (2006) Dietary modulation of *p*-nonylphenol-induced polycystic kidneys in male Sprague-Dawley rats. Toxicol Sci 91(2):631–642

Cunny HC, Mayes BA, Rosica KA et al (1997) Subchronic toxicity (90-day) study with *para*-nonylphenol in rats. Regul Toxicol Pharmacol 26(2):172–178

Daidoji T, Ozawa M, Sakamoto H et al (2006) Slow elimination of nonylphenol from rat intestine. Drug Metab Dispos 34(1):184–190

Danish Environmental Protection Agency (DanishEPA) (2000) Toxicological evaluation and limit values for nonylphenol, nonylphenol ethoxylates, tricresyl, phosphates and benzoic acid,

Nonylphenol and nonylphenol ethoxylates, Section 8 TDI, health based limit values. Danish Environmental Protection Agency. http://www.statensnet.dk/pligtarkiv/fremvis.pl?vaerkid=6944&reprid=0&filid=21&iarkiv=1. Accessed 1 Jan 2014

de Jager C, Bornman MS, Oosthuizen JM (1999a) II. The effect of *p*-nonylphenol on the fertility potential of male rats after gestational, lactational and direct exposure. Andrologia 31(2):107–113

de Jager C, Bornman MS, van der Horst G (1999b) I. The effect of *p*-nonylphenol, an environmental toxicant with oestrogenic properties, on fertility potential in adult male rats. Andrologia 31(2):99–106

Dodson RE, Nishioka M, Standley LJ et al (2012) Endocrine disruptors and asthma-associated chemicals in consumer products. Environ Health Perspect 120(7):935–943

Doerge DR, Twaddle NC, Churchwell MI et al (2002) Mass spectrometric determination of *p*-nonylphenol metabolism and disposition following oral administration to Sprague-Dawley rats. Reprod Toxicol 16(1):45–56

European Commission (EC) (2003) Directive 2003/53/EC of the European Parliment and the Council of 18 June 2003 amending for the 26th time Council Directive 76/769/EEC relating to restrictions on the marketing and use of certain dangerous substances and preparations (nonylphenol, nonylphenol ethoxylate and cement). European Commssion. http://eur-lex.europa.eu/LexUriServ/LexUriServ.do?uri=OJ:L:2003:178:0024:0027:en:PDF. Accessed 8 Jan 2014

European Commission (EC) (2006) Regulation (EC) No 1907/2006 of the European Parliament and of the Council of 18 December 2006 concerning the Registration, Evaluation, Authorization and Restriction of Chemicals (REACH), establishing a European Chemicals Agency amending Directive 1999/45/EC and repealing Council Regulation (EEC) No 793/93 and Commission Regulation (EC) No 1488/94 as well as Council Directive 76/769/EEC and Commission Directives 91/155/EEC, 93/67/EEC, 93/105/EC and 2000/21/EC. European Parliament and the European Union. http://eur-lex.europa.eu/LexUriServ/site/en/oj/2006/l_396/l_39620061230en00010849.pdf. Accessed 8 Jan 2014

Fasano E, Bono-Blay F, Cirillo T et al (2012) Migration of phthalates, alkylphenols, bisphenol A and di(2-ethylhexyl)adipate from food packaging. Food Control 27:132–138

Food and Drug Administration (FDA) (2013) Cumulated Estimated Daily Intake Database. U.S. Food and Drug Administration. http://www.fda.gov/Food/IngredientsPackagingLabeling/PackagingFCS/CEDI/default.htm. Accessed 10 Jan 2014

Food and Drug Administration (FDA) (2014) List of indirect additives used in food contact substances; Doc No. 6025 nonylphenol; Doc No. 6026 nonyphenol, p-; Doc No. 6062 octylphenol; Doc No. 6064 octylphenol,p-; Doc No. 6956 tris(onylphenyl)phosphite. U.S. Food and Drug Administration. http://www.accessdata.fda.gov/scripts/fcn/fcnNavigation.cfm?rpt=iaListing&displayAll=true. Accessed 10 Jan 2014

Fernandes AR, Rose M, Charlton C (2008) 4-Nonylphenol (NP) in food-contact materials: analytical methodology and occurrence. Food Addit Contam Part A Chem Anal Control Expo Risk Assess 25(3):364–372

Ferrara F, Ademollo N, Orru MA et al (2011) Alkylphenols in adipose tissues of Italian population. Chemosphere 82(7):1044–1049

Fukamachi K, Han BS, Kim CK et al (2004) Possible enhancing effects of atrazine and nonylphenol on 7,12-dimethylbenz[*a*]anthracene-induced mammary tumor development in human c-Ha-*ras* proto-oncogene transgenic rats. Cancer Sci 95(5):404–410

Geens T, Neels H, Covaci A (2012) Distribution of bisphenol-A, triclosan and *n*-nonylphenol in human adipose tissue, liver and brain. Chemosphere 87(7):796–802

Gong Y, Han XD (2006) Nonylphenol-induced oxidative stress and cytotoxicity in testicular Sertoli cells. Reprod Toxicol 22(4):623–630

Gong Y, Pan X, Huang Y et al (2008) NP-induced biophysical and biochemical alterations of rat testicular Sertoli cell membranes related to disturbed intracellular Ca^{2+} homeostasis. Toxicol Lett 183(1–3):10–20

Green T, Swain C, Van Miller JP et al (2003) Absorption, bioavailability, and metabolism of *para*-nonylphenol in the rat. Regul Toxicol Pharmacol 38(1):43–51

Guart A, Bono-Blay F, Borrell A et al (2011) Migration of plasticizers phthalates, bisphenol A and alkylphenols from plastic containers and evaluation of risk. Food Addit Contam Part A Chem Anal Control Expo Risk Assess 28(5):676–685

Guenther K, Heinke V, Thiele B et al (2002) Endocrine disrupting nonylphenols are ubiquitous in food. Environ Sci Technol 36(8):1676–1680

Han XD, Tu ZG, Gong Y et al (2004) The toxic effects of nonylphenol on the reproductive system of male rats. Reprod Toxicol 19(2):215–221

Hao CJ, Cheng XJ, Xia HF et al (2012) The endocrine disruptor 4-nonylphenol promotes adipocyte differentiation and induces obesity in mice. Cell Physiol Biochem 30(2):382–394

Hernandez-Rodriguez G, Zumbado M, Luzardo OP et al (2007) Multigenerational study of the hepatic effects exerted by the consumption of nonylphenol- and 4-octylphenol-contaminated drinking water in Sprague-Dawley rats. Environ Toxicol Pharmacol 23(1):73–81

Hill SS, Shaw BR, Wu AH (2003) Plasticizers, antioxidants, and other contaminants found in air delivered by PVC tubing used in respiratory therapy. Biomed Chromatogr 17(4):250–262

Hossaini A, Dalgaard M, Vinggaard AM et al (2001) In utero reproductive study in rats exposed to nonylphenol. Reprod Toxicol 15(5):537–543

Howe SR, Surana P, Jakupca MR et al (2001) Potential dietary exposure to *p*-nonylphenol from food-contact use of tris(nonylphenyl)phosphite (TNPP). Food Addit Contam 18(11): 1021–1039

Hung CH, Yang SN, Kuo PL et al (2010) Modulation of cytokine expression in human myeloid dendritic cells by environmental endocrine-disrupting chemicals involves epigenetic regulation. Environ Health Perspect 118(1):67–72

Inoue K, Kondo S, Yoshie Y et al (2001) Migration of 4-nonylphenol from polyvinyl chloride food packaging films into food simulants and foods. Food Addit Contam 18(2):157–164

Inoue K, Okumura H, Higuchi T et al (2002) Characterization of estrogenic compounds in medical polyvinyl chloride tubing by gas chromatography-mass spectrometry and estrogen receptor binding assay. Clin Chim Acta 325(1–2):157–163

Isidori M, Cangiano M, Palermo FA et al (2010) E-screen and vitellogenin assay for the detection of the estrogenic activity of alkylphenols and trace elements. Comp Biochem Physiol C: Toxicol Pharmacol 152(1):51–56

Iwata M, Eshima Y, Kagechika H et al (2004) The endocrine disruptors nonylphenol and octylphenol exert direct effects on T cells to suppress Th1 development and enhance Th2 development. Immunol Lett 94(1–2):135–139

Jing X, Bing S, Xiaoyan W et al (2011) A study on bisphenol A, nonylphenol, and octylphenol in human urine amples detected by SPE-UPLC-MS. Biomed Environ Sci 24(1):40–46

Jubendradass R, D'Cruz SC, Mathur PP (2012a) Long-term exposure to nonylphenol affects insulin signaling in the liver of adult male rats. Hum Exp Toxicol 31(9):868–876

Jubendradass R, D'Cruz SC, Rani SJ et al (2012b) Nonylphenol induces apoptosis via mitochondria- and Fas-L-mediated pathways in the liver of adult male rat. Regul Toxicol Pharmacol 62(3):405–411

Karrow NA, Guo TL, Delclos KB et al (2004) Nonylphenol alters the activity of splenic NK cells and the numbers of leukocyte subpopulations in Sprague-Dawley rats: a two-generation feeding study. Toxicology 196(3):237–245

Kawaguchi M, Inoue K, Sakui N et al (2004) Stir bar sorptive extraction and thermal desorption-gas chromatography-mass spectrometry for the measurement of 4-nonylphenol and 4-*tert*-octylphenol in human biological samples. J Chromatogr B Analyt Technol Biomed Life Sci 799(1):119–125

Korkmaz A, Ahbab MA, Kolankaya D et al (2010) Influence of vitamin C on bisphenol A, nonylphenol and octylphenol induced oxidative damages in liver of male rats. Food Chem Toxicol 48(10):2865–2871

Korkmaz A, Aydogan M, Kolankaya D et al (2011) Vitamin C coadministration augments bisphenol A, nonylphenol, and octylphenol induced oxidative damage on kidney of rats. Environ Toxicol 26(4):325–337

Kudo C, Wada K, Masuda T et al (2004) Nonylphenol induces the death of neural stem cells due to activation of the caspase cascade and regulation of the cell cycle. J Neurochem 88(6):1416–1423

Kuo C-CD-HK, Haung C-J et al (2010) Nonylphenol-inducted apoptotic pathways in SCM1 human gastric cancer cells. Drug Dev Res 71(2):139–148

Lagos-Cabre R, Moreno RD (2012) Contribution of environmental pollutants to male infertily: a working model of germ cell apoptosis induced by plasticizers. Biol Res 45(1):5–14

Latendresse JR, Newbold RR, Weis CC et al (2001) Polycystic kidney disease induced in F_1 Sprague-Dawley rats fed *para*-nonylphenol in a soy-free, casein-containing diet. Toxicol Sci 62(1):140–147

Lee MJ, Lin H, Liu CW et al (2008) Octylphenol stimulates resistin gene expression in 3T3-L1 adipocytes via the estrogen receptor and extracellular signal-regulated kinase pathways. Am J Physiol Cell Physiol 294(6):C1542–C1551

Li CT, Cheng CY, Ding WH (2008) Determination of alkylphenol residues in baby-food purees by steam distillation extraction and gas chromatography-mass spectrometry. Food Chem Toxicol 46(2):803–807

Li D, Hu Y, Shen X et al (2010a) Combined effects of two environmental endocrine disruptors nonyl phenol and di-*n*-butyl phthalate on rat Sertoli cells in vitro. Reprod Toxicol 30(3):438–445

Li X, Ying GG, Su HC et al (2010b) Simultaneous determination and assessment of 4-nonylphenol, bisphenol A and triclosan in tap water, bottled water and baby bottles. Environ Int 36(6):557–562

Li X, Ying GG, Zhao JL et al (2013) 4-Nonylphenol, bisphenol-A and triclosan levels in human urine of children and students in China, and the effects of drinking these bottled materials on the levels. Environ Int 52:81–86

Lopez-Espinosa MJ, Freire C, Arrebola JP et al (2009) Nonylphenol and octylphenol in adipose tissue of women in Southern Spain. Chemosphere 76(6):847–852

Loyo-Rosales JE, Rosales-Rivera GC, Lynch AM et al (2004) Migration of nonylphenol from plastic containers to water and a milk surrogate. J Agric Food Chem 52(7):2016–2020

Masuo Y, Ishido M, Morita M et al (2004a) Effects of neonatal treatment with 6-hydroxydopamine and endocrine disruptors on motor activity and gene expression in rats. Neural Plast 11(1–2):59–76

Masuo Y, Morita M, Oka S et al (2004b) Motor hyperactivity caused by a deficit in dopaminergic neurons and the effects of endocrine disruptors: a study inspired by the physiological roles of PACAP in the brain. Regul Pept 123(1–3):225–234

McClusky LM, de Jager C, Bornman MS (2007) Stage-related increase in the proportion of apoptotic germ cells and altered frequencies of stages in the spermatogenic cycle following gestational, lactational, and direct exposure of male rats to *p*-nonylphenol. Toxicol Sci 95(1):249–256

Miller D, Wheals BB, Beresford N et al (2001) Estrogenic activity of phenolic additives determined by an in vitro yeast bioassay. Environ Health Perspect 109(2):133–138

Milligan SR, Balasubramanian AV, Kalita JC (1998) Relative potency of xenobiotic estrogens in an acute in vivo mammalian assay. Environ Health Perspect 106(1):23–26

Moon HJ, Han SY, Shin JH et al (2007) Gestational exposure to nonylphenol causes precocious mammary gland development in female rat offspring. J Reprod Dev 53(2):333–344

Muller S, Schmid P, Schlatter C (1998) Pharmacokinetic behavior of 4-nonylphenol in humans. Environ Toxicol Pharmacol 5(4):257–265

Nagao T, Wada K, Marumo H et al (2001) Reproductive effects of nonylphenol in rats after gavage administration: a two-generation study. Reprod Toxicol 15(3):293–315

Narita S, Goldblum RM, Watson CS et al (2007) Environmental estrogens induce mast cell degranulation and enhance IgE-mediated release of allergic mediators. Environ Health Perspect 115(1):48–52

Negishi T, Kawasaki K, Suzaki S et al (2004) Behavioral alterations in response to fear-provoking stimuli and tranylcypromine induced by perinatal exposure to bisphenol A and nonylphenol in male rats. Environ Health Perspect 112(11):1159–1164

Ogawa Y, Kawamura Y, Wakui C et al (2006) Estrogenic activities of chemicals related to food contact plastics and rubbers tested by the yeast two-hybrid assay. Food Addit Contam 23(4):422–430

Olsen CM, Meussen-Elholm ET, Hongslo JK et al (2005) Estrogenic effects of environmental chemicals: an interspecies comparison. Comp Biochem Physiol C: Toxicol Pharmacol 141(3):267–274

Park M-A, Choi K-C (2012) A potential endocrine-disrupting chemical, 4-nonylphenol, stimulated the ovarian cancer cell growth by upregulating cell cycle via an estrogen receptor signaling pathyway in cellular and animal models (abstract). Reprod Fertil Devel 25(1):246

Pirard C, Sagot C, Deville M et al (2012) Urinary levels of bisphenol A, triclosan and 4-nonylphenol in a general Belgian population. Environ Int 48:78–83

Raecker T, Thiele B, Boehme RM et al (2011) Endocrine disrupting nonyl- and octylphenol in infant food in Germany: considerable daily intake of nonylphenol for babies. Chemosphere 82(11):1533–1540

Rudel RA, Camann DE, Spengler JD et al (2003) Phthalates, alkylphenols, pesticides, polybrominated diphenyl ethers, and other endocrine-disrupting compounds in indoor air and dust. Environ Sci Technol 37(20):4543–4553

Rudel RA, Dodson RE, Perovich LJ et al (2010) Semivolatile endocrine-disrupting compounds in paired indoor and outdoor air in two northern California communities. Environ Sci Technol 44(17):6583–6590

Ruehlmann DO, Steinert JR, Valverde MA et al (1998) Environmental estrogenic pollutants induce acute vascular relaxation by inhibiting L-type Ca^{2+} channels in smooth muscle cells. FASEB J 12(7):613–619

Saito I, Onuki A, Seto H (2004) Indoor air pollution by alkylphenols in Tokyo. Indoor Air 14(5):325–332

Schmutzler C, Hamann I, Hofmann PJ et al (2004) Endocrine active compounds affect thyrotropin and thyroid hormone levels in serum as well as endpoints of thyroid hormone action in liver, heart and kidney. Toxicology 205(1–2):95–102

Seike N, Wanibuchi H, Morimura K et al (2003) Enhancement of lung carcinogenesis by nonylphenol and genistein in a F344 rat multiorgan carcinogenesis model. Cancer Lett 192(1):25–36

Shelby MD, Newbold RR, Tully DB et al (1996) Assessing environmental chemicals for estrogenicity using a combination of in vitro and in vivo assays. Environ Health Perspect 104(12):1296–1300

Snedeker SM, Hay AG (2012) Do interactions between gut ecology and environmental chemicals contribute to obesity and diabetes? Environ Health Perspect 120(3):332–339

Song L, Xia W, Zhou Z et al (2012) Low-level phenolic estrogen pollutants impair islet morphology and beta-cell function in isolated rat islets. J Endocrinol 215(2):303–311

Soto AM, Justicia H, Wray JW et al (1991) *p*-Nonyl-phenol: an estrogenic xenobiotic released from "modified" polystyrene. Environ Health Perspect 92:167–173

Suen JL, Hung CH, Yu HS et al (2012) Alkylphenols—potential modulators of the allergic response. Kaohsiung J Med Sci 28(7 Suppl):S43–S48

Talorete TP, Isoda H, Maekawa T (2001) Alkylphenolic compounds and their effect on the injury rate, survival and acetylcholinesterase activity of the rat neuronal cell line PC12. Cytotechnology 36(1–3):163–169

Tan BL, Ali Mohd M (2003) Analysis of selected pesticides and alkylphenols in human cord blood by gas chromatograph-mass spectrometer. Talanta 61(3):385–391

Tanaka T, Kohno H, Tanino M et al (2002) Inhibitory effects of estrogenic compounds, 4-nonylphenol and genistein, on 7,12-dimethylbenz[*a*]anthracene-induced ovarian carcinogenesis in rats. Ecotoxicol Environ Saf 52(1):38–45

Terasaka S, Inoue A, Tanji M et al (2006) Expression profiling of estrogen-responsive genes in breast cancer cells treated with alkylphenols, chlorinated phenols, parabens, or bis- and benzoylphenols for evaluation of estrogenic activity. Toxicol Lett 163(2):130–141

Tyl RW, Myers CB, Marr MC et al (2006) Three-generation evaluation of dietary *para*-nonylphenol in CD (Sprague-Dawley) rats. Toxicol Sci 92(1):295–310

Udoji F, Martin T, Etherton R et al (2010) Immunosuppressive effects of triclosan, nonylphenol, and DDT on human natural killer cells in vitro. J Immunotoxicol 7(3):205–212

USCFR (2013) U.S. Code of Federal Regulations, Title 21 - Food and Drugs, Parts 176.180, 176.200, 176.210, 177.1200, 175.300, 178.210, 175.105, 175.300, 177.2410, and 177. 2600. http://www.gpo.gov/fdsys/pkg/CFR-2013-title21-vol3/pdf/CFR-2013-title21-vol3-chapI.pdf. Accessed 25 Feb 2014

United States Environmental Protection Agency (USEPA) (2010) Nonylphenol and nonyphenol ethoxylates Action Plan summary. U.S. Environmental Protection Agency. http://www.epa.gov/oppt/existingchemicals/pubs/actionplans/np-npe.html. Accessed 8 Jan 2014

Van den Belt K, Berckmans P, Vangenechten C et al (2004) Comparative study on the in vitro/in vivo estrogenic potencies of 17beta-estradiol, estrone, 17alpha-ethynylestradiol and nonylphenol. Aquat Toxicol 66(2):183–195

Wang JL, Liu CS, Lin KL et al (2005) Nonylphenol-induced Ca^{2+} elevation and Ca^{2+} -independent cell death in human osteosarcoma cells. Toxicol Lett 160(1):76–83

White R, Jobling S, Hoare SA et al (1994) Environmentally persistent alkylphenolic compounds are estrogenic. Endocrinology 135(1):175–182

Willoughby KN, Sarkar AJ, Boyadjieva NI et al (2005) Neonatally administered *tert*-octylphenol affects onset of puberty and reproductive development in female rats. Endocrine 26(2):161–168

Woo GH, Shibutani M, Ichiki T et al (2007) A repeated 28-day oral dose toxicity study of nonylphenol in rats, based on the 'Enhanced OECD Test Guideline 407' for screening of endocrine-disrupting chemicals. Arch Toxicol 81(2):77–88

Wu J, Wang F, Gong Y et al (2009) Proteomic analysis of changes induced by nonylphenol in Sprague-Dawley rat Sertoli cells. Chem Res Toxicol 22(4):668–675

Yanagihara N, Toyohira Y, Ueno S et al (2005) Stimulation of catecholamine synthesis by environmental estrogenic pollutants. Endocrinology 146(1):265–272

Yokosuka M, Ohtani-Kaneko R, Yamashita K et al (2008) Estrogen and environmental estrogenic chemicals exert developmental effects on rat hypothalamic neurons and glias. Toxicol In Vitro 22(1):1–9

Zalko D, Costagliola R, Dorio C et al (2003) In vivo metabolic fate of the xeno-estrogen 4-*n*-nonylphenol in Wistar rats. Drug Metab Dispos 31(2):168–178

Zhang HY, Xue WY, Li YY et al (2014) High-fat diet promotes the maternal transgenerational effect of obesity by perinatal exposed to 4-NP in Wistar rats. Toxicol Lett **pii:S0378-4274**(13):01467

Chapter 6
Benzophenone UV-Photoinitiators Used in Food Packaging: Potential for Human Exposure and Health Risk Considerations

Suzanne M. Snedeker

Abstract Chemicals that are used in ultraviolet (UV) print inks include benzophenone-based UV-photoinitiators used in printing the surface of food packaging, especially printed cartonboard. Since these inks are not used up in the printing process, these photoinitiators are available to migrate from the printed surface through porous cartonboard and secondary packaging to food and beverages. The UV-photoinitiators benzophenone (BP) and 4-methylbenzophenone (4-MBP) have been detected in paperboard packaging of a wide variety of foods, and migration studies and analysis of packaged foods indicate they can migrate from the packaging to foods and beverages. BP also has been detected in recycled paper fibers and recycled plastics, and can migrate from packaging both under room temperature and frozen storage conditions even when the food is not in direct contact with the packaging. There is limited information on the toxicology of these chemicals and their metabolism in animals or humans. While neither BP nor 4-MBP appears to be genotoxic, BP administration in rodent cancer bioassays indicates it is a possible human carcinogen. A metabolite of BP, 4-hydroxybenzophenone (4-OH BP), is estrogenic, but the estrogenicity of 4-MBP has not been evaluated. Both BP and 4-MBP have been detected in foods above regulatory specific migration limits and have been the subject of food recalls in Europe. Further research is needed on levels of UV-photoinitiators in food packaging in the global food supply and effective secondary packaging to prevent migration to foodstuffs. More extensive toxicological studies are needed on this class of chemicals to better determine potential risks to human populations.

Keywords UV-photoinitiators · Benzophenone · BP · 4-methylbenzophenone · 4 MBP · 4-hydroxybenzophenone · 4-OH BP · Benzhydrol · Amine synergists · Michler's ketone · 4,4'-bis(dimethylamino)benzophenone · MK · 4,4'-bis(diethylamino)benzophenone · DEAB · 4-(dimethylamino)benzophenone ·

S. M. Snedeker (✉)
Independent Consultant, Ithaca, NY, USA
e-mail: smsnedeker@gmail.com

S. M. Snedeker (ed.), *Toxicants in Food Packaging and Household Plastics*, Molecular and Integrative Toxicology, DOI: 10.1007/978-1-4471-6500-2_6,

DMAB · Print inks · Food packaging · Food contact plastics · Food contact materials · Recycled packaging · Cartonboard · Paperboard · Migration · Set-off · Polyethylene terephthalate plastic · PET · Polyethylene · PE · Polypropylene · PP · Frozen food · Estrogenicity · Endocrine disruption · Carcinogenicity · Genotoxicity · Reproductive toxicity · Sensors · Liver · Hepatic · Kidney · Renal · Hematopoietic system · EFSA Rapid Alert System for Food and Feed · Specific migration limits · SML

6.1 Key Take Home Points

- UV-photoinitiators are commonly used in print inks on the printed surface of food packaging, especially printed cartonboard.
- UV-photoinitiators can be 5 to 10 % percent of the print ink by weight and can migrate through porous cartonboard and commonly used secondary packaging to foods and beverages.
- Benzophenone and 4-methylbenzophenone are UV-photoinitiators that have been detected in printed cartonboard and in fibers of recycled packaging.
- Migration of benzophenone-based photoinitiators from packaging to food has been documented at room- and low-temperature frozen storage conditions, even when the food was not in direct contact with the primary packaging.
- The metabolite of benzophenone, 4-hydroxybenzopheone, is a weak estrogen.
- Animal modeling studies provide evidence that benzophenone is a possible human carcinogen, though neither benzophenone nor 4-methylbenzophenone are genotoxic.
- UV-photoinitiators used in print ink are a class of chemicals that need to be monitored in the global food supply, and additional toxicological risk assessment is needed to better evaluate health effects in human populations.

6.2 Introduction

Chemicals used as photoinitiators in ultraviolet (UV) inks that are applied to food packaging serve as catalysts in the UV-curing process, where the inks dry (cure) rapidly through a photopolymerization process (Papilloud and Baudraz 2002). The advantage of using UV-cured inks is that they can be directly applied to the outer surface of the package without the use of solvents, hence reducing drying time and eliminating solvent exposure to workers. They are most commonly applied to the outer surface of paperboard cartons. Many of these photoinitiators constitute 5 to 10 % of the ink by weight (Anderson and Castle 2003). Since the UV-initiators are not totally used up during the UV-curing, or eliminated from

Table 6.1 Benzophenone UV-photoinitiators and amine synergists detected in food packaging

Chemical name	Common name	Abbreviation	CAS registry number	Detection in food packaging (Reference)
UV-photoinitiators				
Diphenylketone	Benzophenone	BP	119-61-9	(Anderson and Castle 2003)
4-methylbenzophenone		4-MBP	134-84-9	(Castle et al. 1997)
Amine synergists				
4,4'-bis(dimethylamino) benzophenone	Michler's ketone	MK	90-94-8	(Castle et al. 1997)
4,4'-bis(diethylamino)-benzophenone		DEAB	90-93-7	(Castle et al. 1997)
4-(dimethylamino)benzophenone		DMAB	530-44-9	(Castle et al. 1997)

the packaging after the printing process is complete, they are available to migrate through the porous paperboard and secondary packaging to the food or beverage. They also can remain in the paperboard after the outer print layer is removed, and thus UV-initiators can appear in new packaging made with recycled paperboard fibers.

Several amine synergists used in UV-cured inks were identified in cartonboard samples of printed food packaging in the mid-1990s (see Table 6.1), including Michler's ketone 4,4'-bis(dimethylamino)benzophenone (MK), 4,4'-bis(diethylamino)benzophenone (DEAB), and 4-(dimethylamino)benzophenone (DMAB) (Castle et al. 1997). A study has evaluated migration of MK, DEAB, and DMAB from paperboard into food simulants (Ozaki et al. 2006). However, the majority of studies on levels of UV-photoinitiators in food packaging and their migration to food or beverages have focused on the UV-initiators benzophenone (BP) ($C_{13}H_{10}O$) and 4-methylbenzophenone (4-MBP) ($C_{14}H_{12}O$) (see Table 6.1, Fig. 6.1) (Sagratini et al. 2008; Anderson and Castle 2003; Rodriguez-Bernaldo de Quiros et al. 2009; Johns et al. 1995; Bugey et al. 2013; Koivikko et al. 2010; Han et al. 2011; Shen et al. 2009; Jung et al. 2013). BP and 4-MBP have been the subject of several European Food Safety Authority (EFSA) notifications via the *Rapid Alert System for Food and Feed* (RASFF) when food was found to be contaminated with these photoinitiators (Harrington 2010, 2011; EFSA 2009a). Available toxicological risk assessments on UV-initiators are largely limited to studies on BP and 4-MBP. For these reasons, this chapter will focus on: evidence documenting levels of BP and 4-MBP in food packaging and food and beverages; conditions that affect their rate of migration from primary and recycled packaging; adverse health effects that have been identified in animal modeling and cell-based studies; and a summary of regulatory agency responses when the levels of these UV-initiators in food have exceeded regulatory standards.

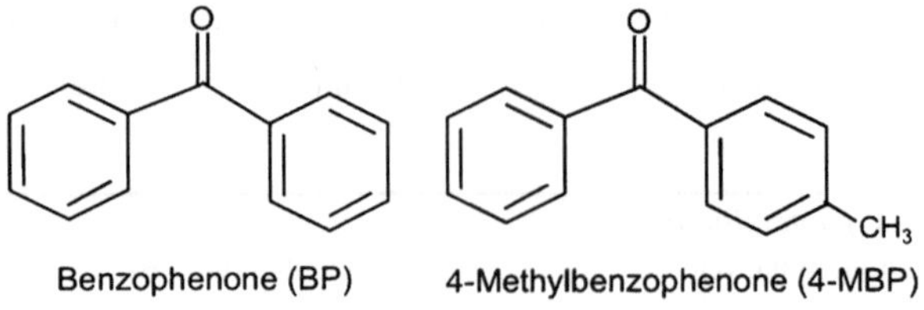

Fig. 6.1 Chemical structures of the UV-initiators benzophenone and 4-methylbenzophenone

6.3 Exposure, Metabolism, and Biomonitoring

6.3.1 Detection in Paperboard Food Packaging and Levels in Food

One of largest studies of the levels of the UV-initiator BP in the printed cartonboard and packaged foods included 350 samples obtained from retail markets in the United Kingdom (Anderson and Castle 2003). BP residues were detected in 41 % of the samples of cartonboard packaging with levels ranging from 0.05 to 3.3 milligrams per square decimeter (mg/dm^2). A random sample of 71 foods from the 143 positive packaging samples were analyzed for BP, and 51 foods (72 %) had detectable levels of BP ranging from 0.01 to 7.3 mg per kilogram (mg/kg). Other studies have focused on the frequency of UV-initiator detection in a specific type of food. In an analysis of 82 cereal products obtained from Swiss retail markets, three samples (3.7 %) had levels in the range of 3.9–7.6 mg/kg (Bugey et al. 2013). These foods exceeded the European *specific migration limit* (SML) for BP of 0.6 mg/kg (EC 2003). The packaging of these three positive food samples had correspondingly high levels BP ranging from 250 to 500 mg/kg. Based on these results, the authors concluded that 1–2 % of the BP in the packaging migrated to the food. For foods where the BP levels were below the SML of 0.6 mg/kg, levels in the corresponding packaging ranged from 5 to 7 mg/kg.

Few studies have analyzed for both BP and 4-MBP in food contact materials or foods. Levels of BP and 4-MBP were determined in 46 paperboard and cartonboard samples obtained from European supermarkets and 19 paper samples obtained directly from suppliers (Koivikko et al. 2010). These two different photoinitiators were not found in the same food packaging sample; one or the other predominated with only traces of the other UV-initiator being present, and these traces were often due contamination from recycled paper fibers. BP was detected more frequently in 59 % of the 46 packaging samples, compared with a 30 % detection frequency for 4-MBP. BP and 4-MBP were present in the paperboard packaging of cakes, chocolate, cookies, and coffee, but were not detected in the packaging of pastry, breakfast cereal, dried fruit, or nuts. The highest level detected for 4-MBP was in the packaging of a cake sample (4.41 mg/dm^2), while the highest levels for BP were in packaging of coffee (3.99 mg/dm^2) and a different cake packaging sample (3.84 mg/dm^2). Levels of 11 different UV-photoinitiators and amine synergists, including BP and 4-MBP, were determined in the packaging (mostly paperboard) of

310 foods obtained in German markets (Jung et al. 2013). The most frequently detected chemical in the samples of food packaging was BP (49 %) followed by 4-MBP (8 %). Subsequently, 99 food samples were analyzed, and 23 were found to have detectable levels of BP with 12 of these foods exceeding the European regulatory SML of 0.6 mg/kg, including rice paper (7.5 mg/kg), chocolate muesli cereal (3.4 mg/kg), fruit muesli cereal (3.7 mg/kg), cocoa-filled cookies (1.8 mg/kg), couscous (1.6 mg/kg and 0.86 mg/kg), and several samples of cinnamon used in a milk powder (range 1.5–50.2 mg/kg). Eight foods had detectable levels of 4-MBP, with the highest levels reported in two samples of chocolate muesli cereal at 8.1 and 0.80 mg/kg.

While most studies evaluating levels of UV-photoinitiators in foodstuff have been conducted on dry food packaged in cartonboard, several studies have determined levels in beverages, including milk, fruit juices, and wine. Levels of different UV-initiators were determined in packaging and food samples of 40 beverages obtained from retail markets in Camerino, Italy (Sagratini et al. 2008). BP was the most frequently detected UV-initiator, being present in all packaging materials in the range of 0.2–387 micrograms per liter (μg/L), and in all beverage samples in the range of 5–217 μg/L. Levels in poly-layered packaging were not directly related to concentrations of BP in the beverage since the BP was not necessarily from the print layer. Polycoupled packaging uses polyethylene (PE) in the layer in direct contact with the beverage, and BP is used as a photoinitiator in the production of the PE film. There was a wide range of concentrations reported in the different types of beverages. While BP levels in most samples of white milk were in the range of 5.25–14 μg/L, chocolate milk samples tended to be higher and ranged from 14.8 to 39 μg/L. Levels of BP in juice samples ranged from 5–41 μg/L. While levels of BP in most wine samples were in the range of 5.5–14.7 μg/L, levels were higher in one red wine sample (217 μg/L) and one white wine sample (87 μg/L). Levels of BP in milk samples obtained from retail markets in China was partially dependent on the type of packaging (Shen et al. 2009). Levels of BP in six milk samples packed in a "carton" (materials and construction not specified) ranged from 2.84 to 18.35 μg/L, while none of the six milk samples packaged in plastic containers (type of plastic unspecified) had detectable levels of BP.

Other studies have investigated levels of BP in the packaging of a variety of dairy products (levels in food were not reported). In an analysis of food packaging of 26 dairy foods, including 16 infant formula samples, two soy drinks, two yogurts, three cheeses, and one ice cream, BP was detected in nine (34.6 %) of the packaging samples (Sanches-Silva et al. 2008). The highest level reported was in ice cream packaging (595 mg/kg). The other positive samples, including packaging from one soy drink sample, one yogurt sample, and six infant formula samples, had BP levels ranging from 0.32 to 2.75 mg/kg packaging.

BP was not detected in retail baby foods obtained in Japanese markets that were packaged in a variety of containers, including glass, plastic (type not specified), metallic laminated with plastic, and multilaminated film packages (Ozaki et al. 2002). This suggests that either these types of packaging were less likely to use

UV-photoinitiators, or that these containers provided effective barriers to the migration of BP-based UV-photoinitiators.

These studies suggest that both BP and 4-MBP are predominately found in printed board packaging, though one type of photoinitiator will predominate over the other. Levels of the UV-photoinitiator migrating from the paperboard to the food is partially dependent the level in the paperboard. More research is needed on levels of BP and 4-MBP used in PE film production and use in and their ability to migrate through multilayered packaging.

6.3.2 Factors that Affect the Migration of UV-Initiators from Paperboard

UV-photoinitiators like BP that are used in print inks are not found in unprinted virgin paperboard (Ozaki et al. 2006). Because the UV-initiator is not consumed in the printing process, it can persist on the printed surface and migrate through the layers of porous paperboard. A variety of factors have been shown to affect migration from the printed surface through the paperboard and to the food itself (direct contact), or from the paperboard through the vapor phase to the food (indirect contact). These factors include set-off, the vapor pressure of the UV-initiator, time and temperature of food storage, porosity and types of materials used for primary and secondary packaging, the fat and moisture content of the food, and whether the UV-initiator is present in recycled materials used to form new packaging. These factors are discussed below.

Set-Off Paperboard is commonly printed by a sheet-fed lithographic process where the use of UV-initiators in the ink results in a rapid cure and drying time (Anderson and Castle 2003). It is common practice to stack sheets of the printed paperboard or to roll it onto reels. The physical transfer of components in the printed inked surface to the interior uninked surface of another stacked sheet is called *set-off* (Johns et al. 2000). Set-off is one the major routes of transfer of UV-initiators in the printed layer to the interior non-printed surface of paperboard used in food packaging.

Vapor-Pressure A model system was developed to evaluate factors affecting the migration of several different types of BP UV-initiators (methyl-2-benzophenone, methyl-2-benzoylbenzoate, BP, 2-hydroxybenzophenone, 4-MBP, 4-benzoylbiphenyl benzophenone, and DEAB) across the vapor phase to selected dry foods of different fat content (Rodriguez-Bernaldo de Quiros et al. 2009). The three UV-initiators with the highest migration into foods were 2-hydroxybenzophenone, BP, and 4-MBP. This was in part due to the higher vapor pressure of these three UV-initiators. Other factors associated with higher migration included a higher fat content and porosity of the foodstuff.

Time and Temperature In a survey of BP levels in cartonboard packaging and retail foods stored at different temperatures, BP migration tended to be highest in foods with direct contact with the packaging stored at room temperature compared to those with indirect contact at room or frozen temperatures (Anderson and Castle 2003). While migration of BP tended to be lower in frozen foods than foods stored at room temperature, cooling and freezing the food did not eliminate BP migration. The average level of BP in food in direct contact with the paperboard packaging held at room temperature was 2.0 mg/kg, compared with 0.58 mg/kg in frozen foods in direct contact, and 0.13 mg/kg in frozen foods with indirect contact with the packaging. Therefore, room temperature storage and direct contact of the packaging with the food enhance BP migration rates, but freezing the food will not prevent migration.

Slow migration of BP has been observed in microwavable packaged foods stored frozen at $-20\ ^{\circ}$C (Johns et al. 2000). The seven frozen foods were purchased in retail markets, and BP was detected in printed packaging of four products: a cheese-egg-bacon flan, Cornish pies, potato waffles, and cheese and onion sticks, in the range 0.4–3.0 mg/dm^2. When levels of BP were determined in the corresponding foods, the percent transfer of BP from the packaging to the food ranged from 0–2 %. Model ink components of chemicals with different volatility were tested for migration to hamburger or potato chips stored frozen for up to 1 year at $-20\ ^{\circ}$C. BP's migration was less than other test substances (9 % transfer for potato chips, 8 % transfer for hamburger) because of BP's relatively low boiling point (305 $^{\circ}$C). While majority of the BP that migrated did so during the 1 week of storage, there was continued transfer of BP during the 1-year period of storage. Neither food was in direct contact with the packaging, and suggests that BP migrated in the vapor phase from the packaging to the frozen food.

In a study of BP migration in ice cream bars, peak migration of BP occurred not in the first week of storage, but after several months of storage. BP migration was examined in wrapped frozen chocolate covered ice cream bars stored for 0, 10, 39, 88, and 242 days (Jickells et al. 2005). The greatest increase in migration occurred between 88 and 242 days of storage. BP levels increased from approximately 900 mg/kg at 88 days to 4,400 mg/kg by day 242 of storage. These are some of the highest levels of BP in food reported in the literature, and the high migration rates may be partially due to the fat content of the ice cream bars.

Moisture Content Modeling studies of BP migration into the food simulant Tenax® have determined that a higher relative humidity increased the rate of BP migration at a constant temperature over a 30-day period (Barnkob and Petersen 2013). The migration of BP from paperboard to the food simulant was 7.3-fold higher at a relative humidity of greater than 73 % compared to a relative humidity of 39–49 %.

Fat Content of Packaged Food Concentrations of the UV-initiator BP were determined in milk samples of different fat content obtained in Chinese retail markets (Shen et al. 2009). BP was detected in five out of six samples from carton-type containers, but was not found in any milk samples packaged in plastic containers. The highest levels of BP (18.35 μg/kg) reported was in the milk sample with the highest fat content [30 grams (g)/L] packaged in a carton. In an investigation of factors that affect migration of UV-cured inks from packaging to dry foods, including cake, bread, cereals, rice and pasta, the level of migration was correlated to both the fat content and porosity of the food (Rodriguez-Bernaldo de Quiros et al. 2009). These factors contributed to cake, with a fat content of 11.6 %, having the highest level of migration of BP. Lower rates of BP migration were reported for low-fat foods such as rice and pasta. In a survey of retail samples of BP in cartonboard packaging and a subset of the foods, the product with the highest level of BP (7.3 mg/kg) was a high fat chocolate product (Chocolate Viennese whirls) that was in direct contact with the packaging and stored at room temperature (Anderson and Castle 2003). Other foods with levels of BP that exceeded 1 mg/kg included four foods stored at room temperature: oatmeal (5.70 mg/kg), Thai jasmine rice (5.6 mg/kg), cranberry orange loaf (1.7 mg/kg), and basmati and wild rice (1.1 mg/kg); and four foods stored at chilled/frozen temperatures including cheese and onion slices (2.10 mg/kg), hot dogs (1.8 mg/kg), pork pies (1.90 mg/kg), and vegetable burgers (1.20 mg/kg). The foods with higher levels of BP tended to fall into several categories: grains and cereals with a high surface to volume ratio stored at room temperature, high fat foods stored at room temperature, and high fat foods stored at chilled/frozen temperatures. It should be noted that other frozen foods that would be expected to have a moderate to high fat and moisture content (grilled lamb, bacon puff pastries, sausages, cheese and ham slices, and lasagna) had BP levels in the range of 0.13–0.84 mg/kg. In one high-fat food that had thicker cartonboard packaging (gravy granules stored at room temperature), BP levels were at the limit of detection at 0.01 mg/kg. Therefore, the thickness of the cartonboard packaging can counter the effect of fat-soluble BP's migration to higher fat foods.

Recycled Paperboard Packaging In recycled paperboard, UV-initiators may be present in the paper fibers even if the food container does not have a printed surface. Many packaged foods do not have any secondary packaging, and the food is in direct contact with the interior surface of the paperboard. In modeling studies of UV-initiator migration from recycled board to solid food, the amount of time the food is in direct contact with the paperboard and the temperature of incubation had a significant positive effect on the migration of test substances (Triantafyllou et al. 2007). When BP-containing paper strips were covered with a food powder (semolina) and incubated for 20–360 min at 70 °C, the percent migration increased with time, especially in the first 120 min. This effect was enhanced when the temperature was increased to 100 °C.

Recycled and Virgin Food Contact Plastics BP is one of the contaminants detected in washed and shredded polyethylene terephthalate plastic (PET) obtained from curbside recycling collection (Konkol et al. 2003b). While 26 chemicals were detected in PET plastic at levels below 215 parts per million (ppm), six contaminants, including BP, were detected at levels above 215 ppm. BP was detected in the curbside recycled PET plastic at the level of 800 ppm, while virgin PET plastic did not contain detectable BP. The source of the BP contamination in the recycled PET plastic was not identified. In optimizing methods to extract and detect contaminants, the same researchers found that using a smaller particle size in the ground PET flakes resulted in higher measured levels of the contaminants, including BP (Konkol et al. 2003a). Some research has been devoted to testing whether coating the interior of PET plastic can prevent BP migration. Coating the inner surface of BP-spiked PET bottles with silicon oxide prevented the migration of BP into food simulants (3 % acetic acid and 10 % ethanol) (Welle and Franz 2008). Therefore, an appropriate interior coating of food containers made with recycled PET may allow for the packaging of water-based beverages without exceeding accepted migration limits for BP.

6.3.3 Effectiveness of Barriers to Prevent Migration of UV-initiators

Studies have evaluated the effectiveness of secondary packaging, coating of primary packaging, and materials used in multi-layered primary packaging, to reduce or eliminate the migration of UV-initiators to food and beverages. A modeling study was designed to determine migration of surrogate contaminants including BP, from paper or paperboard coated with PE to water samples (Choi et al. 2002). The virgin paper and paperboard samples were spiked with BP and coated with PE film (0.012 or 0.03 millimeters), and the migration through the paper and PE layer into water was monitored for 21 days. Migration was monitored both in lower temperature water (10 °C) and in room temperature water (24 °C). BP migrated through the paperboard and thin layer of PE during the first 24 hours (h) at both the lower temperature and room temperature. The thicker layer of PE slowed but did not prevent the migration of BP at either temperature. The authors concluded that a PE coating on paper or paperboard was not an effective barrier to BP migration even at chilled temperatures.

Cakes are commonly marketed in colorful paperboard packages during the Easter and Christmas seasons in Italy, and the shiny finish of the paperboard packaging commonly uses UV-initiator inks that contain BP. Cakes are baked, placed in secondary plastic sleeves, and are then placed in the printed carton. The effectiveness of using different types of plastics in the sleeves to prevent BP migration from the printed paperboard layer to the cake was evaluated (Pastorelli et al. 2008). Polypropylene (PP) is the mostly commonly used plastic for secondary packaging, and two other multi-layer films were tested. PP film was found to be an ineffective barrier to BP migration. This is likely due to BP's migration in the vapor phase both through the porous printed board, and the gas permeable PP. Use of a multilayer film of PP, ethylene vinyl alcohol (EVOH), and PP slowed but did not prevent BP migration from the paperboard to the cake. The secondary packaging that prevented BP migration to the cake was a multilayer film composed of PET, silica oxide, and PP (PET/SiO/PP). The PET/SiO/PP layer was an effective barrier to BP migration at both 40 and 70 °C.

Many foods that have a higher fat content or a high moisture content are packaged in paperboard containers coated with PE (Song et al. 2003). Recycled paperboard is commonly used to manufacture food containers, and can be contaminated with BP, or the recycled paperboard may contain BP applied to cure inks during the printing process. In evaluating whether PP coating provides an effective barrier to BP migration through paperboard packaging, modeling studies have found that PP coated paperboard samples spiked with 50 mg/kg of BP failed to prevent BP migration (Song et al. 2003). BP has been found to migrate from PP coated paperboard to frozen foods obtained in retail markets (Johns et al. 2000). These authors concluded that the migration of BP was partly due to the permeability of PP to low molecular weight substances such as BP. Other researchers also have found that paperboard and PE are ineffective barriers to BP migration in model systems (Guazzotti et al. 2014; Biedermann et al. 2013).

Preliminary research using bio-polymer coatings indicates that some materials are promising for slowing the migration of BP, including waxy starch, starch, gluten, and gelatin coated paper (Guazzotti et al. 2014). Some of these coatings may be useful for recycled packaging contaminated with fibers containing BP. PP has shown to be an effective barrier to prevent 4-MBP migration in food packaged in paperboard over 9 months, but PP was not an effective barrier to BP in these modeling studies (Biedermann et al. 2013).

6.3.4 Development of Sensors to Monitor Levels of Benzophenone Food Packaging

Efforts have begun to use molecular imprinting technology to develop sensors to detect the presence of UV-photoinitiators in packaged food. An amperometric molecularly imprinted polymer sensor has been developed to detect BP in packaging plastics (Li et al. 2012). The sensor method detected BP in over a range of 0.05–5 millimolar (mM), with a detection limit of 10 nM. Levels of BP in different types of food contact packaging were nearly identical using the sensor method as compared to a high-pressure liquid chromatography-UV method.

6.3.5 Use of Benzophenone as Test Compound in Migration Studies

BP is used as a surrogate contaminant in the routine testing of a chemical's ability to migrate from a food contact plastic, including recycled plastics, to a food simulant (Franz and Welle 2008; FDA 2006, 2014). BP is used as a surrogate for non-volatile, polar chemicals (FDA 2006). Food plastics are spiked with BP, and its migration into liquid simulants such 3 % acetic acid or 10–95 % ethanol, is then monitored to determine kinetic properties of the particular polymer plastic being evaluated. This helps determine the type of processing needed for recycled plastics to reduced levels of contaminants in foodstuffs. If possible, it is desirable that a chemical's level in the recycled plastic be the same as those from virgin-PET plastic (Franz and Welle 2008; Franz et al. 2004). Test methods are being developed to examine migration of chemicals from paperboard and plastic-paper materials into solid food simulants (Nerin and Asensio 2007; Sanches-Silva et al. 2009).

6.3.6 Metabolism and Biomonitoring Studies

There is very limited data on the absorption, metabolism, and pharmacokinetics of orally or dermally administered BP in either humans or in animal models, and no published literature could be located on the metabolism of 4-MBP by any route of exposure. Available studies on BP are summarized below.

Several metabolites of BP have been identified using in vitro and in vivo methods. When isolated rat hepatocytes were incubated with BP (0.25 mM) over a 3-h time course, the levels of BP decreased, while levels of its metabolites benzhydrol and the sulfate conjugate of 4-OH BP increased (Nakagawa et al. 2000). There was also a small increase in levels of non-conjugated 4-OH BP. The authors believed the formation of the 4-OH BP was probably due to enzymatic aromatic hydroxylation mediated by cytochrome P-450 enzymes, though the activity of hepatic P-450 enzymes were not reported in this study.

In an in vivo study, BP was administered as a single oral dose (100 or 400 mg/kg) to intact female, Sprague-Dawley rats, and serum levels of BP and metabolites were measured 6 and 24 h later (Nakagawa and Tayama 2002). At 6 h both BP and metabolites benzhydrol, and to a lesser extent 4-OH BP, were present in the serum at both dose levels. At 24 h, serum from 100 mg/kg BP-treated rats had undetectable levels of BP and its metabolites, while in the 400 mg/kg group, BP, 4-OH BP, and benzhydrol were still detectable in the serum. The authors proposed that in rat hepatocytes BP forms benzhydrol by enzymatic ketone reduction, while a mono-oxygenase converts BP to 4-OH BP with subsequent conversion to its sulfate form via a sulfotransferase (Nakagawa and Tayama 2002). Similar results have been reported in male rats treated with BP (Jeon et al. 2008). After an oral administration of BP in corn oil at 100 mg/kg, plasma levels of BP and its metabolites were monitored for 24 h. The two major metabolites detected were benzhydrol and 4-OH BP. Plasma levels of BP and both metabolites peaked at 5 h, and continued to fall during the time course, though levels were still detectable at 24 h. Levels of BP or metabolites in the urine and feces or their tissue distribution were not reported.

The National Toxicology Program (NTP) conducted toxicokinetic studies of BP administered as a single dose in male and female F344/N rats and B6C3F$_1$ mice (Chhabra 2000). BP was monitored in the plasma, and a biphasic pattern in plasma levels suggested enterohepatic circulation. However, there is little data examining BP's or benzhydrol's Phase II metabolism. While the NTP report proposed the BP metabolite being recirculated was a glucoronide of benzhydrol, no published studies have actually evaluated Phase II metabolism of BP and its metabolites.

Percutaneous absorption of BP has been determined in Rhesus monkeys using ^{14}C-radiolabeled BP (Bronaugh et al. 1990). It should be noted that the purpose of this study was to examine the absorption and excretion of dermally-applied compounds used in fragrances, including BP. A single dose of radiolabeled-BP in acetone was applied to abdominal skin of the monkeys, and the level of radioactivity was monitored in the urine for up to 5 days. While 92.6 % of the radiolabel did appear in the urine of the monkeys, there was no attempt to determine the chemical nature the radioactive compounds that were excreted (e.g., parent BP or its metabolites).

The majority of the studies that have determined the extent to which BP and its metabolites are detected in human urine have been small method development studies. In one study that analyzed urine samples obtained from 14 human volunteers (urine collection methods not specified), BP, 2-hydroxybenzophenone, and 3-hydroxybenzophenone, were not detected in any of the urine samples, while the BP metabolite benzhydrol was detected in all urine samples in the range of 0.27–5.16 nanograms per ml (ng/ml), and 4-OH BP was detected in only one urine sample at 0.15 ng/ml (Ito et al. 2009). Levels of urinary benzhydrol have been reported in similar ranges in other studies, including a study that included six human urine samples (0.4–3.7 ng/ml) (Kawaguchi et al. 2008), and a study that included an analysis of ten human urine samples (0.63–5.91 ng/ml) (Kawaguchi et al. 2009).

A liquid chromatography, tandem mass spectrophometry method (LC–MS/MS) has been used to measure a variety of BP compounds in the urine of 625 adult women from California and Utah, including the metabolite of BP, 4-OH BP (Kunisue et al. 2012). The purpose of this study was to determine if there were relationships between these BP compounds, especially those used in personal care products, and the incidence of endometriosis, since several BP compounds, including 4-OH BP have evidence of some estrogenicity. The results of this study suggest that smoking may affect the metabolism of BP compounds. The median levels of 4-OH BP were significantly lower in current smokers ($n = 84$, 0.27 ng/ml) than in non-smokers ($n = 584$, 0.38 ng/ml) (Kunisue et al. 2012). According to the authors, this effect may be due to the induction of cytochrome P450 enzymes in the women who were current smokers. This study did not determine levels of benzhydrol in the urine samples.

While gas chromatography-mass spectrometry methods have been developed to monitor a variety of BP-based compounds in human urine (Kawaguchi et al. 2009; Kunisue et al. 2012; Kawaguchi et al. 2008), most of the biomonitoring studies in human populations have focused on monitoring urinary levels of 2-hydroxy-4-methoxybenzophenone (also called BP-3), a UV-stabilizer commonly used in sunscreen and personal care products (Wolff et al. 2007, 2010; CDC 2013). It should be noted that BP-3 is not used in food contact applications, therefore a discussion of these studies is not relevant to this chapter.

The only other human tissue that has been analyzed for various BP compounds is the human placenta. Using a LC–MS/MS method, levels of several BPs were determined in 16 placental samples of women residing in Granada, Spain (Vela-Soria et al. 2011). The metabolite of BP, 4-OH BP, was detected in 68.8 % of the placenta samples in the range of 0.6–1.2 ng/g of tissue. Further research is needed to determine if the placenta could be used to monitor levels of BP metabolites in human populations.

6.4 Health Effects

Research investigating the toxicity and health effects of BP-related UV-photoinitiators includes studies that have evaluated the general toxicity, carcinogenicity, and reproductive toxicity in rodent feeding studies, as well as studies on the genotoxicity and endocrine disrupting effects of these chemicals. The majority of these studies have evaluated toxicological endpoints for BP; very little toxicological information is currently available for 4-MBP.

6.4.1 General Toxicity Studies

In 14-week subchronic feeding studies conducted in male and female F344/N rats and $B6C3F_1$ mice that received BP in the diet at 0, 1,250, 2,500, 10,000 or 20,000 ppm, the primary organ sites affected were the liver, kidney, and hematopoietic system (Chhabra 2000). The elevated liver weights of BP-exposed rats in the mid- and high-dose groups were due to centrilobular hypertrophy and cytoplasmic vacuolization of hepatocytes, while increases in rat kidney weights were observed in concert with renal tubule dilatation, protein casts, renal tubule epithelial regeneration, and renal papillae necrosis. Increases in liver weights in BP-exposed mice also were accompanied by dose-related hepatocyte hypertrophy. Male mice in the 5,000 and 10,000 ppm groups were anemic, with lower hemoglobin levels and hematocrit values as well as depressed erythrocyte counts. Kidney weights were depressed in mid- and high-dose BP-treated mice (2,500 to 20,000 ppm groups), but unlike the exposed rats, the mice did not show microscopic abnormalities that would explain the increased renal weights. The BP-treated mice also showed changes in weights of reproductive organs, including lower weights of the testis and epididymis in the male 10,000 ppm group, although there were no reports of changes in any sperm indices.

The liver and the hematopoietic system were the primary organ sites affected in a short term 28-day, subchronic feeding study (BP at 100 or 500 mg/kg) conducted in male and female Sprague-Dawley CD rats (Burdock et al. 1991). Both BP-treated groups had increased liver weights accompanied by hepatocellular enlargement. Decreased hematocrit and hemoglobin were observed in both the 100 and 500 mg/kg

BP-treated females, and decreased hemoglobin levels were seen in the high-dose (500 mg/kg) males. There were no renal histological changes reported for BP-treated animals in this short-term study.

6.4.2 Carcinogenicity

In 2-year cancer bioassays conducted by scientists from the National Institute of Environmental Health Sciences (NIEHS) and Batelle (Rhodes et al. 2007), groups of 50 male and 50 female F344 rats and $B6C3F_1$ mice were fed diets containing 0, 312, 625, and 1,250 ppm BP. Diets were started when the animals were 8 weeks of age. Target organs affected included the kidney, liver, hematological systems, and the mammary gland. There was a significantly higher incidence of renal tubular hyperplasia in male rats fed the mid-dose 625 and high-dose 1,250 ppm BP-diets compared to controls, as well as a significantly higher incidence of renal neuropathy in all BP-treated male rats. Evidence of carcinogenicity included a statistically higher incidence of renal adenomas in the 1,250 ppm male rats when analysis of single and step sections of renal tissue were combined. In female rats, while the incidence of renal hyperplasia was significantly increased in both the 625 and 1,250 ppm groups, there were no significant differences for the incidence of renal adenomas or carcinomas between control and BP-treated female animals. There was equivocal evidence of mononuclear-cell leukemia based on increased incidence in the male rats in the 312 and 625 ppm groups, but not in the 1,250 ppm group. However, the authors noted that this leukemia is typically a late developing neoplasm and only two of the 50 animals in the 1,250 ppm group survived to the 105 week time point. Hence, there may have been an insufficient number of surviving animals in the high-dose group to see an effect of BP-treatment. In females, only the mid-dose 625 ppm BP-group had a higher incidence of mononuclear-cell leukemia compared with controls. Several rare histiocytic sarcomas were observed in the 625 and 1,250 ppm BP-treated female rats, while none were observed in the controls or the 315 ppm low-dose group.

In mice, both non-neoplastic and neoplastic hepatic lesions were observed in BP-treatment groups (Rhodes et al. 2007). Both male and female mice had a significantly higher incidence of hepatocyte centrilobular hypertrophy when exposed to any of the three levels of BP. Male mice had a significantly higher incidence of heptocellular adenomas in the 625 and 1,250 ppm groups compared to the controls. In female mice, while there was a higher incidence of hepatocellular adenomas in the 625 and 1,250 ppm groups, it did not reach statistical significance. Rare histiocytic sarcomas in female mice were observed in the 650 and 1,260 ppm groups, but the incidence was significantly higher compared with controls only in the 625 ppm group. The incidence of these types of sarcomas in the female mice exceeded the historical control range for both the 625 and 1,250 ppm BP-treated groups.

The number of mammary fibroadenomas per animal (multiplicity) and incidence was significantly decreased in 625 (15/50) and 1,250 ppm (7/50) BP-exposed female F344 rats compared with controls (27/50 ppm) (Rhodes et al. 2007). In addition, the incidence of mammary fibroadenomas in the 1,250 ppm group was less than half of that predicted by the historical controls from dietary (feed) studies. Mammary fibroadenomas are a common spontaneous neoplasm in female rats. The authors note that other BP-based compounds have been identified as inhibitors of the steroid sulfatase, an enzyme that plays a role in the synthesis of estrone and conversion to estradiol. However, the actual mechanism by which BP or its metabolites may have reduced mammary tumor incidences in these cancer bioassays is not known.

The authors concluded that there is some evidence of carcinogenicity of BP based on increased incidence of renal tubular adenomas in male F344/N rats, hepatocellular adenomas in male B6C3F$_1$ mice, and histiocytic sarcomas in female B6C3F$_1$ mice. The evidence of mononuclear cell leukemia and histiocytic sarcomas in female F344/N rats was considered to be equivocal (Rhodes et al. 2007).

BP does not appear to be carcinogenic when applied dermally in female Swiss mice. BP was dissolved in acetone, and applied on a one-inch square of shaved dorsal skin at concentrations of 0, 5, 25 or 50 % BP twice a week for 120 weeks (Stenback 1974). Skin tumors in BP-treated animals included two squamous cell papillomas of the dorsal skin and one squamous cell carcinoma of the lip. There was not a statistically significant difference in the incidence of tumors in BP-treated animals compared to untreated control animals or to acetone-vehicle controls. However, the survival rate of the BP-treated animals was lower than controls, with less than half of the BP-treated animals surviving at 70 weeks of age (5 % BP-group with 48 % survival; 25 and 50 % BP-groups with 48 % survival) compared with 72 % survival in the acetone-treated vehicle controls and 74 % survival in untreated controls. Hence, there may not have been a sufficient number of viable animals in the BP-treated groups for the development of late stage tumors. This study did not include male mice. No studies could be located that evaluated the carcinogenicity of dermally-applied BP in male or female rats.

There are no published studies on BP or 4-MBP exposure and cancer incidence in humans. Based on sufficient evidence of carcinogenicity in animal cancer bioassays (primarily the evidence cited from the Rhodes et al. 2007 cancer bioassays), the International Agency for Research on Cancer (IARC) classified BP in *Group 2B, possibly carcinogenic in humans* (IARC 2012). The NTP's 12th Report on Carcinogens does not include a substance profile on BP because this chemical has not been nominated for review.

6.4.3 Genotoxicity

The NTP Technical Report on toxicity studies for BP reported a lack of evidence of genotoxicity for this chemical (Chhabra 2000). The NTP report cited mutagenicity studies (Mortelmans et al. 1986) that found no evidence of mutations in *Salmonella typhimurium* strains TA98, TA100, TA1535 or TA1537 treated with BP. These mutagenicity studies were done with and without liver S9 metabolic activation. The NTP conducted micronucleus assays using male B6C3F$_1$ mice, and there were no statistically significant increases in micronuclei from the BP-treated animals compared to controls (Chhabra 2000). Because of concern of BP and 4-MBP's use as UV-initiators in food packaging printing ink and appearance in packaged food, the EFSA asked for additional data on the genotoxicity of these compounds. Two in vivo micronucleus assays in male CBA and NMRI mice were conducted for BP and 4-MBP, and an additional in vitro micronucleus assay using human lymphocytes was conducted for only 4-MBP (Abramsson-Zetterberg and Svensson 2011). These results of these experiments indicated no evidence of genotoxicity for either BP or 4-MBP.

The rec and comet assays have evaluated the genotoxicity of chemicals extracted from the paperboard of virgin and recycled paper products (outer print layer was removed before extraction) (Ozaki et al. 2004). Several BP-based chemicals were identified in the paperboard extracts. They included: BP, MK, DEAB, and DMAB. BP was detected in 8/16 of the extracts of virgin paper products and in 11/12 of the recycled paperboard products. DMAB and DEAB and MK were not detected in virgin paper products, but were detected in 4/12, 8/12, and 9/12 of the extracts of recycled products, respectively. None of the four BP-chemicals produced a killing zone by the rec-assay. In the comet assay BP was weakly positive and DEAB was positive, while MK and DMAB were negative (note, there is an inconsistency in reported results; narrative states DMAB was negative in the comet assay, but results in a table format show it to be positive).

The weight of the evidence from a variety of in vitro and in vivo test indicates a lack of genotoxicity for BP, and no evidence of genotoxicity from micronucleus test for 4-MBP. Additional studies are needed to clarify the presence of and capacity for genotoxicity of other BP-related chemicals used as amine synergists in print ink (including MK, DEAB, and DMAB) that have been identified in recycled paperboard products.

6.4.4 Reproductive Toxicity

The NTP reported few effects on reproductive organs with oral administration of BP via the diet at 0, 1,250, 2,500, 5,000, and 10,000 ppm in F344/N rats and B6C3F$_1$ mice for 14 weeks (Chhabra 2000). In male rats there were no treatment effects on the weights of the testis, epididymis or cauda epididymis, and there were no treatment effects on spermatid or epididymal spermatozoal measurements. In mice, testis weight and epididymis weight were significantly lower in the 10,000 ppm BP-group compared to controls, while in female rats and mice there were no statistically significant effects on the estrous cycle.

In a two-generation reproductive study in Sprague-Dawley rats, the parental (F$_0$) and first generation (F$_1$) were fed diets containing 0, 100, 450 and 2,000 ppm BP (Hoshino et al. 2005). There were no treatment-related effects on serum hormone levels, the estrous cycle, delivery, lactation, sperm indices, viability of pups, anogenital distance, reflexes, physical development or external abnormalities. Decreased body weight gain was observed in both males and females of the F_1 and F_2 2,000 ppm BP-treatment group. As with other toxicology studies in rats, target organs that were affected in BP-treatment groups in this multi-generational reproduction study included the liver and kidney. In both the 450 and 2,000 ppm group the parental animals in the F_1 and F_0 generation showed increase in hepatic weight and centrilobular hepatocyte hypertrophy, elevated renal weights, regeneration of the renal proximal tubular epithelium, and dilation of the renal proximal tubules. Based on these results, the no observed effect level (NOEL) was 2,000 ppm for reproductive effects in the parental generation and 450 ppm in the offspring.

6.4.5 Dermal Effects

There is limited evidence, based on the results of one study, that BP may be an allergen with dermal exposure. In a study of 19 human patients, one patient developed an allergic reaction to a dermally-applied photopatch containing BP (Cook and Freeman 2001). No other studies could be found in the published literature on whether BP, 4-MBP or their metabolites affect immune function.

6.4.6 Endocrine Effects

Estrogenicity Results from in vivo uterotrophic studies in rodents and in vitro estrogenicity assays in a variety of test systems suggest that while the parent chemical BP and its metabolite benzhydrol are not estrogenic, there is some evidence that hydroxylated metabolites of BP, including 4-OH BP, are weak estrogens. The positioning of the hydroxyl group(s) on BP affects the level of estrogenicity of the metabolite. These studies are summarized below.

When BP was given orally at 100 or 400 mg/kg for three days to ovariectomized female Sprague-Dawley rats, at the high-dose BP induced an increase in uterine wet weight (1.9-fold increase), and this uterotrophic effect was accompanied by an increase in uterine luminal epithelial cell height and an increase in thickness of the cornified epithelial cell layer in the vagina (Nakagawa and Tayama 2002). The results of other studies support these findings. When 21-day old juvenile female Sprague-Dawley rats were injected subcutaneously (sc) with BP, benzhydrol, 4-OH BP or 17β-estradiol, only the 4-OH BP and 17β-estradiol induced an increase in uterine weight (Nakagawa and Tayama 2001). There was no uterotrophic effect of BP or its metabolite benzhydrol. In another uterotrophic assay using juvenile female rats, both 4-OH BP, and 3-hydroxylated benzophenone, administrated by sc injection, elicited an estrogenic response while BP did not induce an uterotrophic effect (Hayashi et al. 2006). Results of further experiments indicated that BP has very little binding affinity for the estrogen receptor, but both the 3- and 4-hydroxylated forms of BP showed similar affinities for the estrogen receptor with an IC_{50} of 4.2×10^{-5} M.

In an in vitro estrogenicity assay, incubation of estrogen-dependent MCF-7 breast tumor cell with BP or benzhydrol for six days elicited no cell proliferation response, indicating a lack of estrogenicity of these chemicals (Nakagawa et al. 2000). In contrast, when the MCF-7 cells were incubated with 4-OH BP, there was a significant increase in cell proliferation, indicating estrogenic activity of this metabolite. Other studies using MCF-7 cells have tested a variety of hydroxylated forms of BP for their estrogenic activity (Suzuki et al. 2005). The most potent was 2,4,4'-trihydroxybenzophenone, followed by 2,3',4,4'-tetrahydroxybenzophenone, 4,4'-dihydroxybenzophenone, 2,2',4,4'-tetrahydroxybenzophenone, 4-OH BP, and 2,4-dihydroxybenzophenone, with the parent chemical BP having little estrogen activity. The authors concluded that the presence of a hydroxyl group in the 4-position (*para* position) on the phenyl ring(s) of BP appears to an important determinant of a derivative's estrogenicity, and that the placement of other hydroxyl groups in other positions on one or both of the phenyl rings can modify hormone activity.

A yeast two-hybrid assay was used to evaluate the estrogenicity of BP and 19 hydroxylated derivatives (Kawamura et al. 2003). In general, the presence of a hydroxyl group at the 3- or 4-position on a phenyl ring of BP was associated with a strong estrogenic response. In contrast, the estrogenicity of BP derivatives that had no hydroxyl group at the 3- or 4-position was weakly positive or negative. In derivatives that already had a hydroxyl group at the 4-position, addition of a second hydroxyl group at the 2-position further enhanced the estrogenicity of the compound.

Therefore, the results from these studies support the theory that the placement of hydroxyl groups on the phenyl rings of BP can determine the estrogenicity of the compound, and hydroxylation at the 4-position is essential for determining estrogenicity of a BP derivative. No published studies could be located that evaluated the estrogenicity of 4-MBP.

Other Endocrine Disruption Effects Few studies have examined whether exposure to BP or its metabolites affect other endocrine endpoints. The androgenicity of 4-OH BP has been evaluated using the Hershberger assay, and no agonistic effects were observed (Yamasaki et al. 2003). In a 10-day oral feeding study, administration of 4-OH BP by stomach tube at the highest dose (600 mg/kg), but not at the low- (50 mg/kg) or mid-dose (200 mg/kg) level, caused a significant depression in the weight of the ventral prostate of castrated male Brl Han: WIST Jcl (GALAS) rats compared to controls (Yamasaki et al. 2003). Weights of other male organs, including the seminal vesicle, glans penis, and Cowper's gland, were not affected by exposure to 4-OH BP.

6.5 Regulation and Regulatory Actions

6.5.1 European Union

The Committee of Ministers of the Council of Europe adopted a resolution in September 2002 concerning the transfer of chemicals from paperboard to food. It stated that:

> "paper and board used for all food contact applications...should not transfer constituents to foodstuffs in quantities which would endanger human health" (COE 2002).

General migration limits or limits for specific chemicals were not specified in this resolution. SMLs of non-volatile additives migrating from plastic food contact materials into food are discussed in the *Practical Guide, Food Contact Materials* published by the European Commission in 2003 (EC 2003). Annex A lists a SML of 0.6 mg/kg for BP's migration from plastic food contact materials[1] (page 85). However, this guide did not discuss or set SMLs for volatile substances that can migrate from paperboard packaging though the vapor phase to the food, such as the UV-initiators BP and 4-MBP. The 0.6 mg/kg SML for BP set for migration from food contact plastics is commonly cited as the limit for BP migration from all types food contact materials in the scientific literature since no other standard is available.

In 2009, there were several reports of 4-MBP contaminating European foods. On February 2, 2009 the European Union's (EU's) EFSA RASFF was notified by German authorities that the UV-initiator 4-MBP had migrated from packaging to cereal food products at the concentration of up to 798 μg/kg. Additional data from Belgian authorities was submitted to the RASFF by the end February reporting levels of 4-MBP in cereals was as high as 3,729 μg/kg (EFSA 2009a). The EFSA issued a *Summary of Opinion* on the toxicological evaluation BP on May 14, 2009 (EFSA 2009b), including reassessing the *tolerable daily intake* (TDI) on BP and 4-OH BP and assessing if the 4-MBP should be covered by the TDI for BP and 4-MBP. This document also addressed toxicological concerns and lack of toxicological data on the related UV-initiator 4-MBP (EFSA 2009b). Based on non-neoplastic renal toxicity endpoints, including hyperplasia and nephropathy in the kidneys of BP-treated rats, the TDI for BP was recommended to be reset from 0.01 to 0.03 mg/kg body weight. Other researchers have noted that this new TDI of 0.03 mg/kg body weight for BP has not yet been adopted by the EU (Bugey et al. 2013), hence, the SML of 0.6 mg/kg for BP appears to still be in force. The Summary Opinion of the EFSA panel recommended that a group TDI of 0.01 mg/kg body weight be established for BP and 4-MBP (EFSA 2009b). Although 4-OH BP is one of the primary metabolites of BP, the panel concluded this did not justify

---

[1] The SML is set by multiplying the tolerable daily intake (TDI) by 60 kg (which represents a 60 kg body weight for a reference male). Hence, 60 kg x the BP TDI of 0.01 mg/kg body weight per day results in the SML of 0.6 mg/kg.

including 4-OH BP in same TDI as BP. There was little toxicological data available in 2009 on 4-MBP, hence the calculation of a *Margin of Exposure* (MoE) for 4-MBP had to be based on toxicological and metabolic data for BP. Based on the highest levels of 4-MBP reported in contaminated breakfast cereals, and the highest level of consumption of breakfast cereals in European adults and children, the panel concluded:

> "…for adults, the estimated exposure is unlikely to lead to a health concern, since the estimated MoE is higher than 600. For children, the estimated exposure based on a conservative scenario (high consumption of breakfast cereals, average concentration of 4-methylbenzophenone) is also unlikely to pose a health concern. However, for children, based on the highly conservative scenario (high consumption of breakfast cereals, highest concentration of 4-methylbenzophenone), the estimated MoE is below 600. Therefore a health concern cannot be excluded in this case.
> Based on the limited exposure data available and applying knowledge on the toxicity of a similar substance, benzophenone, EFSA concludes that short term consumption of contaminated breakfast cereals should not pose a risk to most people. However, if the use of 4-methylbenzophenone is to be continued, more data on occurrence of the substance in foods should be provided as well as appropriate toxicity data corresponding to the level of exposure for a full risk assessment." (EFSA 2009a)

In July 2010, Germany authorities notified the RASFF that the UV-initiator BP was detected at levels up to 1,559 μg/kg in imported Italian couscous (Harrington 2010). Over 15,620 cartons of the tainted food were withdrawn from sale. The following year in April 2011, German authorities reported to the RASFF that BP was detected in frozen vermicelli noodles at a concentration of upto 1,746 μg/kg (Harrington 2011). Over 150 kg of the noodle product was withdrawn from sale in Germany.

No other reports of RASFF related to BP or 4-MBP have been located since the 2011 alert for BP. It is not known to what extent manufacturers have adapted or changed the use of these UV-photoinitiators in food packaging, hence, a prudent approach would be to continue to monitor levels of BP and 4-MBP in foods, especially foods packaged in printed cartonboard, as well as seeking additional toxicological data on BP and 4-MBP to better assess potential health risks in human populations, especially children.

6.5.2 Food and Drug Administration

The U.S. FDA does lists BP in its database of *Everything Added to Food in the United States* (EAFUS), which includes a listing of substances directly added to food (FDA 2013a). BP is included in this database because it is allowed as a flavoring in food. BP was not found in FDA's *List of Indirect Additives Used in Food Contact Substances* (LIAUFCS)[2] database. There were no listings for BP in

---

[2] Note there was a LIAUFCS listing for 2-hydroxy-4-*n*-octoxy-benzophenone, CAS Registry No. 1843-05-6, also known as BP-12.

the *Food Contact Substance (FCS) Notifications* database, which lists premarket notifications for food contact substances regarded as safe for their intended use as submitted by the manufacturer (FDA 2013b). BP is not listed in FDA's database of food contact substances that have a *Cumulative Estimated Daily Intake* (CEDI) (FDA 2012). The photoinitiator 4-MBP is not listed in FDA's EAFUS-, LIA-UFCS-, FCS Notifications-, or CEDI database.

6.6 Summary and Conclusions

The UV-initiators BP and 4-MBP which are used in UV-cured print inks, have been detected the cartonboard packaging of a wide number of foods, and have been the subject of EFSA RASFF notifications because of their detection in foods above regulatory migration limits. Modeling studies and studies of packaged foods have shown that BP can migrate from the printed surface of food packaging through primary cartonboard and many types of secondary packaging to food and beverages under a variety of storage conditions, including short and long-term storage at room temperature and low temperature (frozen storage) conditions. There is some evidence that BP may contaminate recycled paper board and recycled plastics including PET, and further studies are needed to determine if 4-MBP also is present in recycled food packaging. Limited information is available on the toxicology of BP, and very little information other than EFSA-mandated genotoxicity studies on 4-MBP, is currently available. While the weight of evidence indicates that neither BP nor 4-MBP are genotoxic, there is evidence that BP is possibly a human carcinogen based on evidence from oral feeding rodent cancer bioassays, especially evidence of renal adenomas in treated male rats, hepatocellular adenomas in male mice, and rare histiocytic sarcomas in female rats and mice. In subacute toxicity rodent feeding studies, the kidney, liver, and hematopoietic system are the primary sites adversely affected by BP-administration. While BP is not estrogenic, there is evidence that the metabolite of BP, 4-OH BP, is a weak estrogen, but to date there has been no evaluation of endocrine disrupting effects of 4-MBP. The detection of these UV-photoinitiators in food packaging and in a wide number of foods, violative levels in foods that exceed migration limits, and toxicological data supporting cancer risk and endocrine disrupting effects, are valid reasons for the need to monitor these UV-photoinitiators in the global food supply and to obtain additional toxicological data on possible adverse health effects.

However, as mentioned in the introduction to this chapter (Sect. 6.2), BP and 4-MBP are only two of the many photoinitiators and related chemicals used in UV-cured inks. The amine synergists MK, DAMB, and DEAB are examples of dialkyamino-benzophenones detected in food packaging (Castle et al. 1997). In addition, other UV-photoinitiators and amine synergists used in food packaging print inks include: 1-hydroxycyclohexyl phenyl ketone (IRGACURE 184), 2-isopropylthioxanthone (ITX) (see Chap. 11, Sect. 11.2.3), and the dialkybenzoates ethyl-4-dimethylbenzoate (EDAB), and 2-ethylhexyl-4-dimethylaminobenzoate

(EHDAB) (Sagratini et al. 2008; Shen et al. 2009; Han et al. 2011). While methodology is available for analyzing levels of these chemicals in packaging and foods, there is limited data available on levels of these photoinitiators, as well as levels of BP and 4-MBP in the global food supply. Given than a food may be packaged in a container manufactured in a different country than where the food originated, this underscores the importance of monitoring UV-photoinitiators and their capacity to contaminate food and beverages.

6.7 Further Research Needs and Recommendations

- As a class of chemicals, BP UV-photoinitiators and amine synergists used in food packaging print inks need to be monitored in the global food supply to identify levels in food packaging and their capacity to migrate to food and beverages.
- UV-photoinitiators that are violative of migration limits should be either prohibited for use in food packaging, or secondary packaging should be in use that effectively prevents the migration of the UV-photoinitiator from the primary packaging to the food or beverage.
- The source of BP found in recycled food contact PET plastic needs to be identified, and more studies are needed to determine the scope and levels of UV-initiator contamination of recycled plastics.
- Since a number of UV-initiators have been identified in food packaging and in food and beverages, more extensive toxicological studies are needed to determine target organs and functional systems that may be adversely affected by exposure to these chemicals.
- Monitoring devices (sensors, see Sect. 6.3.4) should be further developed and used to detect migration of UV-initiators from food packaging to food and beverages.

References

Abramsson-Zetterberg L, Svensson K (2011) 4-Methylbenzophenone and benzophenone are inactive in the micronucleus assay. Toxicol Lett 201(3):235–239

Anderson WA, Castle L (2003) Benzophenone in cartonboard packaging materials and the factors that influence its migration into food. Food Addit Contam 20(6):607–618

Barnkob LL, Petersen JH (2013) Effect of relative humidity on the migration of benzophenone from paperboard into the food simulant Tenax® and modelling hereof. Food Addit Contam Part A Chem Anal Control Expo Risk Assess 30(2):395–402

Biedermann M, Ingenhoff JE, Zurfluh M et al (2013) Migration of mineral oil, photoinitiators and plasticisers from recycled paperboard into dry foods: a study under controlled conditions. Food Addit Contam Part A Chem Anal Control Expo Risk Assess 30(5):885–898

Bronaugh RL, Wester RC, Bucks D et al (1990) In vivo percutaneous absorption of fragrance ingredients in Rhesus monkeys and humans. Food Chem Toxicol 28(5):369–373

Bugey A, Janin Y, Edder P et al (2013) Targeted multidimensional gas chromatography using a heart-cutting device and cryogenic focusing for the determination of benzophenone derivatives in foodstuffs. Anal Bioanal Chem 405(12):4177–4185

Burdock GA, Pence DH, Ford RA (1991) Safety evaluation of benzophenone. Food Chem Toxicol 29(11):741–750

Castle L, Damant AP, Honeybone CA et al (1997) Migration studies from paper and board food packaging materials. Part 2: survey for residues of dialkylamino benzophenone UV-cure ink photoinitiators. Food Addit Contam 14(1):45–52

Centers for Disease Control (CDC) (2013) Fourth national report on human exposure to environmental chemicals, updated tables. Sept 2013, urinary benzophenone-3, downloadable as PDF at: http://www.cdc.gov/exposurereport/pdf/FourthReport_UpdatedTables_Sep2013.pdf. Centers for Disease Control and Prevention, http://www.cdc.gov/exposurereport/. Accessed 18 Feb 2014

Chhabra RS (2000) NTP technical report on the toxicity studies of benzophenone (CAS No. 119-61-9), administered in feed to F344/N rats and $B6C3F_1$ mice. Toxic Rep Ser 61(A51–13):1–53

Choi JO, Jitsunari F, Asakawa F et al (2002) Migration of surrogate contaminants in paper and paperboard into water through polyethylene coating layer. Food Addit Contam 19(12):1200–1206

Council of Europe (COE) (2002) Resolution ResAP (2002)1 on paper and board materials and articles intended to come into contact with foodstuffs, adopted by the committee of ministers on 18 September 2002 at the 808 meeting of the minister's deputies. Council of Europe, committee of ministers.http://www.coe.int/t/e/social_cohesion/socsp/public_health/food_contact/RESOLUTION%202002-1%20PAPER%20AND%20BOARD.pdf Accessed 18 Feb 2014

Cook N, Freeman S (2001) Report of 19 cases of photoallergic contact dermatitis to sunscreens seen at the skin and cancer foundation. Australas J Dermatol 42(4):257–259

European Commission (EC) (2003) Food contact materials, practical guide for users of European directives, updated April 15, 2003, SANCO D3/LR D (04.2003). European Commission, Health and Consumer Protection Directorate-General. http://ec.europa.eu/food/food/chemicalsafety/foodcontact/practical_guide_en.pdf. Accessed 18 Feb 2014

European Food Safety Authority (EFSA) (2009a) EFSA statement on the presence of 4-methylbenzophenone found in breakfast cereals (Question no: EFSA-Q-2009-410) PDF downloadable at: http://www.efsa.europa.eu/en/efsajournal/pub/243r.htm, Accessed 17 Feb 2014 (The EFSA journal RN-243:1-19)

European Food Safety Authority (EFSA) (2009b) Scientfic opinion of European food safety authority (EFSA) prepared by the panel on food contact materials, enzymes, flavourings and processing aids (CEF) on the toxicological evaluation of benzophenone, Adopted 14 May 2009, viewable and PDF downloadable at: http://www.efsa.europa.eu/en/scdocs/scdoc/1104.htm (The EFSA Journal 1104:1-30)

Food and Drug Administration (FDA) (2006) Guidance for industry: use of recycled plastics in food packaging: chemistry considerations. U.S. Food and Drug Administration. http://www.fda.gov/food/guidanceregulation/guidancedocumentsregulatoryinformation/ingredientsadditivesgraspackaging/ucm120762.htm. Accessed 14 Feb 2014

Food and Drug Administration (FDA) (2012) Cumulative estimated daily intake database. U.S. Food and Drug Administration. http://www.accessdata.fda.gov/scripts/sda/sdNavigation.cfm?filter=benzophenone&sortColumn=&sd=edisrev&page=1. Accessed 16 Feb 2014

Food and Drug Administration (FDA) (2013a) Everything added to food in the United States (EAFUS), benzophenone doc no: 0105. U.S. Food and Drug Administration. http://www.accessdata.fda.gov/scripts/fcn/fcnNavigation.cfm?filter=benzophenone&sortColumn=&rpt=eafusListing. Accessed 16 Feb 2014

Food and Drug Administration (FDA) (2013b) Inventory of effective food contact substance (FCS) notifications. U.S. Food and Drug Administration. http://www.fda.gov/food/ingredientspackaginglabeling/packagingfcs/notifications/default.htm. Accessed 16 Feb 2014

Food and Drug Administration (FDA) (2014) Everything added to food in the United States (EAFUS), Benzophenone doc no: 0105. U.S. Food and Drug Administration. http://www.accessdata.fda.gov/scripts/fcn/fcnNavigation.cfm?filter=benzophenone&sortColumn=&rpt=eafusListing

Franz R, Mauer A, Welle F (2004) European survey on post-consumer poly(ethylene terephthalate) (PET) materials to determine contamination levels and maximum consumer exposure from food packages made from recycled PET. Food Addit Contam 21(3):265–286

Franz R, Welle F (2008) Migration measurement and modelling from poly(ethylene terephthalate) (PET) into soft drinks and fruit juices in comparison with food simulants. Food Addit Contam Part A Chem Anal Control Expo Risk Assess 25(8):1033–1046

Guazzotti V, Marti A, Piergiovanni L et al (2014) Bio-based coatings as potential barriers to chemical contaminants from recycled paper and board for food packaging. Food Addit Contam Part A Chem Anal Control Expo Risk Assess. doi:10.1080/19440049.19442013.19869360

Han W, Yu Y, Li N et al (2011) Determination of photoinitiators in printing inks used in food contact materials. Se Pu 29(5):417–421

Harrington R (2010) Benzophenone from packaging taints Italian couscous, 13 July 2010. http://www.foodproductiondaily.com/Safety-Regulation/Benzophenone-from-packaging-taints-Italian-couscous. Accessed 16 Feb 2014

Harrington R (2011) Germany withdraws noodles after benzophenone leaches from packaging, 19 April 2011. http://www.foodproductiondaily.com/Packaging/Germany-withdraws-noodles-after-benzophenone-leaches-from-packaging. Accessed 18 Feb 2014

Hayashi T, Okamoto Y, Ueda K et al (2006) Formation of estrogenic products from benzophenone after exposure to sunlight. Toxicol Lett 167(1):1–7

Hoshino N, Tani E, Wako Y et al (2005) A two-generation reproductive toxicity study of benzophenone in rats. J Toxicol Sci Spec 30:5–20

International Agency for Research on Cancer (IARC) (2012) IARC monographs of carcinogenic risks to humans. In: IARC (ed) Some chemicals in industrial and consumer products, food and drinking-water, benzophenone, vol 101. International Agency for Research on Cancer, Lyon, pp 285–304

Ito R, Kawaguchi M, Koganei Y et al (2009) Development of miniaturized hollow-fiber assisted liquid-phase microextraction with in situ acyl derivatization followed by GC-MS for the determination of benzophenones in human urine samples. Anal Sci 25(8):1033–1037

Jeon HK, Sarma SN, Kim YJ et al (2008) Toxicokinetics and metabolisms of benzophenone-type UV filters in rats. Toxicology 248(2–3):89–95

Jickells SM, Poulin J, Mountfort KA et al (2005) Migration of contaminants by gas phase transfer from carton board and corrugated board box secondary packaging into foods. Food Addit Contam 22(8):768–782

Johns SM, Gramshaw JW, Castle L et al (1995) Studies on functional barriers to migration: transfer of benzophenone from printed paperboard to microwaved food. Dtsch Lebensm-Rundsch 91(3):69–73

Johns SM, Jickells SM, Read WA et al (2000) Studies on functional barriers to migration: migration of benzophenone and model ink components from cartonboard to food during frozen storage and microwave heating. Packag Technol Sci 13:99–104

Jung T, Simat TJ, Altkofer W et al (2013) Survey on the occurrence of photo-initiators and amine synergists in cartonboard packaging on the German market and their migration into the packaged foodstuffs. Food Addit Contam Part A Chem Anal Control Expo Risk Assess 30(11):1993–2016

Kawaguchi M, Ito R, Honda H et al (2008) Measurement of benzophenones in human urine samples by stir bar sorptive extraction and thermal desorption-gas chromatography-mass spectrometry. Anal Sci 24(11):1509–1512

Kawaguchi M, Ito R, Honda H et al (2009) Miniaturized hollow fiber assisted liquid-phase microextraction and gas chromatography-mass spectrometry for determination of benzophenone and derivates in human urine sample. J Chromatogr B Analyt Technol Biomed Life Sci 877(3):298–302

Kawamura Y, Ogawa Y, Nishimura T et al (2003) Estrogenic activities of UV stabilizers used in food contact plastics and benzophenone derivatives tested by the yeast two-hybrid assay. J Health Sci 49(3):205–212

Koivikko R, Pastorelli S, Rodriguez-Bernaldo de Quiros A et al (2010) Rapid multi-analyte quantification of benzophenone, 4-methylbenzophenone and related derivatives from paperboard food packaging. Food Addit Contam Part A Chem Anal Control Expo Risk Assess 27(10):1478–1486

Konkol LM, Cross RF, Harding IH et al (2003a) Contaminants and levels of occurrence in washed and shredded poly(ethylene terephthalate) from curbside collection II: validation of extraction procedures, particle size sampling and crystallinity. Food Addit Contam 20(10):972–984

Konkol LM, Cross RF, Harding IH et al (2003b) Contaminants and levels of occurrence in washed and shredded poly(ethylene terephthalate) from curbside collection. Part 1: extraction conditions. Food Addit Contam 20(9):859–874

Kunisue T, Chen Z, Buck Louis GM et al (2012) Urinary concentrations of benzophenone-type UV filters in U.S. women and their association with endometriosis. Environ Sci Technol 46(8):4624–4632

Li H, Guan H, Dai H et al (2012) An amperometric sensor for the determination of benzophenone in food packaging materials based on the electropolymerized molecularly imprinted poly-o-phenylenediamine film. Talanta 99:811–815

Mortelmans K, Haworth S, Lawlor T et al (1986) *Salmonella* mutagenicity tests, II: results from the testing of 270 chemicals. Environ Mutagen 8(7):1–119

Nakagawa Y, Suzuki T, Tayama S (2000) Metabolism and toxicity of benzophenone in isolated rat hepatocytes and estrogenic activity of its metabolites in MCF-7 cells. Toxicology 156(1):27–36

Nakagawa Y, Tayama K (2001) Estrogenic potency of benzophenone and its metabolites in juvenile female rats. Arch Toxicol 75(2):74–79

Nakagawa Y, Tayama K (2002) Benzophenone-induced estrogenic potency in ovariectomized rats. Arch Toxicol 76(12):727–731

Nerin C, Asensio E (2007) Migration of organic compounds from a multilayer plastic-paper material intended for food packaging. Anal Bioanal Chem 389(2):589–596

Ozaki A, Kawasaki C, Kawamura Y et al (2006) Migration of bisphenol A and benzophenones from paper and paperboard products used in contact with food. Shokuhin Eiseigaku Zasshi 47(3):99–104

Ozaki A, Yamaguchi Y, Fujita T et al (2004) Chemical analysis and genotoxicological safety assessment of paper and paperboard used for food packaging. Food Chem Toxicol 42(8):1323–1337

Ozaki A, Yamaguchi Y, Okamoto A et al (2002) Determination of alkylphenols, bisphenol A, benzophenone and phthalates in containers of baby food, and migration into food simulants. Shokuhin Eiseigaku Zasshi 43(4):260–266

Papilloud S, Baudraz D (2002) Analysis of food packaging UV inks for chemicals with potential to migrate into food simulants. Food Addit Contam 19(2):168–175

Pastorelli S, Sanches-Silva A, Cruz JM et al (2008) Study of the migration of benzophenone from printed paperboard packages to cakes through different plastic films. Eur Food Res Technol 227:1585–1590

Rhodes MC, Bucher JR, Peckham JC et al (2007) Carcinogenesis studies of benzophenone in rats and mice. Food Chem Toxicol 45(5):843–851

Rodriguez-Bernaldo de Quiros A, Paseiro-Cerrato R, Pastorelli S et al (2009) Migration of photoinitiators by gas phase into dry foods. J Agric Food Chem 57(21):10211–10215

Sagratini G, Caprioli G, Cristalli G et al (2008) Determination of ink photoinitiators in packaged beverages by gas chromatography-mass spectrometry and liquid chromatography-mass spectrometry. J Chromatogr A 1194(2):213–220

Sanches-Silva A, Andre C, Castanheira I et al (2009) Study of the migration of photoinitiators used in printed food-packaging materials into food simulants. J Agric Food Chem 57(20):9516–9523

Sanches-Silva A, Pastorelli S, Cruz JM et al (2008) Development of an analytical method for the determination of photoinitiators used for food packaging materials with potential to migrate into milk. J Dairy Sci 91(3):900–909

Shen DX, Lian HZ, Ding T et al (2009) Determination of low-level ink photoinitiator residues in packaged milk by solid-phase extraction and LC-ESI/MS/MS using triple-quadrupole mass analyzer. Anal Bioanal Chem 395(7):2359–2370

Song YS, Begley T, Paquette K et al (2003) Effectiveness of polypropylene film as a barrier to migration from recycled paperboard packaging to fatty and high-moisture food. Food Addit Contam 20(9):875–883

Stenback F, Shubik P (1974) Lack of toxicity and carcinogenicity of some commonly used cutaneous agents. Toxicol Appl Pharmacol 30:7–13

Suzuki T, Kitamura S, Khota R et al (2005) Estrogenic and antiandrogenic activities of 17 benzophenone derivatives used as UV stabilizers and sunscreens. Toxicol Appl Pharmacol 203(1):9–17

Triantafyllou VI, Akrida-Demertzi K, Demertzis (2007) A study on the migration of organic pollutants from recycled paperboard packaging materials to solid food matrices. Food Chem 101(4):1759–1768

Vela-Soria F, Jimenez-Diaz I, Rodriguez-Gomez R et al (2011) Determination of benzophenones in human placental tissue samples by liquid chromatography-tandem mass spectrometry. Talanta 85(4):1848–1855

Welle F, Franz R (2008) SiOx layer as functional barrier in polyethylene terephthalate (PET) bottles against potential contaminants from post-consumer recycled PET. Food Addit Contam Part A Chem Anal Control Expo Risk Assess 25(6):788–794

Wolff MS, Teitelbaum SL, Pinney SM et al (2010) Investigation of relationships between urinary biomarkers of phytoestrogens, phthalates, and phenols and pubertal stages in girls. Environ Health Perspect 118(7):1039–1046

Wolff MS, Teitelbaum SL, Windham G et al (2007) Pilot study of urinary biomarkers of phytoestrogens, phthalates, and phenols in girls. Environ Health Perspect 115(1):116–121

Yamasaki K, Takeyoshi M, Sawaki M et al (2003) Immature rat uterotrophic assay of 18 chemicals and Hershberger assay of 30 chemicals. Toxicology 183(1–3):93–115

Chapter 7
Perfluorinated Compounds in Food Contact Materials

Penelope A. Rice, Omari J. Bandele and Paul Honigfort

Abstract Perfluorinated compounds (PFCs) exhibit unique physical characteristics including hydrophobic and oleophobic properties as well as thermal and chemical stability. Due to these characteristics, PFCs are used in a wide variety of industrial and consumer applications, including automotive, electrical, clothing, and household products. PFCs are ubiquitous in the environment and PFCs greater than seven carbons in length have been found to bioaccumulate in wildlife and humans. Furthermore, laboratory animal studies have raised concerns as to the safety of human exposure to these compounds. Although their use in the manufacture of food contact substances (FCSs) constitutes a small segment of the use of PFCs, their use in food packaging and migration to foodstuffs does represent a potential source of oral exposure to humans. This chapter focuses on aspects of human exposure, and potential health risks associated with two types of long-chain PFCs that historically have been found in food packaging: perfluorinated carboxylic acids (PFCAs) and fluorinated telomer alcohols (FTOHs).

Keywords Perfluorinated · Fluorotelomer · PFOA · Perfluoroalkyl · Perfluorooctanoic · Perfluorononanoic · Perfluorodecanoic · Perfluorododecanoic · Perfluorooctylethanol · FTOH · Perfluorocarboxylates · Food packaging

---

P. A. Rice (✉) · O. J. Bandele · P. Honigfort
Division of Food Contact Notifications, Office of Food Additive Safety, Center for Food Safety and Applied Nutrition, U.S. Food and Drug Administration, College Park, MD, USA
e-mail: Penelope.Rice@fda.hhs.gov

O. J. Bandele
e-mail: Omari.Bandele@fda.hhs.gov

P. Honigfort
e-mail: Paul.Honigfort@fda.hhs.gov

S. M. Snedeker (ed.), *Toxicants in Food Packaging and Household Plastics*, Molecular and Integrative Toxicology, DOI: 10.1007/978-1-4471-6500-2_7,

7.1 Key Take Home Points

- The presence of PFCs in food has been demonstrated to be a major route of exposure to these compounds. However, the origination of these compounds in food, either from the use of these compounds in food packaging or from environmental contamination, has not been definitively elucidated.
- Long-chain PFCAs have clear adverse effects on the liver, the immune system, the testes, and the female mammary gland in rodents, particularly in mice. Most of the observed effects appear to be due to activation of the PPAR-α receptor, although activation of other pathways may also be involved. PFCs as a class have adverse effects on development in rodents, and PFOA is carcinogenic in the liver, pancreas, and testes of male rats.
- The human relevance of the observed effects of PFCAs in rodents is unknown, as the available epidemiology data does not consistently point to similar effects in highly-exposed human population cohorts. Moreover, mechanistic data indicate that the observed effects in the liver and on the developing rodent may be mediated via rodent-specific pathways.
- Little is known about the toxicity of FTOH, as most of the available animal studies have been performed in rats, which is less sensitive to the toxic effects of PFCs than the mouse. However, the available data in the rat indicates that the FTOH are less toxic than the PFCA of comparable chain length.
- Recent regulatory action is expected to limit future exposure to these compounds from food packaging use.

7.2 Overview of Perfluorinated Compounds Used in Food Packaging

7.2.1 Structures and General Uses of Perfluorinated Compounds

Long-chain PFCs contain an extended alkyl chain of eight carbons (C8) or longer where all of the hydrogens have been replaced by fluorine. This perfluorinated alkyl chain is bonded to another functional group, such as an acid (in the case of PFCAs) or an alcohol (FTOHs). The lower cut-off of C8 is based on pharmacological considerations. PFCs with an extended perfluorinated chain length shorter than C8 do not accumulate in mammalian tissues (Chenglis et al. 2009). Unless otherwise specified, all PFCs discussed in this chapter are C8 or longer.

PFCAs are a subset of perfluorinated acids which consist of a non-polar perfluorinated alkyl chain and a polar carboxylic acid end group. Based upon their combination of hydrophobic and hydrophilic properties they are often used as surfactants in emulsion reactions, or as reactants to make low-molecular weight perfluorinated products. Historically the C8 PFCA (perfluorooctanoic acid—

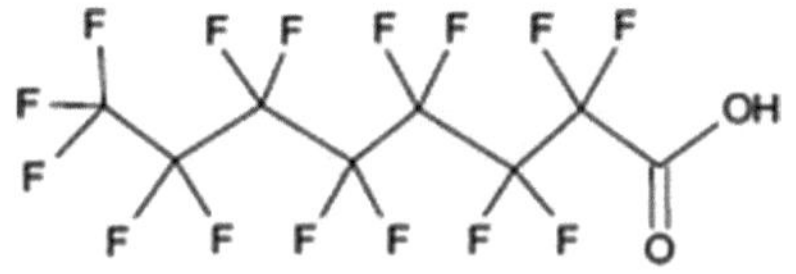

Fig. 7.1 Perfluorooctanoic acid (PFOA)

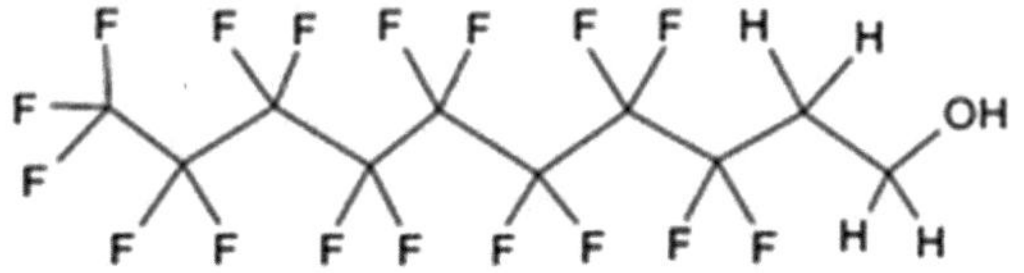

Fig. 7.2 Perfluorooctyl-ethanol (8-2 FTOH)

PFOA, Fig. 7.1) has been the most commonly used, but chain lengths of C9 (perfluorononanoic acid—PFNA), C10 (perfluorodecanoic acid—PFDA), and longer have also been utilized. Although PFCAs can be produced in separate chain lengths, most commercial products consist of mixtures of several chain lengths.

FTOHs are a subset of fluorinated alcohols which consist of a perfluorinated alkyl chain combined with a non-perfluorinated alkyl alcohol. Due to the non-perfluorinated character of the alkyl alcohol, these compounds can participate in condensation reactions which would not be possible if the entire alkyl chain was perfluorinated. In this manner, FTOHs are often incorporated as a side-chain to higher molecular weight, non-perfluorinated polymers. The resultant free perfluorinated alkyl side-chain imparts release properties to the finished polymer. Most commercial products consist of FTOHs with a C2 non-perfluorinated alkyl group coupled to perfluorinated alkyls of various chain lengths. The most common FTOH has a C8 perfluorinated alkyl chain coupled with a C2 non-perfluorinated alkyl alcohol (perfluorooctylethanol—8-2 FTOH, Fig. 7.2). FTOHs used in the production of perfluorinated polymers typically include mixtures of varying chain lengths, with the most prevalent FTOH species being the 8-2 and 10-2 FTOHs.

7.2.2 Use in Food Contact Materials

Since PFCAs and FTOHs are used in the manufacture of FCSs but are not FCSs themselves, they are not specifically regulated by the United States (U.S.) Food and Drug Administration (FDA) as FCSs. However, they are found as impurities, either as residuals from their use in the production or as degradation products of the FCS, in PFCs regulated for food contact use. As such, FDA considers the safety of PFCAs and FTOHs in the regulation of those FCSs where these compounds are present.

PFCs have historically been used in food contact applications prior to 1958. The prior sanctioned use of PFCs manufactured utilizing a subset of perfluorinated acids where the perfluorinated alkyl chain is coupled to a sulfonic acid end group (PFSAs, the most commonly used of which was the C8 perfluorooctane sulfonic acid—PFOS) as coatings for food contact paper can be found in Title 21 of the Code of Federal Regulations part 181.30 (21 CFR 181.30) as well regulations for the same use in 21 CFR 176.170. Based upon information that PFSAs were widespread in the blood of the general population and concerns for bioaccumulation and toxicity, the U.S. Environmental Protection Agency (USEPA) published three significant new use rules (SNURS) to limit any manufacture or importation of 271 related chemicals (US EPA 2002a, b, 2007). These SNURS apply to the manufacture and import of related chemicals for all applications, of which use in the manufacture of FCSs was a small segment. As a result of these SNURS there is no current consumer exposure to the FCSs listed in 21 CFR which utilize PFSAs in their manufacture, and those FCSs will only be discussed in the context of general exposure to PFCs.

Since the 1960s, PFCAs have been used in the manufacture of FCSs as a surfactant in the polymerization of high molecular weight FCS polymers such as polytetrafluoroethylene (PTFE), which is used in non-stick cookware coatings. PTFE also has been used as a reactant to make low-molecular weight paper coatings, where the hydrophobic and lipophobic nature of the extended perfluorinated alkyl chain imparts both water and grease resistance to paper used in food contact applications. PFCAs are indirectly regulated for their use as surfactants and in the polymerization of high molecular weight FCSs under 21 CFR Sections 177.1380, 177.1550, 177.1615, 177.2400 and 177.2510. The use of PFCAs in the manufacture of low molecular weight paper coatings was also considered by FDA in several listings for FCSs in 21 CFR 177.170 and 177.180 as well as in several effective Food Contact Notifications (FCNs) (US FDA 2013).

FTOHs are used as components of paper coatings through the side-chain manipulation of high molecular weight polymeric FCSs. Resultant FCSs have been in use since the early 2000s and are the subject of effective FCNs (US FDA 2013) for their use to impart water and grease resistance to food-contact paper.

7.3 Human Exposure

7.3.1 Levels of PFCs in Serum

Several studies have examined serum levels of PFCs in various populations; most of the available data concern serum levels of PFOA and PFNA. The 2003–2004 National Health and Nutrition Examination Survey (NHANES) (Calafat et al. 2007) reported mean serum PFOA and PFNA levels of 4.5 nanograms per milliliter (ng/ml) and 1.1 ng/ml, respectively, in U.S. males and 3.5 ng/ml and 0.9 ng/ml, respectively, in U.S. females 12 years of age and older. Mean plasma levels of PFOA reported in Denmark are 6.8–6.9 ng/ml for males and 5.4–6 ng/ml for females (Eriksen et al. 2009). Serum PFOA in the U.S. has decreased from levels reported for the 1999–2000 NHANES survey, while serum PFNA levels are increasing (Kato et al. 2011). Serum PFOA levels in occupationally-exposed populations are 691 ng/ml, at least two orders of magnitude higher than the general population (Olsen et al. 2007). Occupationally, highly-exposed segments of the population, such as a cohort of employees from DuPont's Washington Works plant, have reported mean serum PFOA levels of 75.7 ng/ml (Hoffman et al. 2011).

7.3.2 Routes of Exposure

As PFCs are ubiquitous in the environment (Houde et al. 2006), there are multiple routes for human exposure which could potentially account for the blood levels discussed above. The conclusion of several studies which examined monitoring data of likely routes for human exposure is that the diet is a significant source of exposure. A survey of Canadian composite food samples collected between 1992 and 2004 calculated a medium dietary intake of 73 ng/day for perfluoroocatesulfonamides (PFOSAs) and concluded that this was of the same order of magnitude as exposure to these compounds from indoor dust and air inhalation (Tittlemier et al. 2006). Comparisons of the Canadian composite food study using 2004 composites, which took into account PFOS and perfluorocarboxylates, noted a combined exposure of 250 ng/day and concluded that this was slightly higher than exposure resulting from water, dust, treated carpet, and apparel (Tittlemier et al. 2007) or water, dust ingestion/dermal exposure, and indoor/outdoor air inhalation (Lorber and Egeghy 2011). Another study of Norwegian women that correlated serum levels to exposure to PFCs through diet and indoor environmental routes concluded that food was generally the major exposure source, but that the indoor environment was also a significant source of exposure (Haug et al. 2011). Attempts to model comprehensive reviews of available exposure data from various routes have focused on PFOA and PFOS exposure and agree that dietary sources are the major route for human exposure to these compounds, in some scenarios accounting for 90 % or greater of the exposure (Fromme et al. 2009; Trudel et al. 2008; Vestergren and Cousin 2009).

7.3.3 Migration of PFC from Food Contact Materials to Food and Food Simulants

Relatively few studies have been published which directly correlate detected levels of PFCs in foods to the migration of these compounds from food packaging. Interestingly, the Canadian composite food survey (Tittlemier et al. 2006) detected PFOSAs in all foods sampled prior to 2003 with the highest concentration in fast food composites followed by pre-packaged consumer products such as microwave popcorn (both food categories where contact with coated paper packaging would be expected) with the lowest concentrations in less processed foods such as seafood, meat, and dairy, where contact with such packaging would be less likely. However, the same composite survey did not detect PFOSAs in fast food or prepackaged food composites after 2002, a date which coincides with the USEPA SNURs banning the manufacture and import of PFSA related compounds, while the level in foods not commonly packaged in coated paper remained constant. Furthermore, the specific PFOSAs detected in fast food and prepackaged food composites prior to 2003 were used primarily to coat paper, whereas those detected in foods not commonly packaged in coated paper were PFCs primarily used in fabric coatings. These results indicate that, while there is constant, low level exposure to PFCs originating in food from environmental routes, the use of PFCs to coat paper results in exposure to those chemicals when food packaged in those products is consumed. Furthermore, although specific to PFSAs, this study can be taken as generally indicative of the impact of regulatory actions on consumer exposure to PFCs, including PFCAs and FTOHs, from their use in food packaging.

Migration of PFCs into food simulants has been analytically demonstrated (Begley et al. 2005). FDA examined PTFE coated cookware for residual PFCAs, and migration of PFCAs and FTOHs from commercial packaging (microwave popcorn bags) coated with PFC coatings. This work demonstrated that migration of PFCAs from their use as surfactants to make high molecular weight non-stick coatings is negligible due to a low initial residual level, and that further abusive heating did not result in the generation of additional PFCAs. However, this work did detect migration into food simulants of both PFOA and FTOHs (11 micrograms per square decimeter ($\mu g/dm^2$) paper) from PFC coated paper under conditions designed to mimic storage and microwave susceptor use.

In conclusion, serum levels demonstrate that the general population is exposed to PFCs. Examination of potential exposure routes have concluded that exposure through the diet is a significant source of exposure. Potential sources of this dietary exposure may involve environmental contamination of food as well as migration of PFCs to food from packaging. Although few studies have been published, available information demonstrates that migration of PFCs to food from packaging may contribute to dietary exposure.

7.3.4 Pharmacokinetics

PFCAs are readily absorbed via inhalation, ingestion, or dermal exposure (Kennedy 1985; Kennedy et al. 1986; US EPA 2009) and aren't metabolized. Accumulation occurs mainly in lean, well-perfused tissues in order: liver > blood > kidney > lungs > heart, with extremely low levels in other tissues (Benskin et al. 2009). Serum PFCAs are largely bound to serum albumin, with <2 % unbound

PFCA (Ohmori et al. 2003). PFCAs are excreted unchanged in the urine and/or bile; biliary excretion dominates at ≥C9 (Goecke et al. 1992; Ohmori et al. 2003). PFCAs have been shown to cross the human placenta (Apelberg et al. 2007) and enter the milk of the nursing mother (Volkel et al. 2008). Systemic half-lives of PFCA increase with chain length and are sexually-dimorphic in rats, with systemic half-lives for PFOA, PFNA, and PFDA of 2–4 hours (h), 1.4–2.44 days, and 58.6 days in females versus 6–7 days, 29.5–30.6 days, and 39.9–40 days in males (Ohmori et al. 2003; Tatum-Gibbs et al. 2011). The estimated half-life for PFOA in humans does not differ between sexes but is extremely long: 2.3–3.8 years (Olsen et al. 2007). The sexual dimorphism in rats is due to reuptake of PFCAs from the renal filtrate in males via renal organic anion transporters (OATs), whose expression is upregulated by testosterone (Weaver et al. 2010). Comparisons of the pharmacokinetics of PFOA in mice, rats, and humans suggest that mice more closely resemble humans than do rats due to the lack of sexual dimorphism in both mice and humans (US EPA 2009).

Of the FTOHs to which consumers are exposed via the diet, the 8-2 FTOH has been the most extensively studied, with most of the work performed in rats. The 8-2 FTOH (Fasano et al. 2006, 2009) shows sexual-dimorphism in the extent of its absorption from the gastrointestinal-tract under non-fasting conditions, with lower net absorption in males (49 %) versus females (81.2 %) dosed at 5 milligrams per kilogram (mg/kg), while under fasting conditions, there was no gender difference in net absorption. Extent of absorption decreases with increasing dose (49–57 % at 5 mg/kg vs. 27 % at 125 mg/kg) under non-fasted conditions. Most of the absorbed fraction (>90 %) is excreted in bile, with ~6 % excreted in the urine (Fasano et al. 2006). Systemic half-lives of 8-2 FTOH are greater in males (112–217 h) versus females (5.6–17.5 h).

Unlike PFCAs, 8-2 FTOH is preferentially distributed to adipose tissue, whereas its metabolites were located in lean, highly perfused tissues. The 8-2 FTOH is extensively metabolized in vivo to various 8-2 and 7-3 saturated and unsaturated alcohol and aldehyde products, 6-2 fluorotelomer aldehydes, glutathione conjugates, and C5-C9 PFCAs. The most ubiquitous metabolites are the 7-3 acid (both sexes) and PFOA (Fasano et al. 2009). Nabb et al. (2007) found that with rodent hepatocytes metabolized FTOH faster and to a far greater extent than human hepatocytes, and that rodent, but not human, hepatocytes produced PFCAs from FTOHs, indicating that rodents may have significant differences from humans in the metabolism of FTOH. Perfluorinated FCSs, such as the perfluoroalkyl phosphate surfactants, can also be metabolized in vivo to their FTOH components and the corresponding FTOH-metabolites (D'eon and Mabury 2007). This finding implies that direct dietary exposure to these perfluorinated FCSs may produce similar toxic effects as exposure to any FTOH impurities present in the perfluorinated FCS.

7.4 Health Effects and Toxicology

7.4.1 *Systemic Toxicity: Non-neoplastic Effects on Organ Systems*

The toxicity of PFOA has been evaluated in oral studies of varying duration in rats, mice, and monkeys. FTOHs and other PFCAs have been evaluated only in rodents. Oral administration of these compounds result in decreased body weight gain and feed efficiency/intake, with PFCA $\geq$ C10 producing frank body weight loss in rats (Shi et al. 2007; Zhang et al. 2008). Targets organs for that are common to several of the tested compounds are the liver, the immune system, and the male and female reproductive organs. Various other targets of toxicity include the kidney, teeth, and thyroid. Effects on each target organ are discussed separately below.

Effects on the Liver Increased liver weights, hepatocellular hypertrophy, peroxisomal and xenobiotic-metabolizing enzyme induction, and peroxisome proliferation are the signature toxic effects of PFCs in rodents, with hepatotoxicity usually, but not always, manifesting as increases in serum liver enzyme levels (Perkins et al. 2004; Son et al. 2008a; Kudo et al. 2005; Kawashima et al. 1995; Ladics et al. 2005, 2008; Hirata-Koizumi et al. 2012). Dose-related decreases in serum cholesterol, high density lipoprotein (HDL): low density lipoprotein (LDL) ratio, and triglycerides occur in rodents (Borges et al. 1992; Zhang et al. 2008; Minata et al. 2010), but increased triglycerides were noted in cynomolgus monkeys (Butenhoff et al. 2002). Peroxisomal enzyme induction may coexist with peroxisomal enzyme inhibition, such as with PFDA (Borges et al. 1992). Increased liver weights and peroxisomal enzymes appear to be critical effects, occurring at levels as low as 14.1 μg/ml in mice (Wolf et al. 2008) and 70 μg/ml in rats (Perkins et al. 2004). Potency increases with perfluorocarbon chain length and is proportional to the degree of hepatic accumulation (Kudo et al. 2005, 2006), although PFOA and PFNA appear to be the most peroxisome proliferator activated receptor (PPAR)-α agonists (Wolf et al. 2012). Epidemiological studies assessing the association between PFCA exposure and serum lipids and/or serum liver enzymes have found positive associations (Sakr et al. 2007a, b; Lin et al. 2010; Gallo et al. 2012) or no association (Olsen et al. 2000, 2003; Wang et al. 2012; Olsen and Zobel 2007; Gilliland and Mandel 1996). In contrast to laboratory animal studies, some human epidemiology studies have noted positive associations between perfluorochemical exposure and blood total cholesterol and/or triglyceride levels (Olsen et al. 2003; Sakr et al. 2007a, b; Steenland et al. 2009; Frisbee et al. 2010) or only a negative association with serum HDL levels (Wang et al. 2012; Olsen and Zobel 2007; Gilliland and Mandel 1996). No association with liver disease was noted in an occupationally-exposed cohort (Lundin et al. 2009) or in the Washington Works cohort (C8 Science Panel 2012a). The lack of conclusive effects in humans is not surprising, considering that peroxisomal proliferation and enzyme induction is mediated via activation of PPAR-α (Wolf et al. 2008), to which rodents are extremely sensitive. Human liver cells, in contrast, are refractory to peroxisomal proliferation (Bjork and Wallace 2009) even though PFCs are agonists at both the human and rodent PPAR-α receptor with apparently equal potency (Vanden Heuvel et al. 2006). Some of the effects may also be mediated by activation of the Constitutive Androstane Receptor (CAR) and/or PPAR-γ (Cheng and Klaassen 2008), as PFOA produced significant, although qualitatively distinct, hepatic toxicity in both PPAR-α wild-type (WT) and knockout (KO) mice (Wolf et al. 2008).

Effects on the Immune System Most of the studies evaluating the immunotoxicity of PFCs have focused on PFOA and PFNA. Splenic and thymic atrophy with inhibition of cell proliferation and/or induction of apoptosis (Yang et al. 2001; Fang et al. 2008; Hirata-Koizuma et al. 2012), altered T-cell phenotypes (Son et al. 2008b), and decreased cell-mediated and humoral immune responses (Nelson et al. 1992; Yang et al. 2002a) were noted in mice at oral doses of $\geq$10 mg/kg per day PFOA or $\geq$3 mg/kg per day PFNA. Published studies conflict, and have reported that PFOA increased (Loveless et al. 2008) and decreased (Qazi et al. 2009) numbers of circulating neutrophils and monocytes, which may reflect the higher doses used by Loveless and colleagues. PFOA has been reported to have stimulatory (Qazi et al. 2009) or inhibitory (Corsini et al. 2011) effects on inflammatory cytokine production. The observed immunotoxicity may be mediated via PPAR-α activation, as administration of PFOA to PPAR-α KO mice had no effects on splenic or thymic weights or cellularity (Yang et al. 2002b). No immunotoxic effects have been noted in any of the studies conducted with the FTOH, although only immune system organ weights and histopathology were assessed. Serum PFOA was inversely related to serum antibody titer to tetanus and diphtheria in vaccinated children in the Faroe Islands (Grandjean et al. 2012), suggesting that higher PFOA exposure reduced the immune response to vaccination in children. Other epidemiology studies found no association between PFOA exposure and infectious disease morbidity and/or mortality in children from the general population (Fei and Olsen 2011; Okada et al. 2012), the Washington Works cohort (C8 Science Panel 2012b), or in occupationally-exposed populations (Leonard et al. 2008). Dose levels associated with immunotoxicity in animal studies produced significant systemic toxicity and large increases in blood corticosteroid levels. It is possible, therefore, that the discordance between the results of animal and epidemiology studies is related to the 1,000-fold higher blood levels achieved in the animal studies.

Effects on Male Reproductive Organs Effects of PFCs on male reproductive organs have been evaluated for PFOA, PFNA, PFDA, PFDoDA, an FTOH mixture, and the 8-2 FTOH. No effects on fertility or spermatogenesis were noted with any compound, but PFCAs $\geq$ C9 caused marked adverse effects on androgen-production, as measured by serum testosterone; testicular and/or accessory sex organ weights; and testicular morphology in rats (Feng et al. 2009; Bookstaff et al. 1990; Shi et al. 2007). Morphological changes included degenerative and/or apoptotic changes in seminiferous tubules or tubular atrophy. PFOA's effects on male reproductive organs differed between species; rats were the least sensitive to PFOA, with effects limited to decreased testes weights (York et al. 2010). In contrast, PFOA decreased serum testosterone and caused testicular degeneration in mice (Li et al. 2011) and monkeys (Butenhoff et al. 2002), and increased sperm abnormalities in mice. The adverse effects of PFOA in mice were noted in both WT- and humanized-PPAR-α mice, but not in KO mice, indicating the anti-androgenic effects were mediated PPAR-α activation (Li et al. 2011). Direct inhibition of steroidogenesis in Leydig cells may also play a significant role in the observed toxicity, as PFOA, PFDA, and PFDoDA inhibited steroidogenesis in primary Leydig cells derived from adult rats and in Leydig cell tumor lines (Shi et al. 2010; Biegel et al. 1995; Boujrad et al. 2000) via competitive inhibition of steroidogenic enzymes and also possibly downregulation of steroidogenic gene expression. Indicators of cell damage, such as lipid accumulation and reactive oxygen species (ROS) accumulation in mitochondria, were also noted in Leydig cells treated with PFC (Boujrad et al. 2000; Shi et al. 2010). No adverse effects on male reproductive organ weight or histology were noted in rats with the FTOH (Ladics et al. 2005, 2008).

In contrast to the clear evidence of adverse effects of PFCAs on male reproductive organs in animal studies, evidence from epidemiological studies is far less clear-cut. Impaired semen quality was associated with high levels of PFCs in non-occupationally exposed males in a Danish study (Joensen et al. 2009) but not in other studies (Raymer et al. 2012; Toft et al. 2012; Specht et al. 2012). No associations between serum PFOA and reproductive hormone levels were noted in an occupationally-exposed (Olsen et al. 1998) or Northern European (Specht et al. 2012) males. There is no information available regarding PFC activity at the androgen receptor. In summary, ≥C8 PFCAs appear to alter reproductive hormone homeostasis in male rodents and possibly in nonhuman primates without direct effects on fecundity in standard reproductive toxicity studies, possibly because of the high resilience of rodents to anti-fertility effects. The available data indicate that either FTOH may not be toxic to the male reproductive tract or are significantly less potent at these endpoints than PFCAs.

Effects on Female Reproductive Organs and the Mammary Gland Effects of PFCs on the reproductive organs and mammary glands (MGs) on rats and mice have been evaluated for PFOA, PFDoDA, the 8-2 FTOH, and an FTOH mixture. Few adverse effects on fertility or reproductive organ parameters were noted in rats with PFCAs; the only effects noted were increased numbers of estrous cycles with prenatal PFOA exposure (York 2002) and decreased serum estradiol levels with high-dose PFDoDA (Shi et al. 2009). Prenatal, peripubertal, or adult exposure to PFOA had adverse effects on MG development and differentiation in mice. PFOA decreased MG development in prenatally-exposed CD-1 mice (White et al. 2007, 2009, 2011) and delayed differentiation into a lactating phenotype in their dams (White et al. 2007) at doses as low as 3–5 mg/kg per day. At 18 months of age, abnormal histopathological findings were noted in the MGs of prenatally-exposed CD-1 mice, including increased numbers of darkly-staining foci, hyperplasia of the ductal epithelium, increased stromal epithelial densities, and/or inappropriate differentiation of ductal tissue (White et al. 2009). Some studies also report strain-dependent effects of PFOA on MG development when administered around the time of puberty (postnatal day 21), with stimulatory effects noted with low-dose PFOA in C57Bl/6 mice and inhibitory effects noted with higher dose of PFOA in the same strain or with all doses of PFOA in Balb/c mice (Yang et al. 2009). These effects appear to be mediated via estrogen receptor (ER)-signaling, as these effects were abolished with ovariectomy (Zhao et al. 2010). There are significant factors in the design of these studies that complicate interpretation of these results, i.e.: low power ($n < 6$/treatment group) in the Yang et al. study and, in the White et al. studies, use of a subjective, qualitative scoring method instead of quantitative morphometry or 5-bromo-2′-deoxyuridine (BrdU) staining of terminal end buds to evaluate MG cell proliferation and development. As has been previously demonstrated, use of the qualitative method may give disparate results from quantitative morphometry (Hovey et al. 2011). However, the fact that the results of White et al. were also found by Yang et al. using BrdU-staining of terminal end buds lends support to the validity of White et al.'s findings. As such, the available data indicate that PFOA has adverse effects on MG development and differentiation at least in mice. Although the available bioassay data in rats do not indicate treatment-related effects on the MG (Biegel et al. 2001), the pharmacokinetics of PFOA in the female rat may make it less sensitive than the female mouse to the long-term toxicity of PFOA at this endpoint. Moreover, the rats in the bioassay were not dosed beginning in utero, possibly further diminishing the sensitivity of that study to detect toxicity at that endpoint. As such, further studies conducted in mice are needed to discern whether PFCA may have carcinogenic effects in the MG. The FTOH do not appear to have any adverse effects on female reproductive organ or accessory sex organ weights or histopathology in rats (Mylchreest et al. 2005a; Ladics et al. 2005, 2008), although significant estrogenic activity has been reported for the 8-2 FTOH in yeast two-hybrid assays (Ishibashi et al. 2007) and MCF-7 cells (Maras et al. 2006).

Epidemiological evidence of effects of PFCs on female reproductive health is inconclusive. Impaired fecundity, measured as increased time-to-pregnancy, was associated with high serum PFOA levels in non-occupationally exposed Danish women in one study (Fei et al. 2009) but not another (Vestergaard et al. 2012). PFOA levels were associated with increased incidence of pregnancy-induced hypertension and preeclampsia in the Washington Works cohort (C8 Science Panel 2011d). In summary, there is evidence that PFOA may have adverse effects on MG development and differentiation in mice, possibly mediated via estrogenic pathways. No consistent effects of PFC exposure were noted on female reproductive organ health. Further studies are needed to evaluate the relationship between PFC-exposure and potential adverse effects such as preeclampsia and pregnancy-related hypertension in highly-exposed human populations.

Effects on the Thyroid Data on effects of PFCs on the thyroid are comparatively sparse, with the exception of studies conducted in rats with PFDA. PFDA induced marked hypothyroidism in rats after only a single, low-dose injection (Van Rafelghem et al. 1987). PFOA decreased serum thyroxine levels in cynomolgus monkeys (Butenhoff et al. 2002), but not in rats (Perkins et al. 2004); no histopathological correlates were noted for the finding in monkeys. Thyroid follicular cell hypertrophy was noted in rats exposed to mixed FTOH (Ladics et al. 2005), however, the high level of free iodine present as an impurity in the test substance complicates attributing this finding to effects of the FTOH. PFOA's effects on the thyroid are mixed in epidemiological studies. Some noted no association of serum PFOA levels with serum thyroxine levels in occupationally-exposed healthy subjects (Olsen and Zobel 2007; Olsen et al. 2003), with the occurrence of hypothyroxemia during pregnancy (Chan et al. 2011), or with fetal thyroxine levels (Kim et al. 2011). Others found direct associations between serum PFOA levels and the incidence of thyroid disease in the general population (Melzer et al. 2010) and in highly-exposed populations (C8 Science Panel 2012g). Thus, the evidence for effects of PFCs on thyroid parameters is mixed. PFDA appears to have clear adverse effects on thyroid parameters in rodents. However, the effects of PFOA and other PFCs on thyroid function are unclear. Rats are known to be sensitive to secondary effects on thyroid hormone production due to increased hepatic metabolism and clearance of thyroid hormone. This mechanism is thought to not be relevant to humans due to the presence of transthyrethrin which buffers serum thyroxine levels against changes in hepatic metabolic capacity (Wu and Farrelly 2006). While the finding of decreased serum thyroxine in monkeys may indicate an effect of PFOA on the thyroid in humans, interpretation of this finding is complicated by excess mortality. Further studies will be needed to fully-assess the toxic effects of PFCs on the thyroid.

Other Effects Other effects noted with PFCs in rats include: renal tubular hypertrophy and increased kidney weights, decreased red blood cell numbers and hematocrit, and, with FTOHs, ameloblast degeneration with tooth breakage (Ladics et al. 2005, 2008; Cui et al. 2009). The effects on teeth are not apparent in toxicological studies conducted with PFCAs, indicating that these effects may be specific to either the parent compound or one of its non-PFCA metabolites. No effects of PFOA on the incidence of chronic kidney disease were noted in the Washington Works cohort (C8 Science Panel 2012e).

7.4.2 Developmental Toxicity

The effects of perfluorinated compounds on prenatal and postnatal endpoints have been evaluated in rodents for PFOA, PFNA, PFDA, perfluorooctadecanoic acid (PFOcDA), the 8-2 FTOH, and an FTOH mixture and in mice for PFOA. PFOA is the best-characterized of these compounds, having been assessed in both rats and mice. Mice appear to be far more sensitive to the effects of PFOA on development compared to rats, possibly due to the aforementioned species difference in pharmacokinetics. Adverse effects noted prenatally in mice and/or rats with the above-cited compounds include: decreased implantation/number of corpora lutea (mixed FTOH and PFOcDA), increased percent litter loss and/or incidence of full litter resorption (FLR); reduced skeletal ossification and/or increased incidences of skeletal variations; and decreased fetal body weights (Abbott et al. 2007; Lau et al.

2006; Harris and Birnbaum 1989; Mylchreest et al. 2005a, b; Hirata-Koizumi et al. 2012). The effects on early FLR in mice are PPAR-α-independent, as PFOA also increased FLR in KO mice (Abbott et al. 2007), and may be due to direct toxicity of PFOA to the developing placenta (Suh et al. 2011). Dose-related decreased fetal survival was also noted in mice treated with PFOA $\geq$ 20 mg/kg. In contrast, no effects on pregnancy maintenance or fetal survival were observed in rats or rabbits treated with PFOA (Gortner et al. 1981, 1982) or in rats treated with 8-2 FTOH (Mylchreest et al. 2005b). Reduced ossification was noted at doses as low as 1 mg/kg PFOA in mice and at doses of 200 mg/kg in rats with 8-2 FTOH; these doses were associated with maternal toxicity in the FTOH studies but occurred at doses lower than the maternal systemic lowest observed effect level (LOEL) with PFOA, indicating a direct adverse effect of PFOA on the mouse fetus. There also appear to be significant differences in the response of mice to the 8-2 FTOH, as a single oral dose of 30 mg/kg on gestation day (GD) 8 increased the percent nonviable fetuses and litter incidence of anencephaly or exencephaly (Henderson and Smith 2007). These effects don't appear to be due to metabolism of the parent compound to PFOA and PFNA, as neural tube defects were not observed with either compound in mice or with the 8-2-FTOH in rats.

Adverse effects noted on postnatal development include decreased number of pups born (mixed FTOH and PFOcDA), neonatal survival, decreased body weight gain prior to weaning, delayed attainment of eye-opening and hair growth, and changes in the timing of attainment of puberty. Decreased neonatal survival was noted in studies conducted in rats with the mixed FTOH (Mylchreest et al. 2005a) and PFOA (York 2002) and in mice with PFOA (Lau et al. 2006; Wolf et al. 2007; Abbott et al. 2007) and PFNA (Wolf et al. 2010). LOELs ranged from 1–5 mg/kg PFOA in mice to 30 mg/kg PFOA in rats, indicating a 6–30-fold increase in sensitivity of mice to PFOA for this endpoint. Unlike rats, PFOA decreased postnatal weight gain in mice of both sexes, with males slightly more sensitive than females. Brief prenatal exposures to extremely low total doses of PFOA (5 mg/kg on GD 10–17) were sufficient to decrease postnatal growth in male mice. PFOA's effect on postnatal growth was proportional to the total dose received by the offspring, with pups exposed to PFOA prenatally and postnatally having more severe growth deficits than pups exposed during either stage alone. Changes in the timing of attainment of puberty have also been noted in rats and mice after prenatal PFOA exposure, although the direction of the change varied between studies {acceleration in males and delay in females (Lau et al. 2006); delay in males and females (York 2002)} which indicates that the effects of PFOA on this parameter have not been adequately characterized. In contrast to effects of PFOA on FLR, effects of PFNA and PFOA on postnatal endpoints in mice appear to be PPAR-α-dependent (Abbott et al. 2007; Wolf et al. 2010). Interestingly, clofibrate and WY-14,643, highly selective and potent PPAR-α-agonists, had no adverse postnatal effects in either PPAR-α-WT or KO mice fed these compounds throughout gestation (Palkar et al. 2010), indicating that the adverse postnatal effects of PFOA and PFNA may be mediated by additional pathways active during development, such as CAR or PPAR-γ (Abbott et al. 2012).

Several epidemiological studies have examined the association between PFOA exposure and neonatal outcome variables, including birth weight and length, ponderal index, head circumference, gestational age, Apgar scores, and developmental milestones at 6 and 18 months of age. One study reported that cord serum PFOA levels were inversely proportional to head circumference for vaginal births and ponderal index, regardless of method of delivery, in babies born to non-occupationally-exposed women (Apelberg et al. 2007), and non-significant associations of serum PFOA with decreased birth weight also were noted. Another study conducted in the Danish National Birth Cohort noted an inverse association between maternal first trimester serum PFOA levels and birth weight, abdominal circumference, and birth length (Fei et al. 2007, 2008b); non-significant negative associations between maternal PFOA levels and head circumference were also noted. Maternal serum PFOA was similarly inversely-related to body weight and/or body mass index at 12 months old (Andersen et al. 2010) in the same population. Other studies of occupationally-exposed cohorts and the general population have reported no association between maternal or umbilical cord serum PFOA levels and gestational age at birth, birth weight, birth length, head or chest circumference, risk for preterm birth or low birth weight, birth defects, or risk for miscarriage or stillbirth (Nolan et al. 2009; Washino et al. 2009; Chen et al. 2012; C8 Science Panel 2011a, b, c; Hamm et al. 2010).

In summary, the aggregate data from animal studies demonstrate that developmental toxicity is a sensitive endpoint for all PFCs. Mice may be more sensitive to the adverse developmental effects than rats due to pharmacokinetic considerations, and brief in utero exposure alone is sufficient to induce postnatal toxicity into adulthood. Although additional data on the effects of FTOH in mice are needed, comparison of the postnatal effects of PFCAs versus FTOH in rats suggests that FTOH may be slightly less toxic than the PFCAs. However, the conflicting data from epidemiological studies and the results from PPAR-α-KO studies raise the issue of the applicability of the findings from rodent studies to humans.

7.4.3 Carcinogenicity and Genotoxicity

Only PFOA has been evaluated under conditions of a standard rodent bioassay. In one study, Sprague-Dawley rats of both sexes (n = 50/sex/group) were fed PFOA at levels of 0, 30, or 300 parts per million (ppm) in feed (corresponding to intake levels of 0, 1.42, and 14.2 mg/kg per day in males, and 0, 1.61, and 16.1 mg/kg per day in females) for 2 years (Butenhoff et al. 2012). Significantly increased combined incidences of hepatocellular carcinomas and hyperplastic nodules, and incidences of Leydig cell adenomas were noted in high-dose males. No treatment-related neoplastic effects were noted in females. An additional bioassay in which PFOA was also administered in the diet to male CD rats (n = 80/group) for 2 years at levels of 0 or 300 ppm noted significantly increased incidences of hepatocellular adenoma and carcinoma combined, pancreatic acinar cell adenomas and carcinomas combined, and Leydig cell adenomas with PFOA treatment (Biegel et al. 2001). Epidemiological studies conflict, with some studies reporting an association between PFOA-exposure and kidney and testicular cancer in the Washington Works cohort (C8 Science Panel 2012c) or prostate cancer in an occupationally-exposed cohort (Gilliland and Mandel 1993; Lundin et al. 2009). Other studies found no association between PFOA exposure and cancer in either men or women (Leonard et al. 2008; Eriksen et al. 2009) from occupationally-exposed cohorts or cohorts from the general population.

PFOA's neoplastic effects in rats aren't mediated via direct damage to DNA, as PFCs are generally negative in genotoxicity assays (Buhrke et al. 2013; ECHA 2012). PFOA's neoplastic effects may be mediated via rodent-specific pathways, such as PPAR-α activation induction of liver tumors and the disruption of the Hypothalamic-Pituitary-Gonad axis inducing Leydig cell tumors in the testes via increased hepatic androgen clearance (Cook et al. 1992) or aromatase stimulation (Liu et al. 1996). The human liver appears to be refractory to tumor induction by stimulation of PPAR-α, even by extremely potent, specific activators of PPAR-α, such as fibrates (Rosen et al. 2009). The spontaneous incidence of Leydig cell tumors in male rats is 135,500–1,920,000-fold higher incidence than in humans. Several physiological factors make humans refractory to development of this tumor type, including: human sex-hormone binding globulin, which stabilizes blood androgen levels; and increased sensitivity and responsiveness of rodent Leydig cells to proliferatory stimuli compared to human Leydig cells (Cook et al. 1999). While mechanism(s) of action of the induction of pancreatic tumors have not been as thoroughly characterized as those for the liver and testes tumors, sustained elevation of cholecystokinin in rodents is believed to play a role in the development of the observed acinar cell tumors (Klaunig et al. 2012). This mechanism also does not appear to operate in humans, as may be seen by the lack of effect of PFOA exposure on pancreatic cancer incidence in epidemiology studies and by the lack of association between plasma cholecystokinin levels and PFOA exposure in an occupationally-exposed cohort (Olsen et al. 2000).

Since almost all of the long-chain PFCs tested are PPAR-α agonists and/or induce xenobiotic-metabolizing enzymes in the liver, PFCs may be expected to induce some or all of the neoplastic effects seen with PFOA in rodents. There is currently insufficient information to definitively classify long-chain PFCs as human carcinogens. More complete characterization of their mechanisms of action in humans versus rodents will be necessary in order to determine whether this class of compounds poses a carcinogenic hazard to humans at current cumulative levels of exposure.

7.4.4 Emerging Endpoints

Neurobehavioral Effects Epidemiological studies have reported conflicting results concerning associations between PFC levels in mothers and/or children and incidences of Attention Deficit/Hyperactivity Disorder (ADHD) in children or impaired response inhibition, with significant associations reported for cohorts from the general population (Gump et al. 2011; Hoffman et al. 2010). In contrast, no association between PFOA exposure and pre- or postnatal PFOA exposure and neurobehavioral disorders in children was noted for the highly-exposed Washington Works cohort (C8 Science Panel 2012f). Additional studies examining attainment of developmental landmarks and incidences of childhood behavioral or coordination problems found no association between maternal or umbilical cord serum PFOA levels and Apgar scores at birth and the timing of attainment of developmental milestones at 18-months of age (Fei et al. 2008a) or an association between maternal serum PFOA and incidences of behavioral or motor problems in childhood (Fei and Olsen 2011). There is some preliminary evidence of adverse effects of PFOA on habituation and/or activity levels in studies conducted in mice dosed postnatally (Johansson et al. 2008) or in utero (Onishchenko et al. 2011). However, there are significant flaws in the designs of these studies, including: lack of accounting for litter effects in the statistical analyses in the results, low power, and lack of positive controls for the behavioral assays.

Adverse Effects on Body weight and Glucose Homeostasis Conflicting effects of PFCA exposure on glucose homeostasis have been reported, with increased serum glucose levels and hepatic glycogen levels noted in male rats fed PFNA or PFDoDA (Ding et al. 2009; Fang et al. 2012) and no effects on serum glucose reported for PFOA and PFDA (Goecke et al. 1994). Similarly, conflicting results have been noted in epidemiology studies. Both positive effects (Leonard et al. 2008) and no association (Gilliland and Mandel 1993; C8 Science Panel 2012d), between PFOA exposure and Type II diabetes mortality rates or incidence, have been reported for occupationally-populations or the Washington Works cohort.

While studies conducted in animals with PFCs generally note decreased body weight gain and/or weight loss, a recent study reported increased body weight gain and percentage of body fat in middle-aged female mice that had been prenatally-exposed to very low levels of PFOA (Hines et al. 2009). Similarly, an epidemiological study conducted in the general population noted a positive association between maternal serum PFOA levels during pregnancy and average body mass index (BMI), waist circumference, serum leptin and insulin in female offspring (Halldorsson et al. 2012). Hines et al. (2009) used a cross-foster design which has several disadvantages, such as difficulty in subsequently accounting for effect of litter of origin, decreased power, the possibility of sampling errors (Chiarotti et al. 1987), and stress-induced alterations in maternal care (Denenberg et al. 1963). Cross-fostering may also induce epigenetic changes in offspring that influence adult phenotypic markers, such as body weight (Bartolomucci et al. 2004). As such, the results of the Hines study and the epidemiology studies need to be confirmed in an animal study that does not employ a cross-foster design.

7.5 Regulatory Action

Much of the toxicity information described above became available after the regulatory allowances for the use of PFCs in food contact applications. Based upon safety concerns for exposure to PFCs as a class, FDA reached a voluntary agreement with the manufacturers of all long-chain FTOH-based paper coatings, as well as manufacturers of PFCA-based coatings subject to a FCN, to cease manufacture and sale of their products for food-contact applications as of October 1, 2011 (US FDA 2012). Although existing supplies of these products already in the marketplace were allowed to be used until they ran out, FDA estimated that these supplies would be consumed within one year of the cessation of sale; as such consumer exposure to FTOHs from food packaging effectively ceased as of October, 2012.

Based upon concerns for the ubiquitous detection of PFCAs in the environment, their biopersistance, and toxicity, USEPA enacted a *PFOA Stewardship Program* in 2006 with the major manufacturers of PFOA, the intent of which was to reduce global facility emission and product content of PFOA and longer chain homologues by 95 % by 2010, and to eliminate emissions and product content by 2015 (US EPA 2013). In addition, USEPA has posted an action plan (US EPA 2009) to address the use of PFCAs by manufacturers not participating in the stewardship program. As the stewardship program addresses all uses of PFCAs in the U.S., including their use as FCSs, this action also eliminates the use of FCSs listed in 21 CFR that utilize PFCAs in their manufacture.

Although PFCs have been the subject of review of regulatory bodies in other parts of the world, to date these have not resulted in significant restriction on use. In the European Union, restrictions have been limited to PFSAs which are subject to *Directive 2006/122/EC*. This Directive also notes that PFOA and its salts are suspected to have a similar risk profile to PFOS and that these compounds should be the subject of ongoing risk assessment activities to determine what restrictions on marketing and use would be appropriate for these compounds. More recently the European Food Safety Authority has published a scientific report stating that the dietary exposure to PFOS and PFOA is highly unlikely to exceed health-based guidance values (EFSA 2012). In Canada, prohibitions against the manufacture, import, sale, or use of four FTOH-based substances (none of which are used in food packaging) were passed in 2010 (Canada 2010a) and a voluntary *Environmental Performance Agreement* (Canada 2010b), similar to the USEPA stewardship program also went into effect in 2010.

7.6 Summary, Data Gaps, and Recommendations

PFCAs and FTOHs have been used in a variety of applications including food packaging. Human exposure to these compounds has been demonstrated, with diet as a significant contributor although the significance of exposure from food packaging has not been elucidated. Available toxicity information indicates bioaccumulation of long-chain PFCs in well-perfused tissues with toxicity to the liver, immune system, and reproductive organs, as well as concerns for developmental toxicity and carcinogenicity. The applicability of the findings of laboratory animal studies to humans is unclear, as some of the mechanisms of action for these effects may be rodent-specific. Moreover, the toxic effects of FTOHs have not been as well-characterized as those for PFCAs, particularly regarding the endpoints of developmental and immunotoxicity. Additional data from animal studies and human epidemiological studies will be necessary in order to come to firm conclusions regarding the toxicological profile of FTOHs. Given the higher sensitivity of mice versus rats to the toxicity of PFC, any future toxicity studies of PFC should be conducted in mice. Moreover, given the fact that the majority of exposure to PFCs from food packaging involves FTOHs and not PFCAs, future human health risk assessments for long-chain PFCs should focus on fully-delineating the toxicity profile of FTOHs, including exploration of the mechanisms of action and human health relevance of the observed effects. Although data gaps on the toxic effects of PFCs exist, regulatory action by the FDA and USEPA should ensure that future exposure in the U.S. to long-chain PFCs from food packaging is severely limited.

Regulatory agencies will continue to contend with the historical legacy of the use of long-chain PFCs in general, and in food packaging specifically. The effectiveness of the regulatory approach to these compounds described above could be verified through the analysis of composite foods samples gathered prior to and after FDA's 2012 agreement with industry and USEPA's 2015 product stewardship deadline. In addition to the historic use of long-chain PFCs, industry continues to develop replacement products based upon short-chain (C6 and shorter) PFCs. Although these short-chain compounds do not involve concerns for biopersistance, their toxicity continues to be an area of active research, the results of which continue to be taken into account when considering the regulatory allowances for their use in food packaging.

References

Abbott BD, Wolf CJ, Schmid JE et al (2007) Perfluorooctanoic acid-induced developmental toxicity in the mouse is dependent on expression of peroxisome proliferator-activated receptor-alpha. Toxicol Sci 98(2):571–581

Abbott BD, Wood CR, Watkins AM et al (2012) Effects of perfluorooctanoic acid (PFOA) on expression of peroxisome proliferator-activated receptors (PPAR) and nuclear receptor-regulated genes in fetal and postnatal CD-1 mouse tissues. Reprod Toxicol 33(4):491–505

Andersen CS, Fei CY, Gamborg M et al (2010) Prenatal exposures to perfluorinated chemicals and anthropometric measures in infancy. Am J Epidemiol 172(11):1230–1237

Apelberg BJ, Witter FR, Herbstman JB et al (2007) Cord serum concentrations of perfluorooctane sulfonate (PFOS) and perfluorooctanoate (PFOA) in relation to weight and size at birth. Environ Health Perspect 115(11):1670–1676

Bartolomucci A, Gioiosa L, Chirieleison A et al (2004) Cross fostering in mice: behavioral and physiological carry-over effects in adulthood. Genes Brain Behav 3(2):115–122

Begley TH, White K, Honigfort P et al (2005) Perfluorochemicals: potential sources of and migration from food packaging. Food Addit Contam 22(10):1023–1031

Benskin JP, De Silva AO, Martin LJ et al (2009) Disposition of perfluorinated acid isomers in Sprague-Dawley rats; Part 1: single-dose. Environ Toxicol Chem 28(3):542–554

Biegel LB, Hurtt ME, Frame SR et al (2001) Mechanisms of extrahepatic tumor induction by peroxisome proliferators in male CD rats. Toxicol Sci 60(1):44–55

Biegel LB, Liu RCM, Hurtt ME et al (1995) Effects of ammonium perfluorooctanoate on Leydig-cell function - In vitro, in vivo, and ex vivo studies. Toxicol Appl Pharmacol 134(1):18–25

Bjork JA, Wallace KB (2009) Structure-activity relationships and human relevance for perfluoroalkyl acid-induced transcriptional activation of peroxisome proliferation in liver cell cultures. Toxicol Sci 111(1):89–99

Bookstaff RC, Moore RW, Ingall GB et al (1990) Androgenic deficiency in male-rats treated with perfluorodecanoic acid. Toxicol Appl Pharmacol 104(2):322–333

Borges T, Robertson LW, Peterson RE et al (1992) Dose-related effects of perfluorodecanoic acid on growth, feed-intake and hepatic peroxisomal beta-oxidation. Arch Toxicol 66(1):18–22

Boujrad N, Vidic B, Gazouli M et al (2000) The peroxisome proliferator perfluorodecanoic acid inhibits the peripheral-type benzodiazepine receptor (PBR) expression and hormone-stimulated mitochondrial cholesterol transport and steroid formation in Leydig cells. Endocrinology 141(9):3137–3148

Buhrke T, Kibellus A, Lampen A (2013) In vitro toxicological characterization of perfluorinated carboxylic acids with different carbon chain lengths. Toxicol Lett 218(2):97–104

Butenhoff J, Costa G, Elcombe C et al (2002) Toxicity of ammonium perfluorooctanoate in male cynomolgus monkeys after oral dosing for 6 months. Toxicol Sci 69(1):244–257

Butenhoff J, Kennedy G, Chang S et al (2012) Chronic dietary toxicity and carcinogenicity study with ammonium perfluorooctanoate in Sprague-Dawley rats. Toxicology 298:1–13

C8 Science Panel (2011a) Probable link evaluation of birth defects. C8 Probable Link Reports. http://www.c8sciencepanel.org/pdfs/Probable_Link_C8_Birth_Defects_5Dec2011.pdf. Accessed on 17 Jun 2013

C8 Science Panel (2011b) Probable link evaluation of miscarriage and stillbirths. C8 Probable Link Reports. http://www.c8sciencepanel.org/pdfs/Probable_Link_C8_Pregnancy_Loss_5Dec2011.pdf. Accessed on 17 Jun 2013

C8 Science Panel (2011c) Probable link evaluation of perterm birth and low birthweight. C8 Probable Link Reports. http://www.c8sciencepanel.org/pdfs/Probable_Link_C8_Preterm_and_LBW_birth_5Dec2011.pdf. Accessed on 17 Jun 2013

C8 Science Panel (2011d) Probable link evaluation of pregnancy induced hypertension and preeclampsia. C8 Probable Link Reports. http://www.c8sciencepanel.org/pdfs/Probable_Link_C8_PIH_5Dec2011.pdf. Accessed on 17 Jun 2013

C8 Science Panel (2012a) Probable link evaluation for liver diseases. C8 Probable Link Reports. http://www.c8sciencepanel.org/pdfs/Probable_Link_C8_Liver_29Oct2012.pdf. Accessed on 17 Jun 2013

C8 Science Panel (2012b) Probable link evaluation of infectious disease. C8 Probable Link Reports. http://www.c8sciencepanel.org/pdfs/Probable_Link_C8_Infections_30Jul2012.pdf. Accessed on 17 Jun 2013

C8 Science Panel (2012c) Probable link evaluation of cancer. C8 Probable Link Reports. http://www.c8sciencepanel.org/pdfs/Probable_Link_C8_Cancer_16April2012_v2.pdf. Accessed on 17 Jun 2013

C8 Science Panel (2012d) Probable link evaluation of diabetes. C8 Probable Link Reports. http://www.c8sciencepanel.org/pdfs/Probable_Link_C8_Diabetes_16April2012.pdf. Accessed on 17 Jun 2013

C8 Science Panel (2012e) Probable link evaluation of kidney disease. C8 Probable Link Reports. http://www.c8sciencepanel.org/pdfs/Probable_Link_C8_Kidney_29Oct2012.pdf. Accessed on 17 Jun 2013

C8 Science Panel (2012f) Probable link evaluation of neurodevelopmental disorders in children. C8 Probable Link Reports. http://www.c8sciencepanel.org/pdfs/Probable_Link_C8_Neurodevelopment_30Jul2012.pdf. Accessed on 17 Jun 2013

C8 Science Panel (2012g) Probable link evaluation of thyroid disease. C8 Probable Link Reports. http://www.c8sciencepanel.org/pdfs/Probable_Link_C8_Thyroid_30Jul2012.pdf. Accessed on 17 Jun 2013

Calafat AM, Wong LY, Kuklenyik Z et al (2007) Polyfluoroalkyl chemicals in the US population: Data from the National Health and Nutrition Examination Survey (NHANES) 2003–2004 and comparisons with NHANES 1999–2000. Environ Health Perspect 115(11):1596–1602

Canada (2010a) Environmental performance agreement respecting perfluorinated carboxylic acids (PFCAs) and their precursors in perfluorinated products sold in Canada. Canada Gazette. http://www.ec.gc.ca/epe-epa/default.asp?lang=En&n=AE06B51E-1. Accessed on 20 Jun 2013

Canada (2010b) Regulations amending the prohibition of certain toxic substances regulations, 2005 (four new fluorotelomer-based substances). Canada Gazette, Part II, 144 (21):1959–1967

Chan E, Burstyn I, Cherry N et al (2011) Perfluorinated acids and hypothyroxinemia in pregnant women. Environ Res 111(4):559–564

Chen MH, Ha EH, Wen TW et al (2012) Perfluorinated compounds in umbilical cord blood and adverse birth outcomes. PLoS One 7(8):1–8

Cheng XG, Klaassen CD (2008) Perfluorocarboxylic acids induce cytochrome p450 enzymes in mouse liver through activation of PPAR-alpha and CAR transcription factors. Toxicol Sci 106(1):29–36

Chengelis CP, Kirkpatrick JB, Myers NR et al (2009) Comparison of the toxicokinetic behavior of perfluorohexanoic acid (PFHxA) and nonafluorobutane-1-sulfonic acid (PFBS) in cynomolgus monkeys and rats. Reprod Toxicol 27(3–4):400–406

Chiarotti F, Alleva E, Bignami G (1987) Problems of test choice and data-analysis in behavioral teratology: the case of prenatal benzodiazepines. Neurotoxicol Teratol 9(2):179–186

Cook JC, Klinefelter GR, Hardisty JF et al (1999) Rodent Leydig cell tumorigenesis: a review of the physiology, pathology, mechanisms, and relevance to humans. Crit Rev Toxicol 29(2):169–261

Cook JC, Murray SM, Frame SR et al (1992) Induction of Leydig cell adenomas by ammonium perfluorooctanoate—a possible endocrine-related mechanism. Toxicol Appl Pharmacol 113(2):209–217

Corsini E, Avogadro A, Galbiati V et al (2011) In vitro evaluation of the immunotoxic potential of perfluorinated compounds (PFCs). Toxicol Appl Pharmacol 250 (2):108–116

Cui L, Zhou QF, Liao CY et al (2009) Studies on the toxicological effects of PFOA and PFOS on rats using histological observation and chemical analysis. Arch Environ Contam Tox 56(2):338–349

D'Eon JC, Mabury SA (2007) Production of perfluorinated carboxylic acids (PFCAs) from the biotransformation of polyfluoroalkyl phosphate surfactants (PAPS): Exploring routes of human contamination. Environ Sci Technol 41(13):4799–4805

Denenberg VH, Zarrow MX, Grota LJ (1963) Maternal behaviour in rat: analysis of cross-fostering. J Reprod Fertil 5(2):133–141

Ding LN, Hao FH, Shi ZM et al (2009) Systems biological responses to chronic perfluorododecanoic acid exposure by integrated metabonomic and transcriptomic studies. J Proteome Res 8(6):2882–2891

Eriksen KT, Sorensen M, McLaughlin JK et al (2009) Perfluorooctanoate and perfluorooctanesulfonate plasma levels and risk of cancer in the general Danish population. J Natl Cancer Inst 101(8):605–609

European Chemical Agency (ECHA) (2012) CLH Report: proposal for harmonized classification and labeling: 8:2-Fluorotelomer alcohol. Based on regulation (EC) #1272/2008 (CLP Regulation), Annex VI, Part 2. http://echa.europa.eu/documents/10162/570253b4-9e95-4e29-a394-89ece2050905. Accessed on 20 Jun 2013

European Food Safety Authority (EFSA) (2012) Perfluoroalkylated substances in food: occurence and dietary exposure. EFSA J 10(6):2743

Fang XM, Zang L, Feng Y et al (2008) Immunotoxic effects of perfluorononanoic acid in mice. Toxicol Sci 105(2):312–321

Fang XM, Gao GZ, Xue HY et al (2012) Exposure of perfluorononanoic acid suppresses the hepatic insulin signal pathway and increases serum glucose in rats. Toxicology 294(2–3):109–115

Fasano WJ, Carpenter SC, Gannon SA et al (2006) Absorption, distribution, metabolism, and elimination of 8-2 fluorotelomer alcohol in the rat. Toxicol Sci 91(2):341–355

Fasano WJ, Sweeney LM, Mawn MP et al (2009) Kinetics of 8-2 fluorotelomer alcohol and its metabolites, and liver glutathione status following daily oral dosing for 45 days in male and female rats. Chem-Biol Interact 180(2):281–295

Fei CY, McLaughlin JK, Lipworth L et al (2008a) Prenatal exposure to perfluorooctanoate (PFOA) and perfluorooctanesulfonate (PFOS) and maternally reported developmental milestones in infancy. Environ Health Perspect 116(10):1391–1395

Fei CY, McLaughlin JK, Tarone RE et al (2008b) Fetal growth indicators and perfluorinated chemicals: a study in the Danish national birth cohort. Am J Epidemiol 168(1):66–72

Fei CY, McLaughlin JK, Lipworth L et al (2009) Maternal levels of perfluorinated chemicals and subfecundity. Hum Reprod 24(5):1200–1205

Fei CY, McLaughlin JK, Tarone RE et al (2007) Perfluorinated chemicals and fetal growth: a study within the Danish national birth cohort. Environ Health Perspect 115(11):1677–1682

Fei CY, Olsen J (2011) Prenatal exposure to perfluorinated chemicals and behavioral or coordination problems at age 7 years. Environ Health Perspect 119(4):573–578

Feng YX, Shi ZM, Fang XM et al (2009) Perfluorononanoic acid induces apoptosis involving the Fas death receptor signaling pathway in rat testis. Toxicol Lett 190(2):224–230

Frisbee SJ, Shankar A, Knox SS et al (2010) Perfluorooctanoic acid, perfluorooctanesulfonate, and serum lipids in children and adolescents. Archives Pediat Adol Med 164(9):860–869

Fromme H, Tittlemier SA, Volkel W et al (2009) Perfluorinated compounds - Exposure assessment for the general population in western countries. Int J Hyg Environ Health 212(3):239–270

Gallo V, Leonardi G, Genser B et al (2012) Serum perfluorooctanoate (PFOA) and perfluorooctane sulfonate (PFOS) concentrations and liver function biomarkers in a population with elevated PFOA exposure. Environ Health Perspect 120(5):655–660

Gilliland FD, Mandel JS (1993) Mortality among employees of a perfluorooctanoic acid production plant. J Occup Med 35(9):950–954

Gilliland FD, Mandel JS (1996) Serum perfluorooctanoic acid and hepatic enzymes, lipoproteins, and cholesterol: A study of occupationally exposed men. Am J Ind Med 29(5):560–568

Goecke CM, Jarnot BM, Reo NV (1992) A comparative toxicological investigation of perfluorocarboxylic acids in rats by F-19 NMR-spectroscopy. Chem Res Toxicol 5(4):512–519

Goecke CM, Jarnot BM, Reo NV (1994) Effects of the peroxisome proliferator perfluoro-n-decanoic acid on hepatic gluconeogenesis and glycogenesis: A C-13 NMR investigation. Chem Res Toxicol 7(1):15–22

Gortner EG (1981) Oral teratology study of T-2998CoC in rats. Safety Evaluation Laboratory and Riker Laboratories, Inc. Study date Apr 1981

Gortner EG (1982) Oral teratology study of T-3141CoC in rabbits. Safety Evaluation Laboratory and Riker Laboratories, Inc. Study date Sept 1982

Grandjean P, Andersen EW, Budtz-Jorgensen E et al (2012) Serum vaccine antibody concentrations in children exposed to perfluorinated compounds. JAMA 307(4):391–397

Gump BB, Wu Q, Dumas AK et al (2011) Perfluorochemical (PFC) exposure in children: associations with impaired response inhibition. Environ Sci Technol 45(19):8151–8159

Halldorsson TI, Rytter D, Haug LS et al (2012) Prenatal exposure to perfluorooctanoate and risk of overweight at 20 years of age: a prospective cohort study. Environ Health Perspect 120(5):668–673

Hamm MP, Cherry N, Chan E et al (2010) Maternal exposure to perfluorinated acids and fetal growth. J Expo Sci Environ Epidemiol 20(7):589–597

Harris MW, Birnbaum LS (1989) Developmental toxicity of perfluorodecanoic acid in C57BL/6 N mice. Fund Appl Toxicol 12(3):442–448

Haug LS, Huber S, Becher G et al (2011) Characterisation of human exposure pathways to perfluorinated compounds—Comparing exposure estimates with biomarkers of exposure. Environ Int 37(4):687–693

Henderson WM, Smith MA (2007) Perfluorooctanoic acid and perfluorononanoic acid in fetal and neonatal mice following in utero exposure to 8-2 fluorotelomer alcohol. Toxicol Sci 95(2):452–461

Hines EP, White SS, Stanko JP et al (2009) Phenotypic dichotomy following developmental exposure to perfluorooctanoic acid (PFOA) in female CD-1 mice: Low doses induce elevated serum leptin and insulin, and overweight in mid-life. Mol Cell Endocrinol 304(1–2):97–105

Hirata-Koizumi M, Fujii S, Furukawa M et al (2012) Repeated dose and reproductive/developmental toxicity of perfluorooctadecanoic acid in rats. J Toxicol Sci 37(1):63–79

Hoffman K, Webster TF, Weisskopf MG et al (2010) Exposure to polyfluoroalkyl chemicals and attention deficit hyperactivity disorder in US children aged 12–15 years. Environ Health Perspect 118(12):1762–1767

Hoffman K, Webster TF, Bartell SM et al (2011) Private drinking water wells as a source of exposure to perfluorooctanoic acid (PFOA) in communities surrounding a fluoropolymer production facility. Environ Health Perspect 119(1):92–97

Houde M, Martin JW, Letcher RJ et al (2006) Biological monitoring of polyfluoroalkyl substances: a review. Environ Sci Technol 40(11):3463–3473

Hovey RC, Coder PS, Wolf JC et al (2011) Quantitative assessment of mammary gland development in female Long Evans rats following in utero exposure to atrazine. Toxicol Sci 119(2):380–390

Ishibashi H, Ishida H, Matsuoka M et al (2007) Estrogenic effects of fluorotelomer alcohols for human estrogen receptor isoforms alpha and beta in vitro. Biol Pharm Bull 30(7):1358–1359

Joensen UN, Bossi R, Leffers H et al (2009) Do perfluoroalkyl compounds impair human semen quality? Environ Health Perspect 117(6):923–927

Johansson N, Fredriksson A, Eriksson P (2008) Neonatal exposure to perfluorooctane sulfonate (PFOS) and perfluorooctanoic acid (PFOA) causes neurobehavioural defects in adult mice. Neurotoxicology 29(1):160–169

Kato K, Wong LY, Jia LT et al (2011) Trends in exposure to polyfluoroalkyl chemicals in the US population: 1999-2008. Environ Sci Technol 45(19):8037–8045

Kawashima Y, Kobayashi H, Miura H et al (1995) Characterization of hepatic responses of rat to administration of perfluorooctanoic and perfluorodecanoic acids at low-levels. Toxicology 99(3):169–178

Kennedy GL (1985) Dermal toxicity of ammonium perfluorooctanoate. Toxicol Appl Pharmacol 81(2):348–355

Kennedy GL, Hall GT, Brittelli MR et al (1986) Inhalation toxicity of ammonium perfluorooctanoate. Food Chem Toxicol 24(12):1325–1329

Kim S, Choi K, Ji K et al (2011) Trans-placental transfer of thirteen perfluorinated compounds and relations with fetal thyroid hormones. Environ Sci Technol 45(17):7465–7472
Klaunig J, Hocevar B, Kamendulis L (2012) Mode of action of perfluorooctanoic acid (PFOA) tumorigenicity and human relevance. Reprod Toxicol 33(4):410–418
Kudo N, Iwase Y, Okayachi H et al (2005) Induction of hepatic peroxisome proliferation by 8-2 telomer alcohol feeding in mice: Formation of perfluorooctanoic acid in the liver. Toxicol Sci 86(2):231–238
Kudo N, Suzuki-Nakajima E, Mitsumoto A et al (2006) Responses of the liver to perfluorinated fatty acids with different carbon chain length in male and female mice: In relation to induction of hepatomegaly, peroxisomal beta-oxidation and microsomal 1-acylglycerophosphocholine acyltransferase. Biol Pharm Bull 29(9):1952–1957
Ladics GS, Kennedy GL, O'Connor J et al (2008) 90-day oral gavage toxicity study of 8-2 fluorotelomer alcohol in rats. Drug Chem Toxicol 31(2):189–216
Ladics GS, Stadler JC, Makovec GT et al (2005) Subchronic toxicity of a fluoroalkylethanol mixture in rats. Drug Chem Toxicol 28(2):135–158
Lau C, Thibodeaux JR, Hanson RG et al (2006) Effects of perfluorooctanoic acid exposure during pregnancy in the mouse. Toxicol Sci 90(2):510–518
Leonard RC, Kreckmann KH, Sakr CJ et al (2008) Retrospective cohort mortality study of workers in a polymer production plant including a reference population of regional workers. Ann Epidemiol 18(1):15–22
Li YF, Ramdhan DH, Naito H et al (2011) Ammonium perfluorooctanoate may cause testosterone reduction by adversely affecting testis in relation to PPAR alpha. Toxicol Lett 205(3):265–272
Lin CY, Lin LY, Chiang CK et al (2010) Investigation of the associations between low-dose serum perfluorinated chemicals and liver enzymes in US adults. Am J Gastroenterol 105(6):1354–1363
Liu RCM, Hurtt ME, Cook JC et al (1996) Effect of the peroxisome proliferator, ammonium perfluorooctanoate (C8), on hepatic aromatase activity in adult male Crl:CD BR (CD) rats. Fund Appl Toxicol 30(2):220–228
Lorber M, Egeghy PP (2011) Simple intake and pharmacokinetic modeling to characterize exposure of Americans to perfluoroctanoic acid. PFOA. Environ Sci Technol 45(19):8006–8014
Loveless SE, Hoban D, Sykes G et al (2008) Evaluation of the immune system in rats and mice administered linear ammonium perfluorooctanoate. Toxicol Sci 105(1):86–96
Lundin JI, Alexander BH, Olsen GW et al (2009) Ammonium perfluorooctanoate production and occupational mortality. Epidemiology 20(6):921–928
Maras M, Vanparys C, Muylle F et al (2006) Estrogen-like properties of fluorotelomer alcohols as revealed by MCF-7 breast cancer cell proliferation. Environ Health Perspect 114(1):100–105
Melzer D, Rice N, Depledge MH et al (2010) Association between serum perfluorooctanoic acid (PFOA) and thyroid disease in the US National Health and Nutrition Examination Survey. Environ Health Perspect 118(5):686–692
Minata M, Harada KH, Karrman A et al (2010) Role of peroxisome proliferator-activated receptor-alpha in hepatobiliary injury induced by ammonium perfluorooctanoate in mouse liver. Ind Health 48(1):96–107
Mylchreest E, Ladics GS, Munley SM et al (2005a) Evaluation of the reproductive and developmental toxicity of a fluoroalkylethanol mixture. Drug Chem Toxicol 28(2):159–175
Mylchreest E, Munley SM, Kennedy GL (2005b) Evaluation of the developmental toxicity of 8-2 telomer B alcohol. Drug Chem Toxicol 28(3):315–328
Nabb DL, Szostek B, Himmelstein MW et al (2007) In vitro metabolism of 8-2 fluorotelomer alcohol: Interspecies comparisons and metabolic pathway refinement. Toxicol Sci 100(2):333–344
Nelson DL, Frazier DE, Ericson JE et al (1992) The effects of perfluorodecanoic acid (PFDA) on humoral, cellular, and innate immunity in Fischer 344 rats. Immunopharm Immunotoxicol 14(4):925–938

Nolan LA, Nolan JM, Shofer FS et al (2009) The relationship between birth weight, gestational age and perfluorooctanoic acid (PFOA)-contaminated public drinking water. Reprod Toxicol 27(3–4):231–238

Ohmori K, Kudo N, Katayama K et al (2003) Comparison of the toxicokinetics between perfluorocarboxylic acids with different carbon chain length. Toxicology 184(2–3):135–140

Okada E, Sasaki S, Saijo Y et al (2012) Prenatal exposure to perfluorinated chemicals and relationship with allergies and infectious diseases in infants. Environ Res 112:118–125

Olsen GW, Burris JM, Burlew MM et al (2000) Plasma cholecystokinin and hepatic enzymes, cholesterol and lipoproteins in ammonium perfluorooctanoate production workers. Drug Chem Toxicol 23(4):603–620

Olsen GW, Burris JM, Burlew MM et al (2003) Epidemiologic assessment of worker serum perfluorooctanesulfonate (PFOS) and perfluorooctanoate (PFOA) concentrations and medical surveillance examinations. J Occup Environ Med 45(3):260–270

Olsen GW, Burris JM, Ehresman DJ et al (2007) Half-life of serum elimination of perfluorooctanesulfonate, perfluorohexanesulfonate, and perfluorooctanoate in retired fluorochemical production workers. Environ Health Perspect 115(9):1298–1305

Olsen GW, Gillard FD, Burlew MM et al (1998) An epidemiologic investigation of reproductive hormones in men with occupational exposure to perfluorooctanoic acid. J Occup Environ Med 40(7):614–622

Olsen GW, Zobel LR (2007) Assessment of lipid, hepatic, and thyroid parameters with serum perfluorooctanoate (PFOA) concentrations in fluorochemical production workers. Int Arch Occup Environ Health 81(2):231–246

Onishchenko N, Fischer C, Ibrahim WNW et al (2011) Prenatal exposure to PFOS or PFOA alters motor function in mice in a sex-related manner. Neurotox Res 19(3):452–461

Palkar PS, Anderson CR, Ferry CH et al (2010) Effect of prenatal peroxisome proliferator-activated receptor alpha (PPAR alpha) agonism on postnatal development. Toxicology 276(1):79–84

Perkins RG, Butenhoff JL, Kennedy GL et al (2004) 13-week dietary toxicity study of ammonium perfluorooctanoate (APFO) in male rats. Drug Chem Toxicol 27(4):361–378

Qazi MR, Bogdanska J, Butenhoff JL et al (2009) High-dose, short-term exposure of mice to perfluorooctanesulfonate (PFOS) or perfluorooctanoate (PFOA) affects the number of circulating neutrophils differently, but enhances the inflammatory responses of macrophages to lipopolysaccharide (LPS) in a similar fashion. Toxicology 262(3):207–214

Raymer JH, Michael LC, Studabaker WB et al (2012) Concentrations of perfluorooctane sulfonate (PFOS) and perfluorooctanoate (PFOA) and their associations with human semen quality measurements. Reprod Toxicol 33(4):419–427

Rosen MB, Lau C, Corton JC (2009) Does exposure to perfluoroalkyl acids present a risk to human health? Toxicol Sci 111(1):1–3

Sakr CJ, Kreckmann KH, Green JW et al (2007a) Cross-sectional study of lipids and liver enzymes related to a serum biomarker of exposure (ammonium perfluorooctanoate or APFO) as part of a general health survey in a cohort of occupationally exposed workers. J Occup Environ Med 49(10):1086–1096

Sakr CJ, Leonard RC, Kreckmann KH et al (2007b) Longitudinal study of serum lipids and liver enzymes in workers with occupational exposure to ammonium perfluorooctanoate. J Occup Environ Med 49(8):872–879

Shi ZM, Feng YX, Wang JS et al (2010) Perfluorododecanoic acid-induced steroidogenic inhibition is associated with steroidogenic acute regulatory protein and reactive oxygen species in cAMP-stimulated Leydig cells. Toxicol Sci 114(2):285–294

Shi ZM, Zhang HX, Ding LN et al (2009) The effect of perfluorododecanonic acid on endocrine status, sex hormones and expression of steroidogenic genes in pubertal female rats. Reprod Toxicol 27(3–4):352–359

Shi ZM, Zhang HX, Liu Y et al (2007) Alterations in gene expression and testosterone synthesis in the testes of male rats exposed to perfluorododecanoic acid. Toxicol Sci 98(1):206–215

Son HY, Kim SH, Shin HI et al (2008a) Perfluorooctanoic acid-induced hepatic toxicity following 21-day oral exposure in mice. Arch Toxicol 82(4):239–246

Son HY, Lee S, Tak EN et al (2008b) Perfluorooctanoic acid alters T lymphocyte phenotypes and cytokine expression in mice. Environ Toxicol 24(6):580–588

Specht IO, Hougaard KS, Spano M et al (2012) Sperm DNA integrity in relation to exposure to environmental perfluoroalkyl substances - A study of spouses of pregnant women in three geographical regions. Reprod Toxicol 33(4):577–583

Steenland K, Tinker S, Shankar A et al (2009) Association of perfluorooctanoic acid and perfluorooctane sulfonate with serum lipids among adults living near a chemical plant. Am J Epidemiol 170(10):1268–1278

Suh CH, Cho NK, Lee CK et al (2011) Perfluorooctanoic acid-induced inhibition of placental prolactin-family hormone and fetal growth retardation in mice. Mol Cell Endocrinol 337(1–2):7–15

Tatum-Gibbs K, Wambaugh JF, Das KP et al (2011) Comparative pharmacokinetics of perfluorononanoic acid in rat and mouse. Toxicology 281(1–3):48–55

Tittlemier SA, Pepper K, Edwards L (2006) Concentrations of perfluorooctanesulfonamides in Canadian total diet study composite food samples collected between 1992 and 2004. J Agr Food Chem 54(21):8385–8389

Tittlemier SA, Pepper K, Seymour C et al (2007) Dietary exposure of Canadians to perfluorinated carboxylates and perfluorooctane sulfonate via consumption of meat, fish, fast foods, and food items prepared in their packaging. J Agr Food Chem 55(8):3203–3210

Toft G, Jonsson BAG, Lindh CH et al (2012) Exposure to perfluorinated compounds and human semen quality in Arctic and European populations. Hum Reprod 27(8):2532–2540

Trudel D, Horowitz L, Wormuth M et al (2008) Estimating consumer exposure to PFOS and PFOA. Risk Anal 28(2):251–269

United States Environmental Protection Agency (US EPA) (2002a) Perfluoroalkyl sulfonates; significant new rule. Fed Reg 67(236):72854–72867

United States Environmental Protection Agency (US EPA) (2002b) Perfluoroalkyl sulfonates; significant new rule. Fed Reg 67(47):11007–11013

United States Environmental Protection Agency (US EPA) (2007) Perfluoroalkyl sulfonates; significant new rule. Fed Reg 72(194):57222–57235

United States Environmental Protection Agency (US EPA) (2009) Long-chain perfluorinated chemicals (PFCs) action plan. U.S. Environmental Protection Agency. http://www.epa.gov/oppt/existingchemicals/pubs/pfcs_action_plan1230_09.pdf. Accessed on 20 Jun 2013

United States Environmental Protection Agency (US EPA) (2013) 2010/2015 PFOA Stewardship Program. http://www.epa.gov/oppt/pfoa/pubs/stewardship/index.html. Accessed on 20 Jun 2013

United States Food and Drug Administration (US FDA) (2012) Update on perfluorinated grease-proofing agents. http://www.fda.gov/Food/IngredientsPackagingLabeling/PackagingFCS/Notifications/ucm308462.htm. Accessed on 20 Jun 2013

United States Food and Drug Administration (US FDA) (2013) Inventory of effective food contact substance (FCS) notifications. http://www.fda.gov/Food/IngredientsPackagingLabeling/PackagingFCS/Notifications/ucm116567.htm. Accessed on 20 Jun 2013

Vanden Heuvel JP, Thompson JT, Frame SR et al (2006) Differential activation of nuclear receptors by perfluorinated fatty acid analogs and natural fatty acids: A comparison of human, mouse, and rat peroxisome proliferator-activated receptor-alpha, -beta, and -gamma, liver X receptor-beta, and retinoid X receptor-alpha. Toxicol Sci 92(2):476–489

Van Rafelghem MJ, Inhorn SL, Peterson RE (1987) Effects of perfluorodecanoic acid on thyroid status in rats. Toxicol Appl Pharmacol 87(3):430–439

Vestergaard S, Nielsen F, Andersson AM et al (2012) Association between perfluorinated compounds and time to pregnancy in a prospective cohort of Danish couples attempting to conceive. Hum Reprod 27(3):873–880

Vestergren R, Cousins IT (2009) Tracking the pathways of human exposure to perfluorocarboxylates. Environ Sci Technol 43(15):5565–5575

Volkel W, Genzel-Boroviczeny O, Demmelmair H et al (2008) Perfluorooctane sulphonate (PFOS) and perfluorooctanoic acid (PFOA) in human breast milk: results of a pilot study. Int J Hyg Environ Health 211(3–4):440–446

Wang JS, Zhang YT, Zhang W et al (2012) Association of perfluorooctanoic acid with HDL cholesterol and circulating miR-26b and miR-199-3p in workers of a fluorochemical plant and nearby residents. Environ Sci Technol 46(17):9274–9281

Washino N, Saijo Y, Sasaki S et al (2009) Correlations between prenatal exposure to perfluorinated chemicals and reduced fetal growth. Environ Health Perspect 117(4):660–667

Weaver YM, Ehresman DJ, Butenhoff JL et al (2010) Roles of rat renal organic anion transporters in transporting perfluorinated carboxylates with different chain lengths. Toxicol Sci 113(2):305–314

White SS, Calafat AM, Kuklenyik Z et al (2007) Gestational PFOA exposure of mice is associated with altered mammary gland development in dams and female offspring. Toxicol Sci 96(1):133–144

White SS, Kato K, Jia LT et al (2009) Effects of perfluorooctanoic acid on mouse mammary gland development and differentiation resulting from cross-foster and restricted gestational exposures. Reprod Toxicol 27(3–4):289–298

White SS, Stanko JP, Kato K et al (2011) Gestational and chronic low-dose PFOA exposures and mammary gland growth and differentiation in three generations of CD-1 mice. Environ Health Perspect 119(8):1070–1076

Wolf CJ, Fenton SE, Schmid JE et al (2007) Developmental toxicity of perfluorooctanoic acid in the CD-1 mouse after cross-foster and restricted gestational exposures. Toxicol Sci 95(2):462–473

Wolf CJ, Schmid JE, Lau C et al (2012) Activation of mouse and human peroxisome proliferator-activated receptor-alpha (PPAR alpha) by perfluoroalkyl acids (PFAAs): Further investigation of C4-C12 compounds. Reprod Toxicol 33(4):546–551

Wolf CJ, Zehr RD, Schmid JE et al (2010) Developmental effects of perfluorononanoic acid in the mouse are dependent on peroxisome proliferator-activated receptor-alpha. PPAR Res 2010:1–11

Wolf DC, Moore T, Abbott BD et al (2008) Comparative hepatic effects of perfluorooctanoic acid and WY 14,643 in PPAR-alpha knockout and wild-type mice. Toxicol Pathol 36(4):632–639

Wu KM, Farrelly JG (2006) Preclinical development of new drugs that enhance thyroid hormone metabolism and clearance: inadequacy of using rats as an animal model for predicting human risks in an IND and NDA. Am J Ther 13(2):141–144

Yang CF, Tan YS, Harkema JR et al (2009) Differential effects of peripubertal exposure to perfluorooctanoic acid on mammary gland development in C57Bl/6 and Balb/c mouse strains. Reprod Toxicol 27(3–4):299–306

Yang Q, Abedi-Valugerdi M, Xie Y et al (2002a) Potent suppression of the adaptive immune response in mice upon dietary exposure to the potent peroxisome proliferator, perfluorooctanoic acid. Int Immunopharmacol 2(2–3):389–397

Yang Q, Xie Y, Alexson SEH et al (2002b) Involvement of the peroxisome proliferator-activated receptor alpha in the immunomodulation caused by peroxisome proliferators in mice. Biochem Pharmacol 63(10):1893–1900

Yang Q, Xie Y, Eriksson AM et al (2001) Further evidence for the involvement of inhibition of cell proliferation and development in thymic and splenic atrophy induced by the peroxisome proliferator perfluoroctanoic acid in mice. Biochem Pharmacol 62(8):1133–1140

York RG (2002) Oral (gavage) two-generation (one litter per generation) reproduction study of ammonium perfluorooctanoate (APFO) in rats. Argus Research Laboratories. Study date 26 Mar 2002. http://www.epa.gov/opptintr/tsca8e/pubs/8ehq/2002/sep02/fyi_0902_01378am.pdf. Accessed on 17 Jun 2013

York RG, Kennedy GL, Olsen GW et al (2010) Male reproductive system parameters in a two-generation reproduction study of ammonium perfluorooctanoate in rats and human relevance. Toxicology 271(1–2):64–72

Zhang H, Shi Z, Liu Y et al (2008) Lipid homeostasis and oxidative stress in the liver of male rats exposed to perfluorododecanoic acid. Toxicol Appl Pharmacol 227(1):16–25

Zhao Y, Tan YS, Haslam SZ et al (2010) Perfluorooctanoic acid effects on steroid hormone and growth factor levels mediate stimulation of peripubertal mammary gland development in C57Bl/6 mice. Toxicol Sci 115(1):214–224

Chapter 8
Antimony in Food Contact Materials and Household Plastics: Uses, Exposure, and Health Risk Considerations

Suzanne M. Snedeker

Abstract Antimony is a metal that has been used clinically to treat parasitic infections, as synergist in flame retardant materials, and as a catalyst in the manufacturing of plastics. Antimony trioxide is used in the manufacturing of polyethylene terephthalate (PET), a food contact plastic used extensively for single-use water and beverage bottles and food trays. Both storage time and high temperatures are factors that increase the migration of antimony from the food contact plastics to the food or beverage. While polyvinyl chloride (PVC) plastics may be a source of antimony exposure, there is little information on levels of exposure in children from PVC toys, or in those working with recycled electronics. Levels of antimony in drinking water supplies and in food contact materials (FCMs) are regulated in Europe and the United States (U.S.). Studies on the general population of the U.S. indicate that urinary levels of antimony have decreased since 1999. Rodent studies have detected lung tumors with inhalation but not oral exposure to antimony, and increased levels of blood lipids with oral exposure. Additional animal cancer bioassays are needed to evaluate antimony trioxide's potential as a carcinogen. While peripheral artery disease and pre-eclampsia have been reported in antimony-exposed humans, more studies are needed to evaluate these findings. Since antimony trichloride has some estrogenic properties, additional studies are needed to determine if other types of antimony, including antimony trioxide, have the potential to be endocrine disruptors.

Keywords Antimony · Antimony trioxide · Pentavalent antimony · Polyethylene terephthalate · PET · Polyvinyl chloride · PVC · Metals · Food contact materials · FCM · Cancer · Endocrine disruption · Cardiovascular effects · Blood lipids · Immune function · Preeclampsia · Peripheral artery disease

---

S. M. Snedeker (✉)
Independent Consultant, Ithaca, NY, USA
e-mail: smsnedeker@gmail.com

S. M. Snedeker (ed.), *Toxicants in Food Packaging and Household Plastics*, Molecular and Integrative Toxicology, DOI: 10.1007/978-1-4471-6500-2_8,

8.1 Key Take Home Points

- Antimony trioxide is used in the manufacturing of polyethylene terephthalate (PET) , a plastic widely used in food packaging, including single-use water and beverage bottles, and reheatable food trays.
- Length of storage time and high-temperature storage conditions are two factors that increase the release of antimony from PET plastics used in food packaging.
- Antimony is used a synergist for flame retardants in circuit boards and plastic housings of electronics, but little is known about exposure in consumers or electronic waste workers.
- Biomonitoring studies in the U.S. indicate that antimony levels in urine have been decreasing in the general population.
- While there is some evidence of lung tumor induction in laboratory rat inhalation studies, more research is needed to evaluate carcinogenicity of different forms of antimony via oral routes of exposure.
- More research is needed to evaluate whether antimony trioxide used in PET food packaging has the potential to be an endocrine disruptor.
- More research is needed to evaluate whether antimony has the potential to leach from plastics in landfills to surface and drinking water supplies.

8.2 Introduction

While the metal antimony has long been used medically to treat certain parasitic infections, and ecological effects of its use in flame retardant materials are being evaluated by regulatory agencies, its use in food packaging and in other household materials has only recently begun to receive attention. This chapter will trace the widespread use of antimony, especially antimony trioxide (Sb_2O_3), in the manufacturing of certain plastics used for single-use water and beverage bottles and food trays, and emerging research identifying the factors that may enhance migration of this metal from food packaging to foods and beverages consumed by humans. Antimony's use in other household plastics, including use in children's toys and in electronics, and the need for more information on these applications, will be highlighted. Regulation by international, U.S., and European agencies in drinking water and food contact applications will be reviewed. Trends in biomonitoring via urinary excretion patterns over the last decade indicate decreasing levels of antimony, but the reasons for this decline are not known. Much more extensive research is needed on health effects associated with different types of antimony, since some of the health effects observed in animal models and in human populations are limited to certain compounds via specific routes of exposure. Health endpoints of concern include effects on tumor induction, ocular systems, blood lipids and the cardiovascular system, blood pressure and preeclampsia, and endocrine disruption effects.

8.3 Uses of Antimony

8.3.1 Pharmaceutical Use

Antimony-based compounds have been used to treat certain protozoan parasitic diseases. The potassium tartrate and sodium tartrate forms of antimony have been used extensively in the past on a global basis as antischistosomal drugs (Cooper and Harrison 2009; ATSDR 1992), and the pentavalent form of antimony, Sb(V), also has been used to treat leishmaniasis (Cooper and Harrison 2009). Leishmaniasis is caused by protozoan parasites of the Trypanosomatidae family. It has been estimated that leishmaniasis affects 2 million people on a global basis with a mortality rate in excess of 70,000 persons per year (Baiocco et al. 2009). Mechanistically, trivalent antimony, Sb(III), is known to interfere with trypanothion reductase, an enzyme that plays an important role in the survival of the trypanosomatid parasite.

8.3.2 Industrial Use of Antimony in the United States

The U.S. Geological Survey (USGS) compiles information on industrial mineral consumption in the U.S., including the use of antimony (USGS 2012). The USGS lists three major categories of industrial antimony consumption in the U.S.: (1) in alloy metal products, including use in lead batteries (antimony hardens lead), bearings, solder, cable coverings, sheet and pipe metal, and castings; (2) in non-metal products, including use in ammunition primers and fireworks, and in the manufacturing of ceramics, glass, pigments, plastics; and (3) as an oxide as a flame retardant for adhesives, plastics, rubber, paper, and pigments. The major industrial use of antimony in the U.S. was as a flame retardant. Of the 3,520 metric tons of antimony used in flame retardancy in 2011, 3,090 metric tons were used specifically in plastics. China was the major source of imported antimony oxide in 2011, accounting for over 75 % of U.S. imports (15,800 metric tons). Since the focus of this chapter is antimony in household plastics and FCMs, the remainder of this section will give examples of antimony's specific uses in these products.

8.3.3 Use of Antimony in Food Contact Plastics, Electronics, and Other Household Items

Antimony, as antimony trioxide (CAS No. 1309-64-4), is used as a catalyst in the manufacturing of plastics especially PET. PET plastic is used in the production of single-use water bottles (Shotyk and Krachler 2007; Shotyk et al. 2006; Westerhoff et al. 2008), carbonated soft-drink bottles (Tukur et al. 2012), juice containers

(Hansen et al. 2010), and other food containers for liquids (e.g., vinegar and olive oil) (Sanchez-Martinez et al. 2013), as well as PET food trays for ready-to-eat meals (Haldimann et al. 2007; Haldimann et al. 2013).

Antimony has been detected in electronic waste (e-waste) from different uses. This includes use in semi-conductor components as a flame retardant, or as a synergist with other flame retardants in circuit boards (Bi et al. 2011; Lincoln et al. 2007), and in plastic housings of electronic equipment (Santos et al. 2010). Antimony trioxide is used as a flame retardant in hard polymer plastics including acrylonitrile-butadiene-styrene plastic (ABS) commonly used for TV and computer housings (Tostar et al. 2013). Waste electric and electronic equipment (WEEE) exposure to antimony is of concern in Asia where most of the global WEEE is recycled, land-filled, or incinerated (Santos et al. 2010).

Another use of antimony in household plastic includes its use as a flame retardant in PVC covers of crib/cot mattresses (Jenkins et al. 1998a, b; McCallum 2005). While antimony has been detected in children's toys made of PVC, it is not clear if its use was as a pigment in the actual PVC or in paints used on the surface of toys (Kawamura et al. 2006). Antimony is also used in synthetic fabrics as a flame retardant, including use in protective clothing for firefighters (de Perio et al. 2010).

8.4 Regulation of Antimony

8.4.1 Drinking Water Standards for Antimony

World Health Organization A background document on antimony for the development of the *World Health Organization (WHO) Guidelines for Drinking-water Quality* was developed in 2003 (WHO 2003). Because of the lack of genotoxic and carcinogenic effects of antimony by oral routes, WHO drinking water guideline values were based on a *No Observed Adverse Effect Level* (NOAEL) of 6.0 milligrams per kilogram (mg/kg) of body weight (bd wt) per day derived from a sub-chronic 90-day drinking water study in rats. A suggested *Tolerable Daily Intake* (TDI) of 6 micrograms (μg) per kg of bd wt was based on this NOAEL and an uncertainty factor of 1,000; 10 for the use of a short-term study, and 100 to account for intra- and interspecies variation. Based on projected consumption of 10 % of antimony's TDI from drinking water, an adult bd wt of 60 kg, and consumption of 2 liters (L) of water per day, the 2011 WHO guideline for drinking-water quality for antimony was set at 0.02 mg/L in the 2011 guideline (WHO 2011).

European Standard The European Communities' drinking water standard for antimony, based on a 1998 directive, was set at 5.0 μg/L (CEC 1998) The directive document does not specify the parameters used to set this guideline.

United States Environmental Protection Agency The U.S. Environmental Protection Agency (USEPA) issued a *Maximum Contaminate Level* (MCL) of 0.006 mg/L for antimony in their *National Primary Drinking Water Standards* (USEPA 2009). Common sources of antimony contamination listed in this document include: fire retardants, ceramics, solder, electronics, and discharge from petroleum refineries. Release of antimony from degradation of household- and food contact plastics were not included as sources of drinking water contamination.

8.4.2 Regulation of Antimony as a Food Contact Substance

United States Food and Drug Administration The U.S. Food and Drug Administration (FDA) lists antimony oxide (CAS No. 1327-33-9, Doc No. 5093) in their *List of Indirect Additives Used in Food Contact Substances* (LIAUFCS) database (FDA 2011). This listing is a part of the *U.S. Code of Federal Regulations* under Title 21, Part 175 (Regnum 175.10) for indirect food additives. There is no listing for antimony trioxide in the LIAUFCS database. The FDA has estimated the *Cumulated Estimated Daily Intake* (CEDI) of antimony trioxide (CAS No. 1309-64-4) from food contact substances to be 0.0001 mg/kg bd wt per day (FDA 2012). Other types of antimony are approved for use as colorants for polymers and rubber used in making food packaging, including C.I. Pigment Yellow 53 (nickel antimony titanium yellow rutile, CAS No. 8007-18-9) and C.I. Pigment Brown 24 (chrome antimony titanium buff rutile, CAS No. 68186-90-3) (USCF 2010).

European Union (EU) The European Food Safety Authority (EFSA) is given the responsibility to conduct risk assessments on substances used as additives in FCMs, and to set *specific migration limits* (SMLs) for substances migrating from the FCM to the food or beverage. The EFSA's Scientific Panel on food additives, flavourings, processing aids and materials in contact with food (AFC) evaluated limits on the level of antimony trioxide allowed in food in 2004 (EFSA 2004). Antimony is used as an initiator (catalyst) in the manufacturing of PET and other food contact plastics at a maximum percentage (%) of 0.035 % (350 mg antimony/kg of plastic). The EFSA's SML for antimony trioxide (0.04 mg antimony/kg of food) applies to PET-packaged food and food liquids including carbonated beverages, juices, and vinegar. The migration limit was based on the antimony TDI derived for drinking water (0.006 mg antimony/kg bd wt per person per day; 0.36 mg antimony per person per day), with the assumption that up to 10 % of the TDI of antimony could come from FCMs. The panel noted that this migration limit for antimony may be exceeded if PET packaging is exposed to very high temperatures, since antimony is known to migrate from PET plastics under these conditions.

Other Regulatory Considerations It is anticipated that the draft of USEPA's *Antimony Trioxide (ATO) TSCA Chemical Risk Assessment* of will undergo peer-review by a panel of scientific experts in 2013–2014 (https://www.federalregister.

gov/articles/2013/11/08/2013-26846/antimony-trioxide-ato-tsca-chemical-risk-assessment-notice-of-public-meetings-and-opportunity-to). The primary focus of this risk assessment is to evaluate whether antimony trioxide's use as a synergist in halogenated flame retardants poses a potential risk to ecological organisms found in water, sediment, and soil (USEPA 2012). While antimony trioxide is used in the manufacturing of PET-food contact plastics, this use is regulated by other agencies, including the FDA, and hence antimony use and exposure from PET plastics is not covered in the USEPA's ecological risk assessment. However, it should be noted that the degradation of PET plastics in landfills could be a source of antimony trioxide or other forms of antimony in surface water. Future regulatory actions and risk assessments need to more fully consider antimony leachate from degrading PET-plastics since this may affect both ecological organism exposure as well as levels of antimony in human drinking water supplies.

8.5 Exposures to and Biomonitoring of Antimony

8.5.1 Oral Exposure to Antimony from Food Contact Materials

The antimony used as a catalyst in the manufacturing of PET plastics, including single-use bottles and food trays, can migrate from the plastic into the beverage or food during storage. This section will summarize the range of levels detected in published studies, the factors that affect migration (e.g., time and temperature of storage, carbonation, and color of the plastic), and the differences in the speciation of the migrating antimony.

Bottled Water While methods for detecting antimony migration from PET-plastic were published by Japanese researchers in 2004 and 2005, respectively (Kakimoto et al. 2004; Ohkado et al. 2005), the first published report of antimony migration from Canadian and European PET water bottles was published in 2006 by Shotyk and colleagues (Shotyk et al. 2006). The mean level of antimony in 12 brands of PET bottled "natural" water from Canada was reported to be 156 nanograms (ng) per L, compared to 8.2 ng/L in water bottled in polypropylene plastic, and 2.2 ng/L from pristine groundwater. Levels of antimony in PET-bottled water from Germany were substantially higher in the range of 253–546 ng/L. Additional analysis of 35 brands of PET-bottled European water from 11 countries indicated median antimony levels of 343 ng/L, with levels of 725 ng/L detected in one bottled water sample from France. Variations in antimony migration according to brand, and variation from bottle to bottle has been observed in other studies (Keresztes et al. 2009; Reimann et al. 2010). Storage time affected levels of antimony migration from PET water bottles, with levels doubling with room temperature storage over a 3 to 6 month period (Shotyk and Krachler 2007; Shotyk et al. 2006).

A review of conditions that affect migration of PET components concluded that storage temperature is a primary factor affecting the migration of antimony from PET plastics (Bach et al. 2012). Extreme climate conditions (Westerhoff et al. 2008; Greifenstein et al. 2013; Cheng et al. 2010) also can accelerate antimony leaching from PET bottles over time. Using modeling studies, researchers projected that the USEPA MCL for antimony (6 μg/L) could be exceeded if PET water bottles were stored for 176 days at 60 °C, and in only 1.3 days when stored at 85 °C (Westerhoff et al. 2008). The authors noted that temperatures in Arizona in enclosed spaces such as garages and cars regularly exceed 65 °C.

Bottled water is consumed by U.S. military personnel stationed in the Middle East, and in some cases bottled water may serve as the sole source of water for duration of the deployment (Greifenstein et al. 2013). To reduce transportation costs for bottled water, the military developed packaged water systems to bottle purified water on site, with PET bottles used to store this military packaged water (MPW). Studies were conducted to determine if extreme storage and temperature conditions could affect the quality of the MPW. Antimony was detected in MPW bottled in Afghanistan that was stored under laboratory conditions at 60 °C, with a maximum concentration of 3.6 μg/L detected after 28 days of storage; this level is below both military and USEPA MCL of 6 μg/L. In contrast to prediction models (Westerhoff et al. 2008), with increased storage time (62–120 days) levels of antimony decreased over time in the MPW. The authors suggested that calcium, which was not present in the MPW, and is known to influence antimony migration in bottled water, may partially explain why antimony levels in MPW differed from Westerhoff et al.'s prediction models which were based on calcium-containing bottled water stored at 60 °C. More studies are needed under field conditions to determine changes in the levels of antimony in MPW and other types of PET-bottled water under extreme climate conditions/elevated temperatures and prolonged storage conditions.

Other researchers have investigated how boiling, microwaving, or storing PET-water bottles in a hot car can affect antimony migration (Cheng et al. 2010). While all of these extreme conditions resulted in single high values of antimony, there were a wide range of values for the six samples tested for each condition (boiling, range 2.08–7.99 μg/L; microwaving, range 0.261–10.31 μg/L; in-car storage, ND -1.64 μg/L). Studies conducted in France (Bach et al. 2013), Hungary (Keresztes et al. 2009), and Britain (Tukur et al. 2012) support the findings that length of storage time and/or higher storage temperature enhance antimony migration from PET water bottles.

Carbonation appears to affect antimony migration from PET water bottles. Several researchers have reported higher levels of antimony migration in carbonated compared to non-carbonated mineral water (Keresztes et al. 2009) or to ultra pure water, especially at higher storage temperatures (60 °C) (Bach et al. 2013).

In bottle water samples obtained from Boston supermarkets, different types of plastics, levels of carbonation, flavor additives, and storage time affected the levels of antimony leaching into the bottled water samples (Andra et al. 2012). Antimony levels increased with storage time (up to 60 days) at room temperature (23 °C) and tended to be highest in non-carbonated and enriched (NCRE) > carbonated (CR) > non-carbonated (NC) bottled water. The higher antimony levels in NCRE water may be due to the type of plastic used rather than additives in the enriched water, since most of the NCRE water bottles were made of PET plastic, while the NC water bottles were fabricated from a variety of plastics. When only the bottle's plastic type was considered, PET plastic bottles showed the greatest level antimony migration with time, with mean levels increasing from 350 to 400 ng/L from day 1 to day 60 of storage, respectively (levels estimated from Fig. 3 in Andra et al. 2012). There were negligible levels of antimony detected in water samples from polystyrene (PS) or polycarbonate (PC) water bottles (close to null) at both day 1 and day 60; neither type of plastic uses antimony trioxide as a catalyst. The low levels of antimony detected in water from high-density polyethylene (HDPE) bottles (less than 25 ng/L by day 60) may have been due contamination of the HDPE plastics with recycled PET plastic (Andra et al. 2012). Other researchers have reported low levels of antimony migration in non-PET plastics, including PC, PS, HDPE, PVC, polypropylene (PP), and low-density polyethylene (LDPE) (Cheng et al. 2010), and in water samples from glass bottles (Reimann et al. 2010).

While some researchers have noted that in bottled water, antimony levels tend to be higher in PET bottles compared to glass bottles, but in one study of carbonated soft drinks, antimony levels were similar in both glass and PET bottles (Tukur et al. 2012). It is not known whether the antimony detected in carbonated soft drinks is due to antimony leaching from the bottle, or from the actual levels in the drink product. It is conceivable that antimony in the final product also could be due to contamination from PET used in the manufacturing process, or that other ingredients in the carbonated beverages affect factors like pH and oxidation which may affect antinomy migration from PET. It has been suggested that antimony detected in Tetra Pak packaging of juices may be due to a contaminated ingredient or from production equipment (Hansen et al. 2010). In an analysis of antimony concentrations in 41 juices (diluted appropriately) from PET, glass, and Tetra Pak packaging, a wide range of antimony levels were reported, with six of the PET-packaged juices exceeding 5 μg/L (the upper limit for the EU drinking water standard for antimony), and one glass-packaged sample (red cherry juice) exceeding 13 μg/L. It should be noted, however, that it is more appropriate to use SML for antimony in food (40 μg/kg) rather than drinking water guideline levels in interpreting levels of antimony in juice products (Vasami 2010). This is because the water guidelines assume a daily higher consumption of water (2 L), while juice usually makes up a smaller percentage of liquid intake. However, in some at-risk populations, such as children, juice consumption may make up a significant portion of liquid intake, therefore careful consideration needs to be given to evaluating the potential exposure to antimony from juices and other liquids packaged in PET in vulnerable sub-populations. More research is needed to determine influence of manufacturing process, packaging composition, additive ingredients and factors affecting pH or oxidation on the migration of antimony from FCMs.

There is some, but inconsistent evidence, that the color of single-use PET bottles may affect antimony migration. Antimony leaching was reported as four-fold higher from blue-tinted compared to clear PET bottles in one study (Westerhoff et al. 2008), while others reported similar levels of antimony leaching from blue, green or clear PET bottles (Tukur et al. 2012), or have found a range of levels within different colors of PET and glass bottles (Reimann et al. 2010).

Food Trays Because of PET's thermal stability, it is often the choice for packaging frozen food, and hence PET food trays and roasting bags are used for cooking and/or reheating ready-to-eat food in both conventional and microwave ovens. Studies of antimony's ability to leach from PET-food containers suggests that much higher levels can migrate from PET-food trays and roasting bags compared to levels known to leach from PET water bottles (Haldimann et al. 2007). The EFSA SML for antimony from FCM is 0.04 mg antimony/kg of food (40 μg antimony/kg food) (EFSA 2004). Many of the foods prepared in PET containers that were heated to 180 °C exceeded the migration limit for antimony, including dough for pies (mean levels in different brands 111–241 μg/kg), pork (mean level 41.9 μg/kg), and sauerkraut (mean 88.3 μg/kg) heated in PET-roasting bags. Ready-to-eat foods microwaved in PET trays tended to have less antimony migration, with mean antimony levels in the range of 9.7–30.4 μg/kg (median level 17.7 μg/kg). Levels of antimony in uncooked pasta meals (lasagna, pasta gratin, cannelloni, pasta with meat) packaged in PET trays increased four to six fold after they were cooked in a conventional oven, with levels ranging from 9.4–14.9 μg/kg (Haldimann et al. 2013). These results support the general finding that higher cooking temperatures can accelerate the migration of antimony from PET packaging to foodstuffs.

Food Simulant Modeling Studies Using food-simulants, migration models have been developed to predict concentrations of antimony in foods cooked in PET FCMs (Sanchez-Martinez et al. 2013; Welle and Franz 2011; Haldimann et al. 2013). It has been suggested that migration modeling could be used to estimate consumer exposure to antimony (Haldimann et al. 2013). Others have developed migration models based on the activation energy of diffusion for antimony from PET bottles into food simulants/beverages as well as different surface to volume ratios and PET bottle wall concentrations of antimony (Welle and Franz 2011). These researchers from the Fraunhofer Institute for Process Engineering and Packaging in Germany concluded that the migration of antimony from PET bottles cannot reach or exceed the EU SML of 40 μg/kg, and predicted that with room temperature storage for 3 years, antimony levels would not be higher than 2.5 μg/kg, which is well below drinking water limits set for both the U.S. and Europe.

In a study using EU-approved food simulants, antimony migration from PET containers into distilled water, 3 % acetic acid, and 10 % and 20 % ethanol was measured (Sanchez-Martinez et al. 2013). Migration values ranged from 0.5 to 1.2 μg/L, with pentavalent antimony being the only species found in aqueous liquids. Antimony trioxide, the form used as catalyst in PET manufacture, was not detected in the liquid simulants. The authors suggest it is possible that the trioxide form may be oxidized to the pentavalent form of antimony during the polycondensation reactions of PET, and hence be the predominant form that migrates from PET bottles (Sanchez-Martinez et al. 2013). Other researchers also have observed that pentavalent antimony is the predominant form in PET manufactured in Japan in China (Takahashi et al. 2008). In addition to determining the speciation of antimony in food simulants, Sanchez-Marinez and colleagues concluded that the results of their food simulant modeling studies predict that antimony migration from PET would not reach European migration limits for antimony, nor exceed European drinking water standards. However, their storage temperature studies were limited to measuring antimony migration for 10 days at 40 °C, and no data was presented for longer storage times or higher temperatures. Therefore, future studies using food simulants need to include longer storage times and a wider range of temperatures that can better approximate extended storage and extreme climate conditions.

8.5.2 Inhalation and Dermal Exposure to Antimony from Household Products

Inhalation While antimony has been used as a flame retardant in PVC covers of cot/children's mattresses, there is little evidence that the antimony is released from these materials under normal conditions (Jenkins et al. 1998b). There had been concern that contamination of mattresses with the aerobic fungus *Scopulariopsis brevicaulis* could result in the methylation of antimony trioxide and cause the release of trimethyl antimony gas. In testing this hypothesis, biovolatilization was observed only under simulated cot conditions with extreme heat (at 80 or 100 °C), and antimony was not bioavailable from the PVC cot material under room temperature conditions.

Dermal Since a variety of toxic metals have been detected in children's toys, including those made of PVC plastic, more research is needed to determine if toys may be a source of antimony exposure, and if typical hand-to-mouth behavior in children can result in dermal and/or oral exposure to antimony. Analytical tests have used extreme conditions (grinding followed by extraction using acids) to estimate migration of antimony in toys (Kawamura et al. 2006). These methods may not accurately estimate the transfer of antimony from toy surfaces to the hand or mouth. While antimony is used as a synergist for flame retardants in ABS plastics, antimony's migration through the plastic as it ages and the potential for its release from the plastic over the lifetime of the product has not been evaluated. Similarly, while antimony is used in circuit boards of electronic products, the extent to which there is non-occupational exposure to antimony from electronics from dermal contact is not known. More research is needed to examine how processing e-waste, especially in Asia, may lead to occupational and environmental exposure to antimony. Antimony used in flame retardant fabrics for fire fighter clothing (turn-out pants) is not a source of occupational dermal exposure, since there were no differences in urinary levels of fire fighters using or not using antimony-containing clothing in a health assessment conducted by Centers for Disease Control and Prevention (CDC) investigators (de Perio et al. 2010).

8.5.3 Antimony Leaching from Landfills

There is limited information on the leaching of antimony in landfills in the U.S., and when detected as leachate, the source is usually unidentified. The Kim-Stan landfill in Selma, VA, is an USEPA designated Superfund Site that served as a sanitary landfill for over 20 years. Antimony was one of the substances identified by the USEPA in the leachate, but no information was provided on linkages to waste items in the landfill (USEPA 2013). To what extent antimony can leach from the PET bottles or from ABS plastics found in landfills and subsequently migrate to groundwater or drinking water supplies is largely uncharacterized. Given the large volume of single-use PET beverage bottles disposal globally, as well as the increasing volume of e-waste in China, antimony in landfill and e-waste leachate needs to be characterized. More information is needed on the speciation of antimony leachate (trioxide vs. pentavalent forms), since toxicity of antimony is dependent on its species. Dissolved antimony in aquatic environments in China was primarily the pentavalent form, while the trioxide form was a minor fraction (Wu et al. 2011b).

8.5.4 Biomonitoring of Antimony in Humans

Urine Reference ranges for levels of antimony in urine for U.S. residents were first published in 1998 by the National Center for Environmental Health, CDC (Paschal et al. 1998). Antimony was detected in 73.5 % of the 496 urine samples at a mean level of 0.74 μg/L, 0.67 μg/grams (g) of creatinine (geometric mean, all sexes, all ages). Since then, urinary excretion of antimony of the general U.S. population has been monitored as a part of the CDC's National Biomonitoring Program through the National Health and Nutrition Examination Surveys (NHANES). The Fourth Report of the CDC National Biomonitoring Program includes updated summary tables for NHANES 1999–2000, 2001–02, 2002–03, 2003–04 (incomplete for these years), 2005–06, and 2007–08 (CDC 2012). These levels are lower than the first values reported by the CDC in 1998, and for all age groups in both males and females, mean levels of antimony in urine have been falling, from 0.132 μg/L in 1999–00 to 0.061 μg/L in 2007–08. In the 2007–08 NHANES survey, approximately 25 % of the urine samples analyzed were below than the limits of detection (LOD) for antimony (Yorita Christensen 2012), which is consistent with the percentage of subjects with levels lower than the LOD in earlier reports (Paschal et al. 1998).

Other researchers have analyzed 2003–2010 NHANES data for trends, including whether reproductive status, race/ethnicity, smoking or fish/shellfish consumption, or body mass index (BMI) affected urinary antimony excretion patterns in women aged 17 to 39 years (Jain 2013). They reported an overall decline in urinary antimony levels of women for the 2003–10 survey years. While levels rose with parity, urinary antimony excretion was not significantly different in pregnant compared to non-pregnant women. Levels tended to be moderately though significantly higher in Mexican American compared to Non-Hispanic white women. Urinary antimony levels were significantly higher in smokers compared to non-smokers. Neither fish/shellfish consumption nor BMI affected urinary antimony excretion patterns.

Another factor that has been examined in relation to urinary antimony levels is socioeconomic status (SES). Tyrrell and colleagues reported higher urinary antimony levels in those with lower SES (as measured by the Poverty Index Ratio) in NHANES 2001–2010 adult participants who were 18–74 years of age (Tyrrell et al. 2013). Unlike other metals evaluated in this study, the authors could not find any consistent mediator (e.g., shellfish/fish consumption, tobacco use, occupation) that would explain why urinary antimony levels varied by SES.

In European children, urinary antimony levels were not associated with postnatal age in term infants from the United Kingdom in a study conducted in the late 1990s, though antimony levels tended to be higher in pre-term infants compared to urinary levels in older full-term infants (Dezateux et al. 1997). It was not clear why prematurity increased urinary excretion levels of antimony. There was no relationship between urinary antimony and cotinine levels in infants regardless of postnatal age, suggesting that passive smoke exposure did not affect urinary antimony excretion patterns. In Germany, reference values for antimony in urine have been derived from the 2003–06 German Environmental Survey on Children (GerESIV) (Schulz et al. 2009). Based on urine samples from 1,729 children, the reference value for antimony in urine of German children aged 3 to 14 years old was set at 0.3 μg/L (Schulz et al. 2009). This German reference level is similar to the 95th percentile for urinary antimony levels reported for U.S. children ($n = 290$) aged 6 to 11 years in the 2003–04 (0.31 μg/L) NHANES survey (CDC 2012).

Given that one potential source of oral exposure to antimony may be from antimony leaching from PET single-use beverage bottles and other PET food contact plastics, one research group investigated whether drinking water sources and bottled water consumption influenced urinary antimony levels (Makris et al. 2013). This study was limited by the use of a spot urine sample rather than a 24-hour (h) urine collection, and low power with a small sample size of 35 subjects. About 65 % of the water consumption was consumed directly from bottled and tap-water sources, with the remaining 35 % of water consumption from coffee, tea and juices. While the authors reported a significant positive association between "per-capita" water consumption from bottled sources and creatinine-adjusted urinary antimony levels, this effect did not hold when other covariates were entered into their multivariate regression model. A study with a larger sample size is needed to more accurately assess whether PET-bottled water consumption affects levels of urinary antimony in different populations.

Breast Milk Little information has been published on levels of antimony in human breast milk. Antimony was detected in the range of <1.0–49.6 ng/g from 130 samples of breast milk obtained from a small survey of Italian women (21 subjects) (Clemente et al. 1982). No other reports of antimony levels in breast milk were located in the published literature.

Hair To what extent hair levels of antimony are a sensitive measure of exposure or body burden is debatable, but results seem to suggest that urinary levels are a more reliable and sensitive index for antimony biomonitoring than blood or hair levels (Gebel et al. 1998). A method development study reported average hair antimony levels in the range of 0.095–0.226 μg/g based on a small number of samples (Rahman et al. 2000). In areas where antimony exposures are extremely high, such as Xikuangshan, China where some of the largest antimony mines are located, elevated levels of dietary and hair antimony have been reported, with resident's hair levels concentrations averaging 15.7 μg/g dry weight (Wu et al. 2011a). However, the authors point out that hair antimony levels can vary according to sources of external contamination, such as dust and whether contaminated water is used to wash hair. This poses a sampling and analytical problem, since methods to remove external contamination may also affect internal levels of hair antimony. In a health hazard evaluation, investigators from the CDC were asked to investigate a potential toxic exposure to antimony in firefighters that had worn protective clothing containing antimony trioxide and trichloride as flame retardants (de Perio et al. 2010). While a previous chemical analysis of hair samples had reported elevations in hair antimony levels, the CDC risk evaluation could find no difference between urinary levels of antimony in firefighters from departments that did, compared to those that did not use antimony-containing protective clothing. The CDC authors noted that the shortcomings of hair analysis of antimony, stating that urinary levels are the most reliable means for testing for levels of antimony in the human body.

8.6 Health Effects of Antimony

8.6.1 General Toxicity

One of the first chronic (lifetime) toxicity studies on antimony administered 10 parts per million (ppm) antimony as its potassium tartrate salt in the drinking water to male and female Charles River CD mice (Schroeder et al. 1968). Early signs of toxicity included significantly lower mean body weights in treated animals of both sexes compared to controls. Antimony tended to accumulate in the spleen, lung, and kidney, and to a lesser extent in the heart and liver of treated animals, but no specific pathological lesions were identified. In similar lifetime drinking water studies conducted in male and female Long Evans rats, administration of 5 ppm antimony (as potassium antimony tartrate), toxic effects noted included a much shorter lifespan in treated animals, and a suppression of serum glucose levels in non-fasting animals (Schroeder et al. 1970). The antimony-induced decreased longevity was not observed in the chronic drinking water study conducted in mice (Schroeder et al. 1968).

Because of concerns of possible adverse health effects in humans treated with antimony potassium tartrate as an antischistosomal drug, the National Institute of Environmental Health Sciences (NIEHS) conducted preliminary pre-chronic toxicity dosing studies (14- and 90-day studies) using oral (via drinking water) and injection routes in $B6C3F_1$ mice and F344 rats to identify target organs, tissue distribution patterns, and appropriate dosing routes for future studies (Dieter et al. 1991). When administered orally via drinking water, antimony was poorly absorbed and relatively non-toxic compared to the increased mortality, body weight changes, and organ effects seen with administration via intraperitoneal (ip) injection. While treated mice did not exhibit signs of toxicity, antimony did accumulate in both the liver and spleen. In rats, there was a gender-related increased mortality in treated high-dose male but not female rats. While antimony accumulated in the kidney, heart, spleen, liver and the blood of treated rats, the liver was the primary site of antimony's toxic effects. Liver necrosis and heptocellular degeneration were associated with dose-related elevations in the activity of serum levels of the liver enzymes sorbitol dehydrogenase and alanine aminotransferease. There was no evidence of cardiac toxicity and little evidence of renal toxicity in the rats exposed to antimony. The authors concluded that for future chronic studies, the F344 rat would be an appropriate model and that ip administration would be a suitable substitute for the intravenous administration of antimony potassium tartrate used clinically in humans.

In a 90-day subchronic feeding study, the toxicity of antimony trioxide was assessed in male and female Wistar rats (doses: 0, 1,000, 5,000 and 20,000 ppm) (Hext et al. 1999). While there were no histological effects on the liver, in the high-dose animals there were small but statistically significant decreases in plasma alkaline phosphatase in treated male and female rats, and elevated aspartate aminotransferase activity in the high-dose females. In contrast to the NIEHS subchronic ip injection study, there were no antimony trioxide treatment-related effects on the activity of plasma alanine aminotransferase observed in either gender in this oral dosing study. There was a small, but statistically significant increase in the plasma cholesterol levels in high-dose females, and elevated triglyceride levels in high-dose male rats compared to control animals. The authors suggest the changes in plasma liver enzyme activities and blood lipids may be suggestive of minor liver changes in antimony-treated animals.

The authors noted that there were no adverse ocular effects in this oral subchronic antimony trioxide study (Hext et al. 1999). This is in contrast to increased incidence of cataracts in male and female F344 rats exposed to antimony trioxide in a subchronic (13 week exposure, 27 weeks observation) and chronic inhalation study (12 months exposure, 12 months observation) (Newton et al. 1994). Ocular irregularities had been observed as early as 2 weeks after exposure to antimony trioxide dust in the subchronic 90-day inhalation study conducted by Newton and colleagues. Neither the subchronic nor the chronic inhalation study reported any treatment-related effects on blood chemistries, with no reported effects on hemoglobin, hematocrit, serum liver enzyme activities or fasting blood glucose levels (Newton et al. 1994). While lung inflammation and alveolar macrophages were associated with antimony trioxide exposure via inhalation, other than the ocular effects, there were no other clinical effects or histological effects that were related to antimony treatment.

Collectively, these studies suggest that species, route of exposure, and to some extent gender, affect the toxicological responses to antimony administration in rodents.

8.6.2 Carcinogenicity and Genotoxicity

The International Agency for Research on Cancer (IARC) published an assessment of the carcinogenic risk of antimony trioxide in 1989 (IARC 1989). Because occupational exposure to antimony includes inhalation during mining and smelting of the ore, and from the production of pigments, initial rodents cancer bioassays focused on cancer risks via an inhalation route. Antimony trioxide was rated as a 2B carcinogen by IARC (*possibly carcinogenic to humans*) based on the induction of lung tumors in laboratory rats exposed via inhalation. This evidence includes a cancer bioassay demonstrating a significant increase in the incidence of lung neoplasms in adult female, but not male, Wistar rats exposed to antimony trioxide or antimony ore via inhalation (Groth et al. 1986). Dose-response for this study could not be evaluated because only one dose of antimony trioxide was tested. The Groth study's duration (52 weeks of exposure and termination 20 weeks later) was shorter than the standard 104 weeks of exposure for a rodent cancer bioassay.

In contrast to the Groth study, a chronic inhalation study (52 weeks exposure followed by 27 weeks of observation, 0 ppm and three treatment groups) in male and female F344 rats exposed to antimony trioxide as dust failed to observed treatment-related lung neoplasms, though lung inflammation and other clinical lung-related problems were observed in antimony-treated animals (Newton et al. 1994). However, the highest dose of antimony tested in the Newton et al. chronic study was 5 mg per cubic meter (m^3), which was nine-fold lower than the average level of antimony trioxide exposure in the Groth chronic inhalation cancer study (45 mg/m^3). Cancer bioassay inhalation studies on antimony trioxide in mice are being conducted by the National Toxicology Program (NTP) (HCN 2011).

Studies evaluating the cancer risk of antimony via an oral route are limited to lifetime drinking water studies of antimony potassium tartrate conducted in Long Evans rats (Schroeder et al. 1970) and Charles River CD mice (Schroeder et al. 1968). In both studies, a single-dose of antimony was administered during the animal's lifetime at 10 ppm in mice, and 5 ppm in rats, respectively. While some "spontaneous" malignant tumors were noted in the antimony-treated mice, specific incidence information according to tumor site or type was not provided, though the authors noted that lung tumors appeared to be the most prevalent malignancy. In the rat drinking water study, there was no evidence of antimony treatment-related tumorigenicity. Additional studies are needed that evaluate the oral carcinogenicity of antimony over broader dose ranges to assess dose response, as well as to determine if speciation (tri-verses pentavalent antimony) affects carcinogenicity.

Most of the available human studies on the cancer risk of antimony have evaluated cancer mortality rates in occupationally exposed populations, including miners and semi-conductor workers. There is some indication of a significantly higher rate of mortality from lung cancer in male antimony smelter workers from Texas [Standard Mortality Ratio (SMR) 1.39, 90 % Confidence Interval (CI) 1.08–1.88] (Schnorr et al. 1995). Lung cancer risk was positively associated with length of antimony exposure, with risk increasing with 20 or more years of exposure in antimony smelter workers (Jones 1994). Studies of other occupations exposed to antimony, including the semi-conductor industry, have not evaluated mortality from cancer or other aspects of cancer risk in workers (Fowler et al. 1993). It has been noted that cancer risk of antimony occupational exposure is difficult to fully evaluate since co-exposure to arsenic may occur (De Boeck et al. 2003). In general, there is not sufficient evidence to conclude whether or not occupational exposure to antimony affects cancer risk (Hayes 1997).

The genotoxicity of antimony is affected by speciation and the test system. Antimony trioxide and antimony trichloride are not mutagenic to *Salmonella* bacteria (Kuroda et al. 1991; Elliott et al. 1998), while there is evidence of DNA damage the in the Sister Chromatid Exchange (SCE) assay (Kuroda et al. 1991); micronucleus assay using isolated human peripheral lymphocytes (Elliott et al. 1998; Schaumloffel and Gebel 1998), Chinese hamster ovary cells, or human fibroblasts (Huang et al. 1998); and in a modified comet assay using blood samples from antimony trioxide-exposed workers (Cavallo et al. 2002). While some authors suggest that trivalent antimony-induced oxidative damage may partly explain the positive genotoxic effects in the mammalian cell genotoxicity tests (Cavallo et al. 2002), others suggest that antimony may instead primarily affect DNA-repair mechanisms (Grosskopf et al. 2010). Mutagenicity of elemental antimony (99 % purity, form called "metallic") has been demonstrated in strains of *Salmonella* and tested positive in chromosomal aberration tests (Asakura et al. 2009), while other studies indicate a lack of evidence of DNA damage in mammalian cells lines for pentavalent antimony (De Boeck et al. 2003).

In contrast to in vitro tests, most in vivo assays in rodents have failed to show evidence of antimony's genotoxicity. Antimony trioxide tested negative (no evidence of clastogenicity) in the mouse bone marrow micronucleus assay and in the rat liver DNA repair assay (Elliott et al. 1998), and failed to induce micronuclei or chromosomal aberrations in the bone marrow of rats in a subchronic dosing study (Kirkland et al. 2007).

These results support the conclusions drawn by other scientists that while trioxide and trichloride forms of antimony do not appear to be mutagenic in bacterial cells or show evidence of DNA damage in rodent in vivo studies, there is some evidence of genotoxicity in cultured mammalian cell lines for trivalent, but not pentavalent forms of antimony (De Boeck et al. 2003).

8.6.3 Glucose Metabolism, Obesity and Cardiovascular Effects

Glucose Metabolism and Glycolytic Pathways One of the early long-term bioassays evaluated the general toxicity of antimony as potassium antimony tartrate in drinking water (at 5 ppm) in rats (Schroeder et al. 1970). Both fasting and non-fasting levels of glucose were lower in antimony-treated animals compared to controls, and in addition, in the antimony-treated animals the non-fasting levels of glucose were significantly lower than fasting levels. There was no explanation given for this metabolic change.

There is some limited evidence that antimony can affect glycolysis pathways. In vitro studies have demonstrated antimony's ability to inhibit the enzyme phosphofructokinase, an enzyme that affects erythrocyte glycolysis (Poon and Chu 2000). Previous in vivo studies by the same research group had observed increased levels of antimony in the erythrocytes of rats treated with potassium antimony tartrate for 90 days in drinking water compared to untreated controls (Poon et al. 1998). In humans, there are no published studies available that have examined incidence of Type II diabetes and urinary levels of antimony in NHANES survey participants.

Obesity In an analysis of NHANES aggregate data from 1999–2002, and 2001–02, there was no association between urinary levels of antimony with either waist circumference or BMI, suggesting a lack of an obesogenic effect (Padilla et al. 2010).

Cardiovascular Effects Analysis of NHANES 1999–2000 survey data from participants aged 40 years or more identified a positive association of urinary antimony levels and peripheral arterial disease (Navas-Acien et al. 2005). The authors cautioned that this should be considered a preliminary exploratory finding, and more research is needed to explore and define this possible association. Other analyses of NHANES data (1999–2006) have examined whether metals may affect the risk of cardiovascular disease (Agarwal et al. 2011). In this study, urinary antimony levels were significantly positively associated with cardiovascular and cerebrovascular disease; adjusted Odds Ratio (OR) 2.15 (95 % CI: 1.45–3.18). This OR was the second highest (highest was for urinary cadmium) of the 13 metals evaluated.

Cardiovascular problems have been observed clinically in leishmaniasis patients treated with pentavalent antimony (Sundar and Chakravarty 2010). Similar cardiomyopathy effects have been observed in a guinea pig model (Alvarez et al. 2005), and elevated serum cholesterol levels have been reported in long-term drinking water studies of antimony-treated rats (Schroeder et al. 1970).

In a clinical report based on cardiac biopsy samples obtained from surgery patients, antimony levels were found to be greatly elevated (over 10,000 times) in myocardial compared to non-cardiac muscle samples from patients with idiopathic dilated cardiomyopathy (Frustaci et al. 1999). Other studies have shown that antimony may effect cardiac electrolytes and cardiac calcium currents (Kuryshev et al. 2006). More studies are needed to define the etiologic role antimony may play in cardiovascular disease and associated metabolic pathways, and if those with higher exposure to antimony are more at risk for diseases of the cardiovascular system.

8.6.4 Reproductive and Developmental Outcomes

In an analysis of NHANES data, levels of antimony in the urine did not differ appreciably between pregnant and non-pregnant adult women from the U.S. general population (Jain 2013). In a study monitoring urinary antimony levels in 78 pregnant women from Japan, there was no association with urinary levels of this metal and birth outcomes of their infants, including birth weight, birth length and head circumference (Shirai et al. 2010).

There is limited evidence from a single study that blood levels of antimony are associated with preeclampsia (Vigeh et al. 2006). Levels of antimony in whole blood were significantly higher in the 31 women with a diagnosis of preeclampsia (mean 3.69 μg per deciliter (dL)) compared to 365 controls (mean 3.19 μg/dL), and differences were even greater in between antimony levels in umbilical cord blood in preeclampsia (mean 4.16 μg/dL) compared to cord blood from controls (3.17 μg/dL). Levels of other blood metals (lead and manganese) also were higher in women with preeclampsia. This study is observational and does not provide evidence that antimony exposure is causal for preeclampsia. However, since some animal models suggest that antimony trichloride treatment may affect renal function and blood pressure responses in rats (Rossi et al. 1987), more research is needed to evaluate if antimony alone or in conjunction with other heavy or trace metals, affects the risk of preeclampsia in humans (Vigeh et al. 2006). Additionally, 24-h urine samples should be used to evaluate body burdens of antimony in preeclamptic compared to non-preeclamptic pregnant women, since urinary levels of antimony may provide a more accurate estimate of exposure than blood levels.

8.6.5 Dermis

Allergic contact dermatitis associated with exposure to different forms of antimony are limited to several case reports in the literature of occupationally exposed workers in the ceramics (Motolese et al. 1993) and metal smelting industries (White et al. 1993).

8.6.6 Immune System

A small study of workers directly exposed to antimony trioxide as a powder or fumes examined whether immune function was affected by occupational exposure to antimony (Kim et al. 1999). This study's results suggested several indices of immune function, including IgG1, IgE and interferon-gamma levels in serum, were all depressed in antimony-exposed workers compared to workers in the same facility who did not work with antimony, while IgG4 levels were positively related to urinary antimony levels in exposed workers. To what extent there may have been co-exposures to other workplace chemicals is not known. It is not known whether non-occupational antimony exposure in the general population affects immune function.

8.6.7 Endocrine Disruption

While there is some limited evidence that certain types of antimony may affect some endocrine steroid-receptor pathways, there is a lack of information on whether other types of antimony, especially those used in PET manufacture or antimony found in PET bottled water affect estrogenic, androgenic or thyroid-receptor related pathways. For instance, antimony chloride has tested positive for estrogenicity in both the MCF-7 estrogen-receptor (ER) positive breast tumor cell-based assay (*E-screen* assay) and in a estrogen receptor transcriptional expression assay (Choe et al. 2003). The relative efficiency of 1 μM antimony chloride in the ER-transcription assay was 60.9 % compared to 1 nanomolar (nM) 17β-estradiol, and its relative potency (EC_{50}) in the E-screen assay was 3.33×10^{-8} M compared to 17β-estradiol at 1.43×10^{-10} M. Based on this data, others have suggested that antimony migrating from PET plastic used packaging bottled mineral water may be an endocrine disruptor (Sax 2010). These conclusions also were based on studies that have evaluated the estrogenicity (and observed some evidence of) of PET-bottled water, though the components responsible for estrogenic effects were not identified. What is absent from this discussion are data from both cell-based, transcriptional based, and in vivo assays evaluating the estrogenicity of the two most relevant compounds related to antimony use and migration from PET

FCM plastic; antimony trioxide used in the manufacturing of PET plastic and pentavalent antimony, the migratory species of antimony commonly detected in PET bottled water. Similarly, in both the studies cited by Choe's paper, and in more recent studies evaluating endocrine disrupting effects of water bottled in PET plastic (Wagner and Oehlmann 2011), additional data is needed to: 1) identify the chemicals in PET-bottled water responsible for the estrogenic effects, and 2) determine if there are mixture effects, where low levels of several estrogenic components may act additively to illicit an estrogenic response. In contrast, one study has not observed estrogenic (transcriptional activation assay using a HepG2 cell line) or anti-androgenic effects (transcriptional activation assay using the human MDA-MB453-kb2 cell line) from intentional or unintentional additives, including antimony, found in PET-bottled mineral water extracts (10 days at 60 °C) (Bach et al. 2013). These authors noted endocrine-disruption effects observed by other studies, and suggested that differences in endocrine disruption assay systems, storage conditions, extraction procedures, and sampling methods may explain the divergent results.

The only other endocrine disrupting effect that has been evaluated related to antimony in the human population has been an analysis of urinary antimony levels in relation to certain serum thyroid hormones in the U.S. general population as assessed through the NHANES 2007–08 survey (Yorita Christensen 2012). There were no reported significant associations between concentrations of urinary antimony and levels of serum free and total T_3 and T_4 or with serum thyroid stimulating hormone (TSH).

8.7 Conclusions

Antimony trioxide has widespread application in the manufacturing of PET plastics used extensively for single-use water and beverage bottles and PET food trays. There is emerging evidence that antimony can migrate from these food contact plastics to food and beverages consumed by humans, especially during long-term storage and under extreme heat and climatic conditions. Little is known about potential exposure from other sources of household plastics, including PVC used in children's toys and electronics, and use for flame retardancy in circuit boards. Because of these potential exposures, more research is needed to model and assess health effects of antimony trioxide exposure, especially via oral routes of exposure for cancer, cardiovascular, immune response, and endocrine disruption endpoints.

8.8 Data Gaps and Recommendations

- Given the global use of antimony in PET single-use water and beverage bottles and food trays, further research is needed to determine if the degradation of these containers in landfills results in antimony leaching into surface and drinking water supplies. Ecological risks also should be evaluated.
- While antimony is used in PVC products, including children's toys, more information is needed on surface levels in these products, as well as the potential for hand-to-mouth exposure to antimony in children.
- Studies and modeling research suggest that both long-term storage and elevated temperatures may accelerate levels of antimony migration from PET food-contact plastics; parallel studies are needed using EU and FDA-approved food simulants to evaluate antimony migration under a wide range of storage and temperature conditions.
- While various types of antimony, including those used in parasite treatments, have been evaluated for oral carcinogenicity in rodent cancer bioassays, additional research is needed to determine whether orally administered antimony trioxide and pentavalent forms of antimony have any tumorigenic potential.
- More research is needed to determine if antimony exposure in humans affects blood pressure or incidence of preeclampsia in pregnant women, as well as estimation of urinary levels of antimony in preeclamptic compared to non-preeclamptic women.
- Since NHANES data, clinical studies in cardiac patients and antimony-treated leishmaniasis patients suggest antimony may affect cardiovascular pathways, more research is needed to determine if antimony plays an etiologic role in cardiovascular disease and associated metabolic pathways.
- While screening assays indicate that antimony chloride has estrogenic properties, additional studies are needed to determine of antimony trioxide (used in PET food packaging) and other forms of antimony affect estrogen-dependent pathways.

References

Agarwal S, Zaman T, Tuzcu EM et al (2011) Heavy metals and cardiovascular disease: results from the National Health and Nutrition Examination Survey (NHANES) 1999–2006. Angiology 62(5):422–429

Agency for Toxic Substances and Disease Registry (ATSDR) (1992) Toxicological profile for antimony and compounds. Agency for Toxic Substances and Disease Registry, U.S. Public Health Service. http://www.atsdr.cdc.gov/toxprofiles/tp.asp?id=332&tid=58. Accessed 15 Oct 2013

Alvarez M, Malecot CO, Gannier F et al (2005) Antimony-induced cardiomyopathy in guinea-pig and protection by L-carnitine. Br J Pharmacol 144(1):17–27

Andra SS, Makris KC, Shine JP et al (2012) Co-leaching of brominated compounds and antimony from bottled water. Environ Int 38(1):45–53

Asakura K, Satoh H, Chiba M et al (2009) Genotoxicity studies of heavy metals: lead, bismuth, indium, silver and antimony. J Occup Health 51(6):498–512

Bach C, Dauchy X, Chagnon MC et al (2012) Chemical compounds and toxicological assessments of drinking water stored in polyethylene terephthalate (PET) bottles: a source of controversy reviewed. Water Res 46(3):571–583

Bach C, Dauchy X, Severin I et al (2013) Effect of temperature on the release of intentionally and non-intentionally added substances from polyethylene terephthalate (PET) bottles into water: chemical analysis and potential toxicity. Food Chem 139(1–4):672–680

Baiocco P, Colotti G, Franceschini S et al (2009) Molecular basis of antimony treatment in leishmaniasis. J Med Chem 52(8):2603–2612

Bi X, Li Z, Zhuang X et al (2011) High levels of antimony in dust from e-waste recycling in southeastern China. Sci Total Environ 409(23):5126–5128

Cavallo D, Iavicoli I, Setini A et al (2002) Genotoxic risk and oxidative DNA damage in workers exposed to antimony trioxide. Environ Mol Mutagen 40(3):184–189

Centers for Disease Control and Prevention (CDC) (2012) Fourth national report on human exposure to environmental chemicals, updated tables, February 2012. Centers for Disease Control and Prevention, Atlanta, GA. http://www.cdc.gov/exposurereport/. Accessed 15 Aug 2013

CEC (1998) Council Directive 98/83/EC of November 1998 on the quality of water intended for human consumption, L330/32. Official Journal of the European Communities. http://eur-lex.europa.eu/LexUriServ/LexUriServ.do?uri=CONSLEG:1998L0083:20090807:EN:PDF. Accessed 10 Oct 2013

Cheng X, Shi H, Adams CD et al (2010) Assessment of metal contaminations leaching out from recycling plastic bottles upon treatments. Environ Sci Pollut Res Int 17(7):1323–1330

Choe SY, Kim SJ, Kim HG et al (2003) Evaluation of estrogenicity of major heavy metals. Sci Total Environ 312(1–3):15–21

Clemente GF, Ingrao G, Santaroni GP (1982) The concentration of some trace elements in human milk from Italy. Sci Total Environ 24(3):255–265

Cooper RG, Harrison AP (2009) The exposure to and health effects of antimony. Indian J Occup Environ Med 13(1):3–10

De Boeck M, Kirsch-Volders M, Lison D (2003) Cobalt and antimony: genotoxicity and carcinogenicity. Mutat Res 533(1–2):135–152

de Perio MA, Durgam S, Caldwell KL et al (2010) A health hazard evaluation of antimony exposure in fire fighters. J Occup Environ Med 52(1):81–84

Dezateux C, Delves HT, Stocks J et al (1997) Urinary antimony in infancy. Arch Dis Child 76(5):432–436

Dieter MP, Jameson CW, Elwell MR et al (1991) Comparative toxicity and tissue distribution of antimony potassium tartrate in rats and mice dosed by drinking water or intraperitoneal injection. J Toxicol Environ Health 34(1):51–82

Elliott BM, Mackay JM, Clay P et al (1998) An assessment of the genetic toxicology of antimony trioxide. Mutat Res 415(1–2):109–117

European Food Safety Authority (EFSA) (2004) Opinion of the Scientific Panel on food additives, flavourings, processing aids and materials in contact with food (AFC) on a request from the Commission related to a 2nd list of substances for food contact materials. EFSA J 24:1–13

Food and Drug Administration (FDA) (2011) List of indirect additives used in food contact substances; Doc. No. 5093, antimony oxide. U.S. Food and Drug Administration. http://www.accessdata.fda.gov/scripts/fcn/fcnDetailNavigation.cfm?rpt=iaListing&id=151. Accessed 5 Sept 2013

Food and Drug Administration (FDA) (2012) Cumulated Estimated Daily Intake, antimony trioxide. U.S. Food and Drug Administration. http://www.accessdata.fda.gov/scripts/sda/sdNavigation.cfm?sd=edisrev&displayAll=false&page=7. Accessed 4 Sept 2013

Fowler BA, Yamauchi H, Conner EA et al (1993) Cancer risks for humans from exposure to the semiconductor metals. Scand J Work Environ Health 19(Suppl 1):101–103

Frustaci A, Magnavita N, Chimenti C et al (1999) Marked elevation of myocardial trace elements in idiopathic dilated cardiomyopathy compared with secondary cardiac dysfunction. J Am Coll Cardiol 33(6):1578–1583

Gebel T, Claussen K, Dunkelberg H (1998) Human biomonitoring of antimony. Int Arch Occup Environ Health 71(3):221–224

Greifenstein M, White DW, Stubner A et al (2013) Impact of temperature and storage duration on the chemical and odor quality of military packaged water in polyethylene terephthalate bottles. Sci Total Environ 456–457:376–383

Grosskopf C, Schwerdtle T, Mullenders LH et al (2010) Antimony impairs nucleotide excision repair: XPA and XPE as potential molecular targets. Chem Res Toxicol 23(7):1175–1183

Groth DH, Stettler LE, Burg JR et al (1986) Carcinogenic effects of antimony trioxide and antimony ore concentrate in rats. J Toxicol Environ Health 18(4):607–626

Haldimann M, Alt A, Blanc A et al (2013) Migration of antimony from PET trays into food simulant and food: determination of Arrhenius parameters and comparison of predicted and measured migration data. Food Addit Contam Part A Chem Anal Control Expo Risk Assess 30(3):587–598

Haldimann M, Blanc A, Dudler V (2007) Exposure to antimony from polyethylene terephthalate (PET) trays used in ready-to-eat meals. Food Addit Contam 24(8):860–868

Hansen C, Tsirigotaki A, Bak SA et al (2010) Elevated antimony concentrations in commercial juices. J Environ Monit 12(4):822–824

Hayes RB (1997) The carcinogenicity of metals in humans. Cancer Causes Control 8(3):371–385

Health Council of the Netherlands (HCN) (2011) Antimony and antimony compounds, Evaluation of the carcinogenicity and geotoxicity. The Hague: Health Council of the Netherlands, 2011; publication no. 2011/33, ISBN 978-90-5549-877-2. http://www.gezondheidsraad.nl/en/publications/antimony-and-antimony-compounds-evaluation-carconigenicity-and-genotoxicity. Accessed 14 Nov 2012

Hext PM, Pinto PJ, Rimmel BA (1999) Subchronic feeding study of antimony trioxide in rats. J Appl Toxicol 19(3):205–209

Huang H, Shu SC, Shih JH et al (1998) Antimony trichloride induces DNA damage and apoptosis in mammalian cells. Toxicology 129(2–3):113–123

International Agency for Research on Cancer (IARC) (1989). Some organic solvents, resin monomers and related compounds, pigments and occupational exposures in paint manufacture and painting; Section on: Some pigments, antimony trioxide and antimony trisulfide. In: IARC monographs on the evaluation of carcinogenic risks to humans, vol 47. International Agency for Research on Cancer, Lyon, France

Jain RB (2013) Effect of pregnancy on the levels of urinary metals for females aged 17–39 years old: data from National Health and Nutrition Examination Survey 2003–2010. J Toxicol Environ Health A 76(2):86–97

Jenkins RO, Craig PJ, Goessler W et al (1998a) Antimony leaching from cot mattresses and sudden infant death syndrome (SIDS). Hum Exp Toxicol 17(3):138–139

Jenkins RO, Craig PJ, Goessler W et al (1998b) Biovolatilization of antimony and sudden infant death syndrome (SIDS). Hum Exp Toxicol 17(4):231–238

Jones RD (1994) Survey of antimony workers: mortality 1961–1992. Occup Environ Med 51(11):772–776

Kakimoto S, Ikebe K, Hori S (2004) Improved test for the migration of antimony and germanium from polyethylene terephthalate. Shokuhin Eiseigaku Zasshi 45(5):264–269

Kawamura Y, Kawasaki C, Mine S et al (2006) Contents of eight harmful elements in baby toys and their migration tests. Shokuhin Eiseigaku Zasshi 47(2):51–57

Keresztes S, Tatar E, Mihucz VG et al (2009) Leaching of antimony from polyethylene terephthalate (PET) bottles into mineral water. Sci Total Environ 407(16):4731–4735

Kim HA, Heo Y, Oh SY et al (1999) Altered serum cytokine and immunoglobulin levels in the workers exposed to antimony. Hum Exp Toxicol 18(10):607–613

Kirkland D, Whitwell J, Deyo J et al (2007) Failure of antimony trioxide to induce micronuclei or chromosomal aberrations in rat bone-marrow after sub-chronic oral dosing. Mutat Res 627(2):119–128

Kuroda K, Endo G, Okamoto A et al (1991) Genotoxicity of beryllium, gallium and antimony in short-term assays. Mutat Res 264(4):163–170

Kuryshev YA, Wang L, Wible BA et al (2006) Antimony-based antileishmanial compounds prolong the cardiac action potential by an increase in cardiac calcium currents. Mol Pharmacol 69(4):1216–1225

Lincoln JD, Ogunseitan OA, Shapiro AA et al (2007) Leaching assessments of hazardous materials in cellular telephones. Environ Sci Technol 41(7):2572–2578

Makris KC, Andra SS, Herrick L et al (2013) Association of drinking-water source and use characteristics with urinary antimony concentrations. J Expo Sci Environ Epidemiol 23(2):120–127

McCallum RI (2005) Occupational exposure to antimony compounds. J Environ Monit 7(12):1245–1250

Motolese A, Truzzi M, Giannini A et al (1993) Contact dermatitis and contact sensitization among enamellers and decorators in the ceramics industry. Contact Dermatitis 28(2):59–62

Navas-Acien A, Silbergeld EK, Sharrett R et al (2005) Metals in urine and peripheral arterial disease. Environ Health Perspect 113(2):164–169

Newton PE, Bolte HF, Daly IW et al (1994) Subchronic and chronic inhalation toxicity of antimony trioxide in the rat. Fundam Appl Toxicol 22(4):561–576

Ohkado Y, Kawamura Y, Mutsuga M et al (2005) Metals in recycled polyethylene terephthalate and discrimination method for its use. Shokuhin Eiseigaku Zasshi 46(3):109–115

Padilla MA, Elobeid M, Ruden DM et al (2010) An examination of the association of selected toxic metals with total and central obesity indices: NHANES 99-02. Int J Environ Res Public Health 7(9):3332–3347

Paschal DC, Ting BG, Morrow JC et al (1998) Trace metals in urine of United States residents: reference range concentrations. Environ Res 76(1):53–59

Poon R, Chu I (2000) Effects of trivalent antimony on human erythrocyte glutathione-S-transferases. J Biochem Mol Toxicol 14(3):169–176

Poon R, Chu I, Lecavalier P et al (1998) Effects of antimony on rats following 90-day exposure via drinking water. Food Chem Toxicol 36(1):21–35

Rahman L, Corns WT, Bryce DW et al (2000) Determination of mercury, selenium, bismuth, arsenic and antimony in human hair by microwave digestion atomic fluorescence spectrometry. Talanta 52(5):833–843

Reimann C, Birke M, Filzomser (2010) Bottled drinking water: water contamination from bottle materials (glass, hard PET, soft PET), the influence of colour and acidification. Appl Geochem 25(7):1030–1046

Rossi F, Acampora R, Vacca C et al (1987) Prenatal and postnatal antimony exposure in rats: effect on vasomotor reactivity development of pups. Teratog Carcinog Mutagen 7(5):491–496

Sanchez-Martinez M, Perez-Corona T, Camara C et al (2013) Migration of antimony from PET containers into regulated EU food simulants. Food Chem 141(2):816–822

Santos MC, Nobrega JA, Baccan N et al (2010) Determination of toxic elements in plastics from waste electrical and electronic equipment by slurry sampling electrothermal atomic absorption spectrometry. Talanta 81(4–5):1781–1787

Sax L (2010) Polyethylene terephthalate may yield endocrine disruptors. Environ Health Perspect 118(4):445–448

Schaumloffel N, Gebel T (1998) Heterogeneity of the DNA damage provoked by antimony and arsenic. Mutagenesis 13(3):281–286

Schnorr TM, Steenland K, Thun MJ et al (1995) Mortality in a cohort of antimony smelter workers. Am J Ind Med 27(5):759–770

Schroeder HA, Mitchener M, Balassa JJ et al (1968) Zirconium, niobium, antimony and fluorine in mice: effects on growth, survival and tissue levels. J Nutr 95(1):95–101

Schroeder HA, Mitchener M, Nason AP (1970) Zirconium, niobium, antimony, vanadium and lead in rats: life term studies. J Nutr 100(1):59–68

Schulz C, Angerer J, Ewers U et al (2009) Revised and new reference values for environmental pollutants in urine or blood of children in Germany derived from the German environmental survey on children 2003–2006 (GerES IV). Int J Hyg Environ Health 212(6):637–647

Shirai S, Suzuki Y, Yoshinaga J et al (2010) Maternal exposure to low-level heavy metals during pregnancy and birth size. J Environ Sci Health A Tox Hazard Subst Environ Eng 45(11):1468–1474

Shotyk W, Krachler M (2007) Contamination of bottled waters with antimony leaching from polyethylene terephthalate (PET) increases upon storage. Environ Sci Technol 41(5):1560–1563

Shotyk W, Krachler M, Chen B (2006) Contamination of Canadian and European bottled waters with antimony from PET containers. J Environ Monit 8(2):288–292

Sundar S, Chakravarty J (2010) Antimony toxicity. Int J Environ Res Public Health 7(12):4267–4277

Takahashi Y, Sakuma K, Itai T et al (2008) Speciation of antimony in PET bottles produced in Japan and China by X-ray absorption fine structure spectroscopy. Environ Sci Technol 42(24):9045–9050

Tostar S, Stenvall E, Boldizar A et al (2013) Antimony leaching in plastics from waste electrical and electronic equipment (WEEE) with various acids and gamma irradiation. Waste Manag 33(6):1478–1482

Tukur A, Sharp L, Stern B et al (2012) PET bottle use patterns and antimony migration into bottled water and soft drinks: the case of British and Nigerian bottles. J Environ Monit 14(4):1237–1247

Tyrrell J, Melzer D, Henley W et al (2013) Associations between socioeconomic status and environmental toxicant concentrations in adults in the USA: NHANES 2001–2010. Environ Int 59C:328–335

USCFR (2010) U.S. Code of Federal Regulations, Title 21: Food and Drugs, vol 3, Chapter 1, subchapter B food for human consumption, Title: Section 178.3297—colorants for polymers; original date 2009-04-01; revised 2010-01-01. U.S. Government. http://www.gpo.gov/fdsys/pkg/CFR-2010-title21-vol3/xml/CFR-2010-title21-vol3-sec178-3297.xml. Accessed 4 Sept 2013

U.S. Environmental Protection Agency (USEPA) (2009) National primary drinking water regulations, EPA 816-F-09-004. United States Environmental Protection Agency. http://water.epa.gov/drink/contaminants/index.cfm. Accessed 2 Sept 2013

U.S. Environmental Protection Agency (USEPA) (2012) TSCA workplan chemical risk assessment antimony trioxide CASRN: 1309-64-4 (Draft). United States Environmental Protection Agency, Office of Chemical Safety and Pollution Prevention, December 2012. http://www.epa.gov/oppt/existingchemicals/pubs/workplans.html and http://www.epa.gov/oppt/existingchemicals/pubs/TSCA_Workplan_Chemical_Risk_Assessment_of_ATO.pdf. Accessed 17 Oct 2013

U.S Environmental Protection Agency (USEPA) (2013) Kim-Stan Landfill, EPA ID: VAS077923449, Selma, VA. U.S. Environmental Protection Agency, Mid-Atlantic Superfund. http://www.epa.gov/reg3hwmd/npl/VAD077923449.htm. Accessed 6 Sept 2013

U.S. Geological Survey (USGS) (2012) 2012 Minerals yearbook, antimony [advance release]. U.S. Department of the Interior, U.S. Geological Survey. http://minerals.usgs.gov/minerals/pubs/commodity/antimony/myb1-2012-antim.pdf. Accessed 14 Sept 2013

Vasami R (2010) Comments on 'Elevated antimony concentrations in commercial juices'. J Environ Monit 12(12):2307 (author reply 2308)

Vigeh M, Yokoyama K, Ramezanzadeh F et al (2006) Lead and other trace metals in preeclampsia: a case-control study in Tehran. Iran Environ Res 100(2):268–275

Wagner M, Oehlmann J (2011) Endocrine disruptors in bottled mineral water: estrogenic activity in the e-screen. J Steroid Biochem Mol Biol 127(1–2):128–135

Welle F, Franz R (2011) Migration of antimony from PET bottles into beverages: determination of the activation energy of diffusion and migration modelling compared with literature data. Food Addit Contam Part A Chem Anal Control Expo Risk Assess 28(1):115–126

Westerhoff P, Prapaipong P, Shock E et al (2008) Antimony leaching from polyethylene terephthalate (PET) plastic used for bottled drinking water. Water Res 42(3):551–556

White GP Jr, Mathias CG, Davin JS (1993) Dermatitis in workers exposed to antimony in a melting process. J Occup Med 35(4):392–395

World Health Organization (WHO) (2003) Antimony in drinking-water, Background document for development of WHO guidelines for drinking-water quality, WHO/SDE/WSH/03.04/74. World Health Organization. http://www.who.int/water_sanitation_health/dwq/chemicals/0304_74/en/index.html. Accessed 26 Nov 2013

World Health Organization (WHO) (2011) Guidelines for drinking-water quality, 4th edn, Chapter 12, Chemical fact sheets, Section 12.1 Chemical contaminants in drinking-water, antimony. World Health Organization. http://www.who.int/water_sanitation_health/publications/2011/dwq_chapters/en/index.html. Accessed 26 Nov 2013

Wu F, Fu Z, Liu B et al (2011a) Health risk associated with dietary co-exposure to high levels of antimony and arsenic in the world's largest antimony mine area. Sci Total Environ 409(18):3344–3351

Wu XD, Song JM, Li XG et al (2011b) Behaviors of dissolved antimony in the Yangtze River Estuary and its adjacent waters. J Environ Monit 13(8):2292–2303

Yorita Christensen KL (2012) Metals in blood and urine, and thyroid function among adults in the United States 2007–2008. Int J Hyg Environ Health

Chapter 9
Lead in Household Products

Joseph Laquatra

Abstract Humans have used lead for various purposes for thousands of years. Despite awareness of the dangers associated with lead, this element continues to appear in a wide range of household products and poses hazards through different exposure pathways. Sources of household lead exposure discussed in this chapter include: paint dust, drinking water, solder, candle wicks, wood finishes and brass fittings, ceramics, shot and bullets, food and spices, toys and jewelry, lead as a stabilizer in vinyl and polyvinyl chloride-based plastics and wiring, cosmetics, electronic equipment and electronic waste, contaminated soil, and lead batteries. Despite awareness of the dangers of lead exposure, lead in consumer products continues to be the reason for recalls, and the majority of those recalled products originate in China. Consumer education on this topic is ongoing, but should be expanded. Targeted audiences should include new parents, medical professionals, teachers, and others. Important messages should stress that any exposure to lead, especially by children, is harmful.

Keywords Lead poisoning · Environmental exposure · Lead in household products · Lead dust · Vinyl products · Polyvinyl chloride · Lead solder · Electronic waste · Toys · Lead shot and bullets · Glass and ceramics · Cosmetics · Batteries · Contaminated soil

9.1 Key Take Home Points

- Lead is present in many types of household products, including leaded paint and dust, solder, wood finishes and brass products, vinyl plastics, toys and

J. Laquatra (✉)
Department of Design and Environmental Analysis, Cornell University, Ithaca, NY, USA
e-mail: JL27@cornell.edu

S. M. Snedeker (ed.), *Toxicants in Food Packaging and Household Plastics*, Molecular and Integrative Toxicology, DOI: 10.1007/978-1-4471-6500-2_9,

jewelry, food and beverages stored in leaded ceramic and crystal, and is present in electronic waste and lead batteries.
- Lead can contaminate the soil, drinking water, and the air we breathe.
- Lead-painted surfaces can produce a fine dust that is poisonous, especially to infants and children.
- Dermatological contact with lead from paint dust and certain vinyl products can lead to ingestion through hand-to-mouth contact.
- The current level of concern set by the Centers for Disease Control and Prevention (CDC) is 5 micrograms per deciliter (μg/dL) blood.
- There is no safe level of exposure to lead.

9.2 Introduction

Human beings have used lead for various purposes for at least 7,000 years (Cochran 2006). The Ancient Egyptians, Chinese, Romans, and others used lead for medicinal and cosmetic purposes, roofing, plumbing pipes, goblets, vases, pots, coins, stationery and pottery glazing, among other uses. Although awareness of lead poisoning has existed for over 2,000 years (Chauhan et al. 2010), lead is still present in our environment.

As is illustrated in Fig. 9.1, lead exposure occurs through ingestion, inhalation, and dermatological contact; and lead poisoning can affect nearly every organ in the body (The Lead Group 2013). According to the World Health Organization (2010), lead's adverse health effects include cognitive deficits, attention deficit disorder, behavior problems, dyslexia, hypertension, immunotoxicity, reproductive system damage, convulsions, coma, and death. While children are at higher risk of problems associated with lead poisoning, adults are affected as well. Although the United States's (U.S.'s) CDC has set 5 μg of lead per dL of blood as a reference level for public health actions, research has demonstrated that there is no threshold for health problems associated with lead exposure (American Academy of Pediatrics 2012). In other words, there is no safe level of contact with lead.

In 1909 France, Belgium, and Austria banned the use of white-lead paint (The Lead Group 2013). Lead was banned in the U.S. as an ingredient in residential paint in 1978, and from gasoline in 1986, but there are an estimated 50 million homes and apartments in the U.S. with lead-based paint (Heimlich 1997), and soil throughout the country is contaminated with lead from car exhaust emissions that occurred before the leaded gasoline ban. Beyond these issues, lead continues to be used in products that are used and consumed on a daily basis around the world. Lead exposure also occurs when processing discarded consumer goods. This includes an emerging risk of extensive lead poisoning in China, where crude methods are used to harvest metals from recycled electronic waste (e-waste), resulting in extensive soil and water contamination and elevated blood lead levels in children (Huo et al. 2007). To examine the extent of this issue, a literature

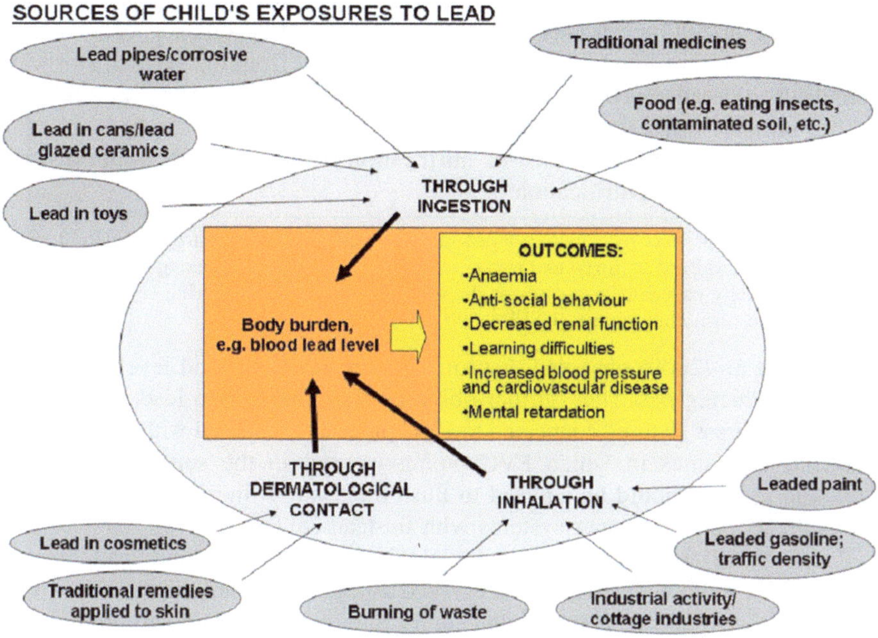

Fig. 9.1 Lead exposure sources. Credit: Permission granted to reproduce this figure by the World Health Organization (WHO) from Childhood Lead Poisoning (ISBN 978 92 4 150033 3), Fig. 1, pg. 19, Copyright 2010, http://www.who.int/ceh/publications/childhoodpoisoning/en/, Accessed 16 April 2013

search was conducted of academic journals, health-related websites, and other sources. Findings are organized through the exposure pathways (oral, inhalation and dermatological) with a separate section on drinking water in the oral exposure section.

9.3 Oral Exposure

9.3.1 Drinking Water

Drinking water can become contaminated with lead in numerous ways. Lead emissions from coal-fired power plants can contaminate water supply sources (Fischetti 2013). Lead-tin solder was used in water supply pipes until 1986. This solder was used to seal joints in copper pipes. In homes that have these pipes that were installed before 1986, lead levels can build up in water when it is not flushed from faucets. For example, when water is stationary in pipes overnight, the first draw of water in the morning may have detectable lead levels. Public health officials recommend that the first use of water from these systems be done carefully. Water should be run during the first draw until it becomes as cold as it can get. At that point, residual lead will have been flushed from the system. Hot water from lead-soldered pipes should never be used for consumption, as more lead leaches into hot water than cold.

Under certain conditions such as nitrification, lead can leach from polyvinyl chloride (PVC) pipe. Nitrification is:

> …a microbial process by which reduced nitrogen compounds (primarily ammonia) are sequentially oxidized to nitrite and nitrate. Ammonia is present in drinking water through either naturally-occurring processes or through ammonia addition during secondary disinfection to form chloramines (USEPA 2002).

Zhang et al. (2009) observed that nitrification increased lead levels in potable water by reducing pH. They further observed that 45 % more lead was released from leaded brass fixtures connected to PVC pipes as compared with copper pipes. Residents of homes in which PVC piping is used in the supply lines of the plumbing system should be advised to flush their water line in the same manner that is prescribed for copper systems with tin-lead solder.

Rabin (2008) relates how old municipal water systems used lead pipes to deliver water to homes. This was recognized as a public health threat in the 1800s, and in the early part of the twentieth century municipalities began to prohibit lead service lines. This movement was countered by the Lead Industry Association (LIA), which began a public campaign to commend the benefits of using lead pipes. Problems remain to this day, as the federal *Lead and Copper Rule* (LCR) requires water companies to initiate lead water pipe replacement when lead levels in water exceed 15 parts per billion (ppb). But the LCR allows water utilities to replace only the public portion of lead pipes. The private portion of these pipes, from the main service line to a house, can be left in place. Renner (2007) describes that when partial replacement is undertaken, that disturbance results in a rise in lead levels in water. Millions of homes in the U.S. have lead service lines as part of their water supply systems (McCartney 2010).

Drinking water can be contaminated with lead from brass or chrome-plated brass faucets. Most of these faucets that were purchased before 1997 contain up to 8 % lead (Massachusetts Water Resources Authority 2012). Federal legislation enacted in 2010 mandates that all faucets purchased after January 14, 2014 have no more than a weighted average of 0.25 % lead per wetted surface area.

Lead solder was used in electronic equipment, but that practice ended with a European Union ban in 2006. Black (2005) reported that when older electronic devices that were disposed of in landfills break down, lead could leach from those landfills and contaminate drinking water. Electronic waste, or e-waste, consists of unwanted electronic devices or Cathode Ray Tubes (CRT). These devices frequently contain hazardous materials, including lead. To prevent groundwater contamination from e-waste, these materials should be properly recycled. E-waste disposal in landfills is illegal in some states (Jarnot 2013).

Lead exposure can occur during manual harvesting of metals from recycled e-waste. In China, where crude methods are used to harvest precious metals from electronics, housings are dismantled manually (Huo et al. 2007). Older television sets with CRTs contain 8 to 9 pounds of lead, and workers can be exposed dermatologically during the manual demolition of the electronics, with discarded materials potentially leaching lead into soil and waterways. This may result in other routes of exposure including ingestion from contaminated drinking water obtained from polluted streams and river beds.

9.3.2 Ingestion

Lead can also enter the body through ingestion. The case of crawling toddlers was mentioned above as an example of lead ingestion through contaminated hand-to-mouth activity from paint and household dust contaminated with lead. But lead can also be ingested directly through contaminated food and other items. In 1991 U.S. food canners voluntarily stopped using lead/tin solder to seal seams in cans used to preserve food. In 1995 the U.S. Food and Drug Administration (FDA) banned the use of lead to solder the seams of such cans. Not all countries have done this, however. Lead solder has been found in cans of rice pudding, lotus-nut paste, and bamboo shoots from China and in canned ham from Denmark (Knight-Ridder News Service 1997).

Lead has been found in venison harvested with lead shot or bullets. An estimated 10 million hunters and their families, as well as those who obtain venison from food pantries, are considered to be at risk for lead ingestion from this source (Hunt et al. 2009). Lead shot and bullets also affect wildlife populations.

Slow cookers, or crock-pots, are another source of lead in the human diet (Insightful Nana 2008). Lead is used in the glazed ceramic insert that is in direct contact with the food that is being cooked. Glazed ceramic dishes, cups, and other tableware can contain lead. Glazed ceramic pots from Latin American countries, such as Mexican bean pots, are likely to have lead in the glazing (Contra Costa Health Services n.d.). Another way that lead can get into food is by growing it in contaminated soil. A study of this issue found that leafy vegetables grown in such soil had higher concentrations of lead than did beans, fruits, and root vegetables (Karnpanit et al. 2010).

Lead-contaminated soil is of particular concern for urban gardens used for growing food for human consumption. Clark et al. (2008) tested 141 backyard gardens in Roxbury and Dorchester, Massachusetts, and found that 81 % of these plots had levels of lead above the U.S. Environmental Protection Agency (USEPA) hazard level of 400 μg per gram (g). While a recommendation to avoid growing vegetables in contaminated soil is to use raised beds lined with landscaping fabric and filled with compost, these authors found that this practice requires maintenance. They observed that lead levels in raised beds increased from

150 μg/g to an average of 336 μg/g over a 4-year period and theorized that the contamination originated from lead particles that were transported by wind.

Lead is also used in glass making to produce lead crystal, which has traditionally been favored because of its radiance and durability (Barbee and Constantine 1994). Barbee and Constantine (1994) found that lead levels in sherry stored in lead crystal decanters were 50 μg per liter (L) after storage for 2 months in a 20-year old decanter, 163 μg/L in a 10-year old decanter, and 1,410 μg/L in a new decanter. They concluded that lead leaching decreases with the age of a decanter. Other researchers have found that lead levels in 4 % acetic acid, white port, and a synthetic alcoholic beverage stored in lead crystal decanters for 1-, 2-, and 10-day periods at room temperature ranged from 100 to 1,800 μg/L, which are above the level that a warning under California's Proposition 65 is required (Appel et al. 1992). This proposition requires that consumer products with known hazards must carry a warning label about such hazards.

High levels of lead have been found in candy imported from Mexico. Some of the candy ingredients, including chilies and tamarind, are dried in the sun. Lead emissions from gasoline and factories can be deposited on the drying foods that are then used in candy making. Some candies are made in ceramic pots that can leach lead (Center for Environmental Health 2006). Mexican candies can also become contaminated with lead when lead ink is used in the candy wrappers.

Lead has been found in baby food, especially baby food with carrots, peaches, pears, and sweet potatoes (Dearen 2013). This particular issue involves a lawsuit filed by an environmental group that is demanding that contaminated baby food must carry a label for California's Proposition 65. Grape juice and fruit cocktail have also been known to have detectable levels of lead (Bolokhova 2013). The sources of lead appear to be contaminated soil and older processing equipment. Some Indian spices and cultural powders have been found to have lead levels that resulted in elevated blood lead levels in children (Gurgel 2010).

In 1994 the Consumer Product Safety Commission (CPSC) recalled crayons imported from China because of high lead levels (CPSC 1994). There was concern that the children could be exposed to lead by chewing on or eating pieces of the lead-contaminated crayons over an extended period of time.

Ayurveda, a traditional medical practice followed in India and other South Asian countries, is associated with medications that can contain lead. This is also the case with traditional or folk medications used in Middle Eastern, West Asian, and Hispanic cultures. The CDC has tracked specific cases of lead poisoning in the U.S. that involved Americans who consumed folk medications (CDC 2004). Lead is deliberately added to some of these medications because of the mistaken belief that it has curative properties. In some cases the source of lead is contaminated soil or older processing equipment.

9.4 Inhalation

Lead inhalation can occur when particle sizes are below 10 μm (micrometers) in diameter, as is the case with fumes and fine dust (World Health Organization 2010). Lead-polluted air results from lead smelter emissions and car emissions in countries that still use leaded gasoline. Lead fumes are also produced when heat guns are used to remove lead-based paint.

Before it was banned, lead-based paint was the paint of choice for double-hung windows because of its superior adhesive properties (Park and Hicks 1995). But as sashes rub against each other when the lower sash is opened and closed, a fine dust is produced (CDC 2012). This is the source of much lead in dust in older homes (CDC 2012). The dust can be inhaled and ingested by toddlers who accumulate the lead on their hands as they crawl (CDC 2012). Hand-to-mouth activity then results in ingestion (World Health Organization 2010). Dust on floors below windows is covered under the clearance standards set by the U.S. Department of Housing and Urban Development (HUD). After work in a house by a paid contractor that disturbs lead paint, the amount of lead on a floor can be no more than 40 μg per square foot (μg/ft^2); interior windowsills can have no more than 250 μg/ft^2; and window troughs can have no more than 400 μg/ft^2.

Candles represent another source of lead fumes and dust. Some candlewicks contain lead as a stiffening agent. Wasson et al. (2002) observed that burning a single candle in a room can raise the amount of lead in the air of that room above the ambient air lead concentration limit set by the USEPA of 1.5 μg per cubic meter (m^3). This lead is in the air as fine particulates that can settle on furniture and floors. Although the Consumer Product Safety Commission (CPSC) banned the manufacture of lead-containing candle wicks in 2003, imported candles, and those purchased at yard sales and from thrift stores, may contain them.

Lead is also present in varnish, shellac, and wood stains in older homes. Sanding surfaces with these coatings can generate high amounts of lead-containing dust. Precautions to be taken during sanding include isolating the work area from the rest of the house and attaching a high efficiency particulate air (HEPA) vacuum to the sanding machine (Hardwood Floors 2012).

9.5 Dermatological Contact

A third pathway for lead into the human body is through dermatological contact. Note that this pathway can also lead to ingestion if the affected area is touched by a hand and is followed up with hand-to-mouth contact. Children's toys are in this category because of lead paint and lead-contaminated vinyl. Bounce houses are the large, inflatable jump houses and bouncy castles that are made of vinyl. The Center for Environmental Health (2010) began testing for lead in these products in 2010. In nearly every test, high levels of lead were detected. Lead is added to the vinyl to stabilize colors. Some vinyl mini-blinds can contain lead. The CPSC recommends that these blinds should be removed from homes and replaced with blinds that are labeled "No Lead Added" or "Non-leaded formula" (Contra Costa Health Services n.d.a 2013). Lead has been found in toy and adult jewelry and in Disney® charms (Center for Environmental Health 2013).

Lead in amounts that exceed safety standards for children's toys has been found in pet toys (Dale 2013). No safety standards exist for lead in pet toys. Lead was found in the ink of tennis balls for dogs. In addition to safety concerns for dogs, children can pick up a wet ball after it has been in a dog's mouth and get lead on their hands.

Another plastic household product that can contain lead is the garden hose (Hickman 2012). The study that examined this issue focused on PVC garden hoses and found lead levels in 30 % of those tested to contain lead levels over 100 parts per million (ppm), which is the CPSC maximum level for lead in children's products. Water from one of the tested hoses contained lead at 0.280 mg/L (280 ppm); the USEPA action level for lead in drinking water is 0.015 mg/L (15 ppm). Apart from handling the hoses and potential dermatological contact, watering a garden with PVC hoses can pollute the soil with lead. Of course, drinking water from a PVC garden hose can result in lead exposure from ingestion. These hazards can be avoided by choosing polyurethane or rubber hoses. In addition to lead in PVC, lead is also a component of brass garden hose connectors and has been found in excess of 2,500 ppm (Hickman 2012). Garden gloves with raised PVC dots on the fingers also contain lead in excess of 2,000 ppm (Hickman 2012).

Lead has also been found on the wiring of Christmas lights and appliance cords (Laquatra et al. 2008). Lead is used as a stabilizer in the PVC coating that covers the electrical conductors. Lead can be transferred to hands when lights or cords are handled. An additional concern is degradation of Christmas light strings that are installed outdoors. Exposure to sunlight can result in lead being released from the PVC and then contaminating soil.

Cell phone cases have been found to contain lead (Sauler 2013). Those most likely to contain lead are brightly colored and made from synthetic materials, including PVC. To avoid lead exposure from these cases, consumers should select cases made of fabric or non-PVC cases.

The use of lead as a stabilizer in vinyl has been mentioned as a component in bounce houses. This is also the case for vinyl tile flooring (Main 2013). Lead dust can be released from vinyl tiles over time. Removal of vinyl tiles should be undertaken with lead-safe work practices; or they can be covered with linoleum, cork, or hardwood floors.

Vinyl lunch boxes have been found to contain lead (NRDC 2012). Lead is added to vinyl as a stabilizer. Exposure to sunlight and air breaks the chemical bond between lead and vinyl and causes lead dust to form. Children can get the dust on their hands and then transfer it to food, or food can become directly contaminated with the dust. Safer alternatives for carrying lunch are metal lunch boxes, canvas sacks, or paper bags.

Lead in amounts as high as 500 ppm has been found in faux leather purses and sandals at 700 ppm (Nguyen et al. 2013), and wallets at 58,700 ppm (Main 2013). These exposures can be avoided by choosing real leather, canvas, or cotton products.

The FDA conducted studies of lipstick in 2009 and 2012 and found lead levels of up to 7.19 ppm in this cosmetic (Severns 2013). For average and high use of lipstick, Liu et al. (2013) estimated an acceptable daily intake of lead to be less than 20 %. The Environmental Working Group maintains *Skin Deep,* a website with safety information on cosmetics and personal care products (EWG's Skin Deep 2013).

Lead is also present in brass house and car keys (Lucas 1999). Health officials recommend that children should not be given keys to play with and that people wash their hands after handling keys (Rother 2000).

About half the content of a type of eye makeup known as kohl, kajal, al-kahl, or surma can be comprised of lead (FDA 2011a). Kohl has been used since ancient times in Africa, the Middle East, Iran, Pakistan, and India. Although illegal in the U.S., kohl has been found in Europe and North America, particularly in Asian and Middle Eastern specialty stores, and for sale on websites (FDA 2011a).

Progressive hair dye products contain lead in the form of lead acetate (FDA 2011b). These dyes are applied over a period of time to achieve a gradual hair color change. People using such dyes have been monitored and no increase in blood lead level has been found from such use. However, a warning is required on these products that states they are for external use only, should not be applied in areas where there are scalp abrasions, and the product should not be allowed to get into eyes.

Lead is in some types of batteries: automobiles and other vehicles use lead-acid batteries, as do some alarm and emergency lighting systems (Wieman 2013). These batteries do not threaten human health when they are used. Problems occur when they are disposed of improperly. If they are landfilled or incinerated, lead can be released into the environment. Recycling is the best option for disposal of lead-acid batteries.

Table 9.1 provides a summary of lead-contaminated items commonly found in household settings and their exposure pathway(s).

9.6 Conclusions and Future Directions

Wherever possible manufacturers should be encouraged or required to use substitutes for lead in their products. Markowitz and Rosner (2013) describe the difficulty of this work and provide a detailed history of the U.S. effort to ban lead as an ingredient in gasoline and paint. They explain the role of the LIA to discredit the science of lead poisoning and the scientists involved in its research. The LIA deliberately misled the public about adverse health effects from lead and enlisted the assistance of politicians to protect their interests. The bans on lead in gasoline

Table 9.1 Lead-contaminated items and exposure pathway(s)

Item(s)	Source(s) of lead	Exposure pathway(s)		
		Inhalation	Ingestion	Dermal contact
Drinking water	Pipes, solder, faucets		✔	
Walls, windows, doors	Paint	✔	✔	
Candles	Wicks	✔		
Wood surfaces	Varnish, shellac, stains	✔		✔
Canned food	Solder		✔	
Game	Shot or bullets		✔	
Slow cookers (crock pots)	Glazing		✔	
Ceramic dishes, cups, tableware	Glazing		✔	
Garden vegetables	Contaminated soil		✔	
Stored beverages	Crystal decanters		✔	
Mexican candy	Sun-dried chilis and tamarind		✔	
Mexican candy	Glazing in pots used for production		✔	
Mexican candy	Ink on wrappers		✔	
Baby food	Soil, processing equipment		✔	
Grape juice, fruit cocktail	Soil, processing equipment		✔	
Indian spices	Soil, processing equipment		✔	
Folk medications	Deliberate ingredient, soil, processing equipment		✔	
Bounce houses, lunch boxes	Vinyl		✔	✔
Toy and adult jewelry, charms	Deliberate ingredient		✔	✔
Art materials	Crayons		✔	✔
Mini-blinds	Vinyl		✔	✔
Garden hoses	PVC, brass connectors		✔	✔
Garden gloves	Raised dots		✔	✔
Purses, sandals, wallets	Faux leather, vinyl			✔
Lipstick	Deliberate ingredient		✔	✔
Tile flooring	Vinyl	✔	✔	✔
House and car keys	Brass		✔	✔
Kohl	Deliberate ingredient			✔
Progressive hair dyes	Lead acetate			✔
Pet toys	Ink		✔	✔
Christmas lights, appliance cords	PVC			✔
Cell phone cases	Synthetic materials, PVC			✔
Electronic equipment	Solder		✔	
Batteries	Lead-acid		✔	

and paint were ultimately successful. In the case of paint, this was only because dedicated scientists, public health professionals, and others persisted in their efforts to convince the CDC and the USEPA of the seriousness of the issue. In the case of gasoline, the USEPA had ordered a reduction in sulfur emissions from car exhaust. In response, General Motors announced that it would install catalytic converters in cars. These converters were fouled by lead and required the use of unleaded gasoline (Markowitz and Rosner 2013).

Despite awareness of the dangers of lead exposure, lead in consumer products continues to be the reason for recalls, and the majority of those recalled products originate in China (Gips 2009–2010). Even if products that contain lead could be kept from entering markets in the U.S., there is no mechanism in place to prevent their sale at thrift shops, yard sales, and eBay auctions. Consumer education on this topic is ongoing, but should be expanded. Targeted audiences should include new parents, medical professionals, teachers, and others. Important messages should stress that any exposure to lead, especially by children, is harmful. Much work is necessary to raise awareness of the dangers posed by exporting e-waste and its resulting manual recycling, especially by children, in China. USEPA (2012) reported that there have been attempts to institute a federal law regarding e-waste recycling, but these have so far been unsuccessful.

References

American Academy of Pediatrics (2012) AAP commends CDC for recognizing that for children, there is no safe level of lead exposure. http://www.aap.org/en-us/about-the-aap/aap-press-room/pages/AAP-Statement-CDC-Revised-Lead-Exposure-Guidelines.aspx?nfstatus=401&nftoken=00000000-0000-0000-0000-000000000000&nfstatusdescription=ERROR%3a+No+local+token. Accessed 28 Jun 2013

Appel BR, Kahlon JK, Ferguson J et al (1992) Potential lead exposures from lead crystal decanters. Am J Public Health 82:1671–1673

Barbee SJ, Constantine LA (1994) Release of lead from crystal decanters under conditions of normal use. Food Chem Toxicol 32:285–288

Black H (2005) Getting the lead out of electronics. Environ Health Perspect 113(10):A682–A685

Bolokhova E (2013) Baby food companies sued over lead. Parenting, April 8

Center for Environmental Health (2010) Lead in bounce houses. http://www.ceh.org/what-we-do/eliminating-toxics/current-work/lead-current-work/lead-in-bounce-houses. Accessed 28 Jun 2013

Center for Environmental Health (2006) Lead in candy. http://www.ceh.org/what-we-do/eliminating-toxics/current-work/lead-current-work/lead-in-candy. Accessed 7 Jun 2013

Center for Environmental Health (2013) Lead in jewelry. http://www.ceh.org/what-we-do/eliminating-toxics/current-work/lead-current-work/lead-in-jewelry-current-work. Accessed 7 Jun 2013

Centers for Disease Control and Prevention (CDC) (2004) Lead poisoning associated with Ayurvedic medications, 2000–2003. Morb Mortal Wkly Rep 53:582–584

Chauhan AS, Bhadauria R, Singh AK et al (2010) Determination of lead and cadmium in cosmetic products. J Chem Pharm Res 2:92–97

Clark HF, Hausladen DM, Brabander DJ (2008) Urban gardens: lead exposure, recontamination mechanisms, and implications for remediation design. Environ Res 107(3):312–319

Cochran D (2006) Household products with lead, Health News From Dr. Yu Thurston County (Washington) Health and Social Services Department. http://www.co.thurston.wa.us/health/ehkids/pdf/Lead.pdf. Accessed 16 Apr 2013

Consumer Product Safety Commission (CPSC) (1994) CPSC announcers recalls of imported crayons because of lead poisoning hazard. Release # 94-055, April 5

Contra Costa Health Services (n.d.) Questions and answers about lead in ceramic tableware. http://cchealth.org/lead-poison/pdf/ceramics.pdf. Accessed 10 Jun 2013

Contra Costa Health Services (n.d.a) Questions and answers about lead in older vinyl mini-blinds. http://cchealth.org/lead-poison/pdf/miniblinds.pdf. Accessed 7 Jun 2013

Dale S (2013) First children's toys, are pet toys next? The Internet Home of Steve Dale. http://www.stevedalepetworld.com/home/46-archived-features/310-first-childrens-toys-are-pet-toys-next. Accessed 29 Jun 2013

Dearen J (2013) Lead in baby food: trial begins over warnings in California. Denver Post, April 9

EWG's Skin Deep (2013) Cosmetics Database http://www.ewg.org/skindeep/site/about.php. Accessed 10 Jun 2013

Fischetti M (2013) Lead exposure on the rise despite decline in poisoning cases. Sci Am, Energy & Sustainability News February 17, http://www.scientificamerican.com/article.cfm?id=lead-exposure-on-the-rise&page=2. Accessed 16 Apr 2013

Gips RT (2009-2010) From China with lead: the hasty reform of the Consumer Product Safety Commission. Hous L Rev 46:545-583

Gurgel C (2010) Pediatric lead exposure from imported Indian spices and cultural powders. Pediatrics 125:827–835

Hardwood Floors (2012) Refinishing old floors leads to high levels of lead exposure. http://hardwoodfloorsmag.com/editors/blog/default.aspx?id=1146. Accessed 10 Jun 2013

Heimlich JE (1997) Lead. Ohio State University Extension Fact Sheet. http://ohioline.osu.edu/cd-fact/0193.html. Accessed 16 Apr 2013

Hickman M (2012) Lead in garden hoses? Study finds high levels of toxic chemicals in gardening gear. Mother Nature Network http://www.mnn.com/your-home/organic-farming-gardening/blogs/lead-in-your-garden-hose-study-finds-high-levels-of-toxic-. Accessed 8 Jun 2013

Hunt WG, Watson RT, Oaks JL et al (2009) Lead bullet fragments in venison from rifle-killed deer: potential for human dietary exposure. PLos ONE, 4; doi: 10.1371/journal.pone.0005330

Huo X, Peng L, Xu X et al (2007) Elevated blood lead levels of children in Guiyu, an electronic waste recycling town in China. Environ Health Perspect 115(7):1113-1117 http://www.ncbi.nlm.nih.gov/pmc/articles/PMC1913570/#__ffn_sectitle. Accessed 28 Jun 2013

Insightful Nana (2008) Lead poisoning and crock pots. http://insightfulnana.com/home-garden/housekeeping-home-garden/lead-poisoning-and-crock-pots/html. Accessed 10 Jun 2013

Janot J (2013) E-waste, Central Minnesota Water Education Alliance. http://www.mnwaterconnection.com/e-waste. Accessed 28 Jun 2013

Karnpanit W, Beniapong W, Srianuiata S et al (2010) Nitrate, lead and cadmium contents in beans, leafy, fruit and root vegetables from conventional, good agricultural practice and organic cultivations. Toxicol Lett 196(Supplement 17):S342

Knight-Ridder News Service (1997) State, federal officials alerted by 6th-grader's report. Boy documents use of lead in food cans. http://articles.baltimoresun.com/1997-03-23/news/1997082044_1_lead-soldered-canned-foods-lead-in-food. Accessed 10 Jun 2013

Laquatra J, Coyne L, Pierce M (2008) Lead in Christmas lights. J Environ Health 71:8–11

Liu S, Hammond K, Rojas-Cheatham A (2013) Concentrations and potential health risks of metals in lip products. Environ Health Perspect 121(6):705–710

Lucas G (1999) State files suit over lead in keys – attorney general wants warning labels on brass variety. San Francisco Chronicle, October 13

Main E (2013) 5 sneaky sources of lead. Prevention. http://www.prevention.com/health/healthy-living/lead-hiding-households. Accessed 10 Jun 2013

Markowitz G, Rosner D (2013) Lead wars: the politics of science and the fate of America's children. University of California Press, Berkeley

Massachusetts Water Resources Authority (2012) Lead in faucets – questions and answers. http://www.mwra.state.ma.us/04water/html/Lead_Faucets.htm. Accessed 20 Jun 2013
McCartney R (2010) Debacle over drinking water deals a blow to CDC and EPA. The Washington Post, December 5 http://www.washingtonpost.com/wp-dyn/content/article/2010/12/04/AR2010120400223.html. Accessed 10 Jun 2013
Natural Resources Defense Council (NRDC) (2012) Lead in her lunch box? Smarter Living. http://www.nrdc.org/living/healthreports/lead-lunchbox.asp. Accessed 8 Jun 2013
Nguyen V, Putnam J, Carroll J (2013) Lead lurking in purses, wallets, sandals. The Investigative Unit. http://www.nbcbayarea.com/investigations/Lead-Lurking-in-Purses-Wallets-and-Sandals-164285156.html. Accessed 9 Jun 2013
Park SC, Hicks DC (1995) Appropriate methods for reducing lead-paint hazards in historic housing. About.com Architecture http://architecture.about.com/library/bl-preservationbrief-leadpaint.htm#Lead. Accessed 29 Jun 20
Rabin R (2008) The lead industry and lead water pipes: A MODEST CAMPAIGN. Am J Public Health 98:1584–1592
Renner R (2007) Lead pipe replacement should go all the way. Env Sci Technol 41(19): 416637–416638
Rother C (2000) Danger unlocked – keys found to leave behind unsafe amounts of lead. San Diego Union-Tribune. http://www.ayurveda-florida.com/Research_non_ayurvedic_topics_health_disease_related/lead_in_keys_dangerous.htm. Accessed 10 Jun 2013
Sauler E (2013) Cell phone cases test positive for lead. Inquirer News http://newsinfo.inquirer.net/368079/cell-phone-cases-test-positive-for-lead. Accessed 10 Jun 2013
Severns M (2013) Which 20 lipsticks contain the most lead? Mother Jones. http://www.motherjones.com/environment/2013/05/study-lead-metals-lipstick-top-20. Accessed 10 Jun 2013
The Lead Group (2013) Facts and firsts of lead. http://www.lead.org.au/fs/fst29.html. Accessed 16 Apr 2013
U.S. Centers for Disease Control and Prevention (CDC) (2012) Lead. Prevention tips. http://www.cdc.gov/nceh/lead/tips.htm. Accessed 29 Jun 2013
U.S. Environmental Protection Agency (USEPA) (2002) Nitrification. http://water.epa.gov/lawsregs/rulesregs/sdwa/tcr/upload/nitrification.pdf. Accessed 4 June 2014
U.S. Environmental Protection Agency (USEPA) (2012) Legislative recycling mandates. http://www.epa.gov/osw/conserve/materials/ecycling/rules.htm. Accessed 29 Jun 2013
U.S. Food and Drug Administration (FDA) (2011a) Kohl, kajal, al-kahal, or surma: by any name, a source of lead poisoning. http://www.fda.gov/Cosmetics/ProductandIngredientSafety/ProductInformation/ucm137250.htm. Accessed 10 Jun 2013
U.S. Food and Drug Administration (FDA) (2011b) Lead acetate in "progressive" hair dye products. http://www.fda.gov/Cosmetics/ProductandIngredientSafety/ProductInformation/ucm143075.htm. Accessed 10 Jun 2013
Wasson SJ, Guo Z, McBrian JA et al (2002) Lead in candle emissions. Sci Total Environ 296:159–174
Wieman B (2013) Importance of recycling batteries. National Geographic: green living. http://greenliving.nationalgeographic.com/importance-recycling-batteries-2966.html. Accessed 10 Jun 2013
World Health Organization (WHO) (2010) Childhood Lead Poisoning (ISBN 978 92 4 150033 3). http://www.who.int/ceh/publications/childhoodpoisoning/en/. Accessed 16 Apr 2013
Zhang Z, Griffin A, Rahman M et al (2009) Lead contamination of potable water due to nitrification. Environ Sci Technol 43(6):1890–1895

Chapter 10
Methylnaphthalene in Food Packaging and Cadmium in Food Packaging and Household Items: Overview of Exposure, Toxicology, Regulatory Aspects, and Research Needs

Suzanne M. Snedeker

Abstract This chapter provides an overview of two toxicants, methylnaphthalene and cadmium. Both of these chemicals are examples of toxicants for which there are significant research data gaps. Reports of an off-odor and off-flavor in certain breakfast cereals led to a recall of certain Kellogg's®-brand breakfast cereals in 2010, with subsequent identification of methylnaphthalene as the chemical used in the wax liners of these cereal boxes. While orally- and dermally-treated laboratory rodents have displayed pulmonary alveolar proteinosis in response to methylnaphthalene exposure, very few other toxicological endpoints have been evaluated for exposures to either 1- or 2-methylnaphthalene. While cadmium is an established heavy metal toxicant, with the lung, kidney, and immune system as primary target organs, recent studies of non-occupationally exposed populations suggest that the risk of cardiovascular and peripheral artery disease may be associated with higher levels of urinary cadmium. While diet and tobacco use are known sources of non-occupational exposure to cadmium, there is emerging evidence that consumer products also may be a source of exposure to general populations. There are reports of cadmium leaching from glazes and decorations used in ceramics and glassware, and detection in plastic housings of cell phones. There is limited data characterizing exposures from cadmium use as a pigment in household or food contact plastics. More research is needed to characterize exposures as affected by the type of cadmium pigment, type of plastic, effects of wear and age, as well as the leaching of cadmium from end-of-use products in landfills.

Keywords Methylnaphthalene · 1-methylnaphthalene · 2-methylnaphthalene · Hydrocarbon · Polycyclic aromatic hydrocarbon · PAH · Food packaging · Food contact material · Breakfast cereal packaging · Waxes · Pulmonary alveolar

---

S. M. Snedeker (✉)
Independent Consultant, Ithaca, NY, USA
e-mail: smsnedeker@gmail.com

S. M. Snedeker (ed.), *Toxicants in Food Packaging and Household Plastics*, Molecular and Integrative Toxicology, DOI: 10.1007/978-1-4471-6500-2_10,

proteinosis · Cadmium · Heavy metal · Cadmium pigments · Household plastics · Food contact plastic · Electronic waste · Ceramic glazes · Post-consumer recycled plastic · High-density polyethylene · HDPE · Polyvinyl chloride · PVC · Acrylonitrile-butadiene styrene · ABS · Carcinogenicity · Lung cancer · Renal damage · Osteoporosis · Endocrine disruption · Estrogenicity

10.1 Key Take Home Points

10.1.1 Methylnaphthalene

- The chemical methylnaphthalene was detected in the wax paper liners of certain Kellogg's®-brand breakfast cereals in 2010.
- The Kellogg® Company issued a voluntary recall of 28 million boxes of breakfast cereals that used methylnaphthalene in the secondary food packaging.
- There is limited information available on the toxicology of 1- and 2-methylnaphthalene.
- More research is needed to assess the toxicology of methylnaphthalenes and the potential for human exposure through their use in food contact applications.

10.1.2 Cadmium

- Cadmium is a heavy metal and is a known human carcinogen.
- Cadmium exposure can cause permanent damage to the lung and kidney, affect bone health and the immune system, and may be an endocrine disruptor.
- While primary non-occupational exposures to cadmium are via the diet and tobacco products, exposure may also occur via use of consumer products.
- Cadmium, used as a pigment in plastics and ceramic glazes, has been detected in some types of household plastics, food packaging, and from ceramics and glassware.
- More research is needed to determine the ability of cadmium used in pigments to migrate out of household plastics and food contact materials.

10.2 Introduction

The chemicals in this chapter, methylnaphthalene[1] and cadmium, are not presented because they are related to each other, but rather because the scope of the subject matter for each chemical is limited, hence they are presented in a combined

[1] When possible, information on 1-methylnaphthalene and 2-methylnaphthalene is presented.

chapter. The primary purpose of this chapter is to provide a framework to justify research needed to better characterize human exposures from their use in food packaging and household products, and to better characterize toxicological and potential health effects in general, non-occupationally exposed populations.

Reports of an off-flavor and off-odor from the secondary food packaging, wax liners of breakfast cereals, resulted in a large recall of affected cereals by the Kellogg® Company in 2010, and the chemical responsible was identified as methylnaphthalene (Kellogg 2010; Schroeder 2010). However, very little toxicological information is available on this chemical as noted by several federal agencies (USEPA 2003; ATSDR 2005). Cadmium, though its toxicological effects from occupational exposures have been well characterized, is an example of a chemical for which we have limited information on exposures though consumer products, especially from use in pigments used in ceramic and glass glazes, household plastics, and food packaging plastics, as well as use in electronics.

The brief overview format used in this chapter focuses on several aspects: available information on exposure through the chemical's use in food contact materials (food packaging, food containers, and utensils) and household products, a summary of known toxicological endpoints, regulations on use in food contact materials and household products, and further research needs.

10.3 Methylnaphthalene

10.3.1 Methylnaphthalenes in Food Packaging

The first brief in this chapter will include the history of the detection and subsequent recall of cereal boxes with liners containing the polycyclic aromatic hydrocarbon (PAH) methylnaphthalene, and present an overview of the limited data currently available on the toxicology of 1- and 2-methylnaphthalene. Significant data gaps on the toxicology and potential human health effects of methylnaphthalenes are noted, and recommendations for further needed research are made.

10.3.1.1 Methylnaphthalene Released from Cereal Box Liners

The Kellogg Company issued a voluntary recall of certain breakfast cereals on June 25, 2010, in response to consumer reports of off-flavor and off-odor (Kellogg 2010). The recall also was posted on the United States (U.S.) Food and Drug Administration (FDA) website (FDA 2010). The cereal recall included of 28 million boxes of Kellogg's® Corn Pops®, Kellogg's® Honey Smacks®, Kellogg's® Froot Loops® and Kellogg's® Apple Jacks®. The press release issued by the Kellogg Company stated that some consumers may be sensitive to the off-flavor

Fig. 10.1 Chemical structures of 1-methylnaphthalene and 2-methylnaphthalene

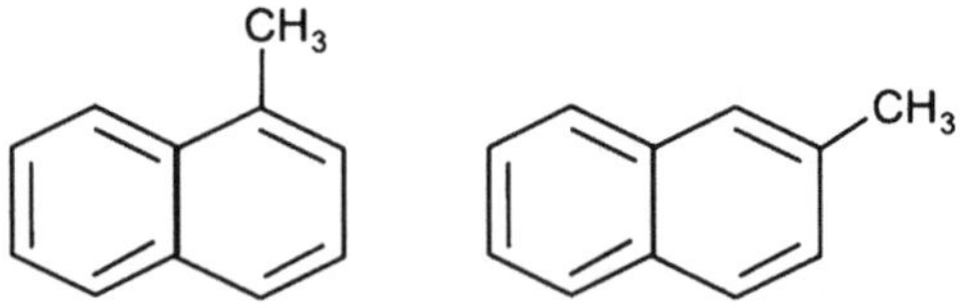

and odor, and recommended that the recalled products not be eaten since consumers could experience temporary symptoms, including nausea and/or vomiting. *Food and Business News* released an article on July 13, 2010, stating that the Kellogg Company had identified the chemicals responsible for the off-flavor and off-odor as hydrocarbons, including methylnaphthalene ($C_{11}H_{10}$), which is used in the food industry in protective wax coatings of food and in the film liners of the cereal boxes (Schroeder 2010). While this article quoted a Kellogg's spokesperson that the wax is approved for use by the FDA, the name and chemical composition of the wax was not given.

Media reports differ on whether the chemical identified was methylnaphthalene (Harrington 2010) or 2-methylnaphthalene (Layton 2010). It appears that some form of methylnaphthalene was used in the wax paper liners (see Fig. 10.1 for chemical structures of 1- and 2-methylnaphthalene), Kellogg identified suspect samples of the wax paper liner, reserved tested samples, and then destroyed the remaining inventory of the wax paper to keep it out of future production (Harrington 2010). Kellogg does not use either 1-methylnaphthalene or 2-methylnaphthalene for the manufacture of cereal box liners in Europe (Harrington 2010). The levels of 1- or 2-methylnaphthalene found in the tainted cereals in the U.S. have not been made public by either the FDA or the Kellogg Company.

10.3.2 Toxicology of 1-Methylnaphthalene and 2-Methylnaphthalene

There is relatively limited information on the toxicology of 1- and 2-methylnaphthalene. Available information has largely been published in two reports. A Toxicological Review of 2-methylnaphthalene (CAS No. 91-57-6) was released by the U.S. Environmental Protection Agency (USEPA) in 2003 (USEPA 2003), and a Toxicological Profile on 1-methylnaphthalene (CAS No. 90-12-0) and 2-methylnaphthalene (CAS No. 91-57-6) was released by the Agency for Toxic Substances and Disease Registry (ATSDR) in 2005 (ATSDR 2005). Discussion on the available toxicological data cited by these reports includes comments on the deficiencies in the data.

The USEPA deemed the existing data on the carcinogenicity of 2-methylnaphthalene from a cancer bioassay conducted in mice to be inadequate to assess the potential for carcinogenicity in humans (USEPA 2003). There were indications of lung adenomas and carcinomas when combined in B6C3F_1 mice in low-dose males [54.3 milligrams per kilogram per day (mg/kg per day)] in the 81-week oral toxicology study cited (Murata et al. 1997), however, there was no data supporting a dose-response effect, since there was no evidence of carcinogenicity in treated high-dose males, nor in the low- or high-dose treated female mice compared to controls.

Non-cancer effects noted in the USEPA report from the Murata et al. (1997) mouse oral-feeding study included increased incidence (43–55 % incidence) of pulmonary alveolar proteinosis in both male and female 2-methylnaphthalene treatment groups at both low- and high-dose levels (USEPA 2003). There was no evidence of 2-methylnaphthalene treatment-related histopathological changes to the liver or kidney, indicating that the lung appeared to be the primary target organ. An earlier study using a dermally applied mixture of 1- and 2-methylnaphthalene showed 100 % incidence of pulmonary alveolar proteinosis in female B6C3F_1 mice (Murata et al. 1992). The USEPA report's authors suggested that either the acetone-applied dermal route may have resulted in more methylnaphthalene being absorbed than administration via the oral route, or that 2-methylnaphthalene may have been metabolized initially in the liver in the dietary study, leading to decreased pulmonary toxicity compared to laboratory mice exposed to the methylnaphthalene mixture via the dermal route (USEPA 2003).

While the USEPA report cited some evidence of T-cell-independent and -dependent immunity in mice injected with a mixture of PAHs that included 2-methylnaphthalene, is was not clear if the effects were due to 2-methylnaphthalene or other PAHs in the mixture. The USEPA report noted there was no evidence of 2-methylnaphthalene's mutagenicity in *Salmonella typhimurium* bacterial mutation tests. The USEPA report also stated that there was a lack of information on many toxicological endpoints for 2-methylnaphthalene, including little information on its developmental, reproductive, or neurological toxicity, as well as no available toxicology data on humans exposed to this chemical via an oral route (USEPA 2003). Overall, the USEPA reported concluded that the primary effect of 2-methylnaphthalene exposure is pulmonary toxicity (from studies in mice), though it is not known if the parent compound or metabolites are responsible for the observed pulmonary alveolar proteinosis.

The ATSDR report cited a lack of available data in humans or animal models on effects from inhalation exposure to either 2-methylnaphthalene or 1-methylnaphthalene (including acute toxicity, mortality incidence, and systemic effects), and noted that there was no available information on respiratory, cardiovascular, gastrointestinal, musculoskeletal, renal, hepatic, ocular, immunological, neurological, reproductive, developmental, or cancer effects (ATSDR 2005). Data presented in the ATSDR report on toxicological effects with oral exposure to methylnaphthalene compounds were largely based on the oral toxicological studies of Murata and colleagues on 1-methylnaphthalene (Murata et al. 1993) and 2-methylnaphthalene (Murata et al. 1997). This included the effects on pulmonary toxicity and lung tumor incidence, which were previously noted in the 2003 USEPA report. The ATSDR Toxicological Profile stated that the Murata et al. oral feeding studies in mice indicated some hematological effects from methylnaphthalene exposure, but these effects were not consistent across gender, dose, or 1- versus 2-methylnaphthalene exposure. The authors of the report concluded that the hematological system is probably not a target for methylnaphthalene compounds (ATSDR 2005). As with the USEPA report, the ATSDR Toxicological Profile noted a lack of evidence for the genotoxicity of 1- and 2-methylnaphthalene.

Very little is known about the metabolism of methylnaphthalene, though the ATSDR (2005) report suggested that for 2-methylnaphthalene, the methyl group and various points on the ring structure could undergo oxidation by CYP monooxygenases. There is some evidence from rodent studies that over 50 % of 2-methylnaphthalene is oxidized to produce 2-hydroxymethylnaphthalene, and that ring epoxidation of 2-methylnaphthalene can occur (15–20 %) at various positions on the ring structure via CYP isozymes. The ATSDR reported noted there is little information from any species on the mechanism of absorption (via any route) and the excretion of methylnaphthalene compounds.

Since these toxicity profiles were published by the USEPA and the ATSDR, other toxicological effects of methylnaphthalene exposure have been reported. This includes effects on neurological systems, with increased pain sensitivity and depressed respiration rate with inhalation exposure to 1- and 2-methylnaphthalene in rats and mice (Korsak et al. 1998). Other information on these compounds includes some data on the toxicokinetics of inhaled 2-methylnaphthalene in male Wistar rats. This rodent inhalation study demonstrated that with exposure at 200 or 400 mg 2-methylnaphthalene per cubic meter, blood levels in either species increased during the first 2 hours (h) of exposure, and then plateaued (Swiercz et al. 2010). No published toxicokinetic data in humans for 1- or 2-methylnaphthalene could be located.

10.3.3 Food Regulations

FDA Methylnaphthalene is not listed in the FDA's *Generally Regarded as Safe* (GRAS) on-line inventory (FDA 2014a), or in FDA's *List of Indirect Additives Used in Food Contact Substances* (FDA 2011). There is a listing for methylnaphthalene as 1-methylnaphthalene (CAS No. 90-12-0, Doc. No. 0995) in FDA's *Everything Added to Food in the United States* (EAFUS) database (FDA 2013b), with the document type listed as "ASP" (fully up-do-date toxicology information has been sought). No Regnum numbers for links to the U.S. Code of Federal Regulations, Title 21, where approved uses of food contact substances are listed, was provided for 1-methylnaphthalene entry in the EAFUS database.

10.3.4 Summary and Research Needs for Methylnaphthalene

Consumers reported off-flavors and off-odors from certain Kellogg's-brand breakfast cereals in 2010. After a product recall, the chemicals responsible were identified and appeared to be hydrocarbons. This included the PAH methylnaphthalene[2] that was used in the wax paper liner of the cereal boxes. The actual concentration of methylnaphthalene detected in the recalled cereals was not made public by the Kellogg Company or the FDA.

Toxicological reviews published by the USEPA and the ATSDR have noted that the primary target organ for oral or inhalation exposure to methylnaphthalene compounds is the lung, and in chronic exposure studies conducted in laboratory mice the primary adverse effect appears to be induction of pulmonary alveolar proteinosis. Because of limitations in existing animal cancer bioassays on methylnaphthalene compounds, the USEPA deemed available carcinogenicity data insufficient to make conclusions on the potential for human carcinogenicity. Both the USEPA and ATSDR reviews cited a large number of toxicological endpoints for which there is a lack of data, including acute toxicity, mortality incidence, systemic effects, respiratory, cardiovascular, gastrointestinal, musculoskeletal, renal, hepatic, ocular, immunological, neurological, reproductive, and developmental effects from inhalation or oral routes of exposure to 1- or 2-methylnaphthalene. There is also a lack of data on absorption, excretion, metabolism or pharmacokinetics in humans. While methylnaphthalene is used in food contact materials, and has been the subject of a large recall of breakfast cereals, there is a lack of data on many toxicological endpoints in animal models and little information on exposure, health effects or pharmacokinetics in human populations. Therefore, more research is needed to fill the significant gaps in current knowledge of the toxicology of 1- and 2-methylnaphthalene, as well addressing the lack of information on the potential for exposure to these chemicals through their use as food contact substances.

10.4 Cadmium

10.4.1 Cadmium Exposure

10.4.1.1 Overview of Non-occupational Sources of Cadmium Exposure

Cadmium is a heavy metal found in agricultural soils. Source include natural levels in the soil, as well cadmium as from phosphate-based fertilizers, atmospheric deposition from incinerated municipal waste, and from municipal waste that may be used in composting (ATSDR 2012). The U.S. Consumer Product Safety Commission (CPSC) has noted that while cadmium is widely found in the environment and primary non-occupational exposures in humans are via the diet and through use of tobacco products, exposure may also occur via contact with

[2] Some uncertainty if the identified compound was 1-methylnaphthalene or 2-methylnaphthalene, or a combination of both chemicals.

consumer products (CPSC 2010a). Exposure to cadmium in consumer products can occur due to cadmium's use in metal alloys used in soldering and electroplating and in rechargeable nickel-cadmium batteries (use in decline due to replacement by nickel-metal hydride batteries), as well as use in pigments for paints and plastics. The CPSC has expressed concern that heavy metals, including cadmium and lead, have been detected in children's jewelry, and exposure in children can occur via mouthing objects and hand-to-mouth behavior (CPSC 2010a). However, exposure to cadmium from other consumer products, including food contact and household plastics, and glass and ceramic products, has received less attention. Some of these potential exposures are discussed below.

10.4.1.2 Cadmium in Household Plastics, Food Contact Plastics, and Glass and Ceramics

Historically, cadmium has been used extensively both as a pigment in plastics, especially in high-density polyethylene (HDPE) and acrylonitrile-butadiene styrene (ABS) (Wilson et al. 1982). In the early 1980s about 20–40 % of cadmium's total usage was as pigments, and 80 % of cadmium pigments were used in plastics. Cadmium has been used in artist's paints since the 1840s, though this represents a small percentage (5–7 %) of its pigment usage. Because of health concerns, cadmium's use in pigments has declined in the U.S. from 2,375 tons in 1987 to 550 tons in 1995 (JustPaint 2014). Cadmium, like the heavy metal lead (see Chap. 9) has been used as a stabilizer in polyvinyl chloride (PVC) plastics (Batzer 1983; Wilson et al. 1982). Most of the cadmium in the municipal waste stream is from its use in plastics and pigments, with estimates that 36 % of cadmium is discarded as materials that are combustible, appearing as bottom ash in municipal incinerators or as atmospheric particulates from fly ash (Korzun and Heck 1990). Household waste, including red, orange, and yellow plastics, have been identified as sources of cadmium, with levels ranging from less than 50 mg/kg to over 1,000 mg/kg (Bode et al. 1990).

There is relatively little information on historic or current levels of cadmium in food packaging. Cadmium has been detected in paper products, including tea bag tissue and unbleached craft paper at 0.3 mg/kg (Castle et al. 1997), and in silver food grade foils used as coloring agents in India at 97 micrograms per gram (μg/g) (Das et al. 2005). A study conducted in 1983 detected cadmium in plastic plates at 0.45–5.6 μg per 100 square centimeters (cm^2) (Preda et al. 1983). A 2009 methods development study determined levels of cadmium in different types of food packaging plastics (Martinis et al. 2009). The food contact plastic with the highest level of cadmium was PVC ranging from 5.1 to 10.4 mg/kg. Cadmium levels for other plastics ranged from non-detectable (n.d.) to 0.2 mg/kg for polyethylene terephthalate (PET); n.d. to 0.12 mg/kg for HDPE; n.d. to 0.11 mg/kg for low-density polyethylene (LDPE); n.d. to 0.14 mg/kg for polypropylene (PP); and n.d. to 0.22 mg/kg for polystyrene (PS). These levels were higher than those detected in food packaging in a 2011 methods development study (Perring et al. 2001).

Perring and colleagues reported concentrations of cadmium in food packaging tenfold lower, in the range of 0.01–0.03 mg/kg for PET, PP, and PS. Only polyethylene (PE) had levels as high as 69.7 mg/kg, while cadmium in HDPE was undetectable in this study.

Only a few published studies have evaluated the migration of cadmium from food contact plastics. Cadmium was not found to migrate from plastic tableware (treated with heat and 4 % acetic acid) that contained 6,863 parts per million (ppm) cadmium (Hosogai et al. 1992). Glass baby bottles had no evidence of cadmium migration (Kubwabo et al. 2009). However, cadmium has been the subject of a CPSC recall due to its detection in decorative designs used on collectable movie-themed drinking glasses sold through McDonalds restaurants (CPSC 2010b). These glasses were voluntarily recalled by the CPSC with cooperation from McDonalds in June 2010. No other reports of product recalls due to cadmium's use in food contact materials were located on the CPSC product recall site (CPSC 2014).

There is relatively little information on cadmium's use in household plastic consumer products. While the CPSC did conduct tests on 11 PVC children's products that Greenpeace had alleged to contain hazardous lead and cadmium levels, the CPSC found eight products had no or trace levels of cadmium, and one other product was not considered hazardous because exposure from this product was deemed unlikely (CPSC 1997a, b). It should be noted that in the U.S. cadmium is not banned from children's products, though the CPSC has suggested that based on the USEPA's reference dose for cadmium of 1 μg/kg body weight per day, chronic ingestion of cadmium should have the following limits: 1 year old, not to exceed 9.2 μg/day; 3 year old, not to exceed 13.5 μg/day; 6 year old, not to exceed 20.2 μg/day (CPSC 1997b).

Other researchers have conducted modeling migration tests to determine if cadmium can be released from vinyl or acrylic resin enamel (used in paint films for children's toys); release was dependent on the method and solvents used, though cadmium migrated more readily from the acrylic than vinyl enamel (Kawamura et al. 2009). This and a previous study (Kawamura et al. 2006) called for additional research to determine release of cadmium from children's toys.

There is emerging data indicating that cadmium can be detected in the waste plastic housings of mobile phones (Nnorom and Osibanjo 2009). The concentrations of cadmium in the plastic phone housings were 69.9 ± 145 mg/kg (mean ± SD), with the high standard deviation indicating a wide range of levels in the samples and between brands of phones. The maximum level of cadmium detected was 1,005 mg/kg. Since levels of cadmium were determined via acid digestion of the plastic, it is not clear to what extent consumers would be exposed to cadmium via handling plastic phones. It is possible that those working in waste electronic recycling, where low temperatures are used to melt plastic housings and open burning is used to recover precious metals from mobile phones, could be exposed to cadmium and other metals via inhalation or dermally. Given the short life of mobile phones and other electronics such as personal computers, additional studies are needed to determine the level of consumer, occupational, and environmental releases of cadmium from waste electronics and plastic housings of electronics (Santos et al. 2010).

While there is some information on cadmium pigments leaching from plastics in landfill sites, this information is over 30 years old (Wilson et al. 1982). Since there have been large increases in the use of plastics in household and food packaging applications over the last several decades, it is imperative that more research be done to better characterize exposure to cadmium from food packaging and household plastics. One area where we are lacking data is the level of cadmium in the painted surfaces of toys and from handling plastic toys. Some of the questions that need to be resolved include: (1) is cadmium transferable from the surface of household plastic items to the skin via handling, (2) can cadmium be ingested from hand-to-mouth behavior via use of household plastics, (3) what is the extent of cadmium migration from pigmented food contact plastics to foodstuffs, (4) can release rates of cadmium from plastic items be affected by wear and tear and the age of the plastic, (5) is cadmium released at different rates depending on the type of cadmium pigment, type of plastic, or type of food packaging, and (6) when household items and food packaging are discarded in landfills, how do environmental conditions (e.g., time, temperature) affect leaching of a cadmium pigment from the plastic to the environment (e.g., soil, atmosphere, and drinking water sources)?

10.4.2 Toxicology of Cadmium

The toxicology of cadmium has been extensively reviewed by the Agency for Toxic Substances and Disease Registry (ATSDR) (2012), and by Bertin and Averbeck (2006). Target organs for cadmium toxicity depend on the route of exposure, with the kidney and bone being the most sensitive organs with oral exposure, and the lung and kidney being the primary target organs with inhalation exposure (ATSDR 2012). Adverse effects of chronic exposure to cadmium in humans include irreversible renal damage (kidney tubular damage, reduced glomerular filtration rate, and renal failure), lung disorders (obstructive airway disease and emphysema), bone disorders (Itai–Itai disease, osteoporosis, and changes in bone mineralization), and immune-suppression (Bertin and Averbeck 2006; IARC 2012; Borchers et al. 2010). There is some evidence from estrogen-dependent tumor cells lines (MCF-7 cells, E-screen assay) and estrogen receptor (ER) dependent transcriptional assays that cadmium chloride ($CdCl_2$) is a weak estrogen (Choe et al. 2003; Isidori et al. 2010), though the magnitude of estrogenicity differed in the E-screen assay between laboratories. In contrast to estrogenicity demonstrated when using in vitro tests, cadmium nitrate $Cd(NO_3)_2$ did not show estrogenic effects in an in vivo test of vitellogenin induction in juvenile goldfish (Isidori et al. 2010), suggesting different sensitivities due to different test systems or the different types of cadmium.

The cancer risks of cadmium to humans have been evaluated by the International Agency for Research on Cancer (IARC) in monographs published in 1973, 1976, 1987, 1993, and most recently in 2012 (IARC 2012), and by the National Toxicology Program (NTP) in their *12th Report on Carcinogens* (NTP 2011). Inhalation of cadmium has been associated with lung cancer in metal smelting and nickel-cadmium battery workers, with evidence that this dose-related effect was not explained by exposure to other metals (such as arsenic in smelter workers) or tobacco use (Boffetta 1993; IARC 2012). Evidence for occupational exposure to cadmium and risks of prostate and kidney cancer show some, but limited evidence of carcinogenicity. Other studies of cadmium exposure and cancer mortality or cancer incidence for other cancer sites, including the pancreas and breast, have been suggestive, but do not provide sufficient evidence of a causal relationship (IARC 2012). The evidence that cadmium compounds cause lung cancer in humans and the association of cadmium exposure with prostate and kidney cancer were cited by IARC as sufficient evidence that cadmium and cadmium compounds are carcinogenic to humans (rated as GROUP 1 carcinogens) (IARC 2012). Similarly, the *12th Report on Carcinogens* lists cadmium compounds as *known to be human carcinogens* (NTP 2011).

Mechanistically, several explanations have been proposed for the toxic and carcinogenic effects of cadmium on cellular systems. They include: increased oxidative stress via inhibition of anti-oxidant enzymes that may induce oxidative DNA damage; inhibition of DNA repair; induction of genetic damage including DNA strand breaks and chromosomal damage; and effects on cell proliferation mediated by effects on protein kinases; increased expression of proto-oncogenes; and inactivation of tumor suppressor protein p53 (Beyersmann and Hartwig 2008; NTP 2011).

While many of the toxic and cancer-related effects of cadmium compounds have been documented in occupationally or in highly exposed human populations, there also is evidence supporting adverse health effects with lower exposures in the general population in the U.S. and in Asia. Several of these studies are summarized below.

The National Health and Nutrition Examination Surveys (NHANES) have examined the association between cadmium levels in urine and health endpoints in the U.S. general population. These studies point to possible adverse health effects from non-occupational exposures to cadmium. This includes a positive association of urinary and serum cadmium levels with the incidence of cardiovascular disease using pooled data from NHANES 1999–2006 (Agarwal et al. 2011). The authors suggest that some of the association of smoking with cardiovascular disease may partially be mediated by cadmium exposure in smokers. Urinary cadmium levels were positively associated with peripheral artery disease (Shirai et al. 2010) in a cross-section of male and female adults over 40 years of age in NHANES 1999–2000, suggesting that cadmium may have a role in the risk of developing atherosclerosis (Navas-Acien et al. 2005). In this study, urinary cadmium levels were 36 % higher in those with peripheral artery disease, and analysis indicated this association was not explained by smoking status. Other associations observed in NHANES studies (2007–2008) include a positive association between urinary cadmium excretion and serum levels of the thyroid hormones triiodothyronine (T_3) and thyroxine (T_4) (Yorita Christensen 2013). Other associations between urinary cadmium in general populations include a significant negative association of cadmium in the urine of 78 pregnant women with their infant's birth weight ($r = -0.271$) in a small study conducted in Japan. In contrast, body mass index and waist circumference were found to be inversely associated with urinary cadmium level in NHANES 1999–2000 and 2001–2002 participants (Padilla et al. 2010). There was no explanation for why urinary cadmium was inversely associated with central obesity.

10.4.3 Regulations Restricting Cadmium in Household Plastics and Food Contact Materials

10.4.3.1 Household Plastics

Underwriter's Laboratory (UL) A commonly use for disposable plastic includes bags used for groceries, retail shopping, garbage, and other uses. Many of these plastic bags are marked as recyclable. The UL released a standard in 2012 called UL 126. This standard has testing requirements for biodegradation ability of plastic bags, and states that certified plastic film should not intentionally contain heavy metals, including cadmium (UL 2012).

European Union (EU) The EU set several directives limiting levels of cadmium in pigments in plastics and paints, as a stabilizer, and in packaging in the early to mid-1990s. The 1991 EU Directive 91/338/EEC restricted the use of cadmium in pigments and stabilizers[3] (CoEC 1991). This included setting the maximum level of cadmium used as a pigment in many types of plastics and in paint to 0.01 % by mass of cadmium (as the metal), and the level of cadmium used as a stabilizer in PVC plastics also was restricted to 0.01 % by mass (0.01 % is equivalent to 100 mg cadmium per kg of plastic). EU legislation set in 1994 specified that the combined levels of the heavy metals lead, cadmium, mercury, and hexavalent chromium in packaging or packaging components could not exceed 100 mg/kg as set forth in the European Parliament and Council Directive 94/62/EC (EPCEU 1994). The EU restricted the use of cadmium in a number of plastic materials as a part of amendments to the Registration, Evaluation, and Authorisation and Restriction of Chemicals (REACH) regulations in 2012 (EU 2012). The EU's regulation No. 835/2012 amends REACH specifying that cadmium must be below 0.01 % by weight of the plastics listed in this regulation (plastics listed include: PVC, polyurethane, LDPE, cellulose acetate, cellulose acetate butyrate, epoxy resins, melamine-formaldehyde, urea-formaldehyde,

[3] These restrictions in Directive 91/338/EC did not apply when cadmium was used as a colorant in the specified plastics or as a stabilizer in PVC for "safety reasons" (CoEC 1991).

unsaturated polyesters, PET, polybutylene terephthalate, general purpose PS, acrylonitrile methyl methacrylate, cross-lined PE, high-impact PS, and PP). This amendment does not limit the cadmium content of other commonly used plastics, including HDPE, ABS, or poly(methyl methacrylate) .

In 2012, the European Chemicals Agency (ECHA) called for more information on the use of cadmium and cadmium compounds in plasticsplastics (ECHA 2012). The ECHA was considering whether to increase the scope of the EU's current restrictions of cadmium to include all plastic materials. The identification of alternatives and assessment of risks of the alternatives to human health and the environment were to be part of the ECHA review. During a similar timeframe, legal action has been taken against the EU's European Commission by several cadmium industry-based groups that challenged the risk assessments used as a basis for earlier 2011 REACH legislation restricting the levels of cadmium in plastics, including questioning the scope and applicability of studies assessing the leaching of cadmium from different types of plastics in landfills (CURIA 2013). Part of the argument was that studies were largely based on cadmium leaching from PVC plastics, and more extensive risk assessments needed to be extended to determine the extent of cadmium leaching from other types of plastics.

10.4.3.2 Food Contact Materials

European Union The Council of the European Communities issued directives limiting cadmium levels in ceramics in 1984. Directive 84/500/EEC recognized that cadmium can leach from cadmium glazes or decorations used for ceramic vessels and cookware, and specified maximum allowable extractable levels dependent on the article: flat or non-fillable articles, 0.07 mg per square decimeters (dm^2); fillable articles, 0.3 mg per liter (L); and cookware with a capacity larger than 3 L, 0.1 mg/L (CoEC 2007). This directive specified tests to be conducted for determining the level of cadmium migration from ceramic articles, using 4 % acetic acid as the food simulant.

The EU's Directive 91/338/EC of 1991 (CoEC 1991), and amendments to REACH legislation (EU 2012) restricted the use of cadmium in some but not all plastics to 0.01 % by weight as noted previously in the *Household Plastics* regulation section (Sect. 10.4.3.1). Directive 91/338/EC also restricted the use of cadmium as a stabilizer in PVC products used as "packaging materials (bags, containers, bottles, lids)" to levels less than 0.01 % (by weight) of the product, and prohibited the use of cadmium in plating any metallic equipment or machinery used in food production. Directive 282/2008 regulates recycled plastics that come in contact with food, but specific limits on the levels of cadmium in recycled food contact plastics are not stated in this directive (CoEC 2008).

While directive 94/62/EC restricts the combined levels of the heavy metals lead, cadmium, mercury, and hexavalent chromium in packaging or packaging components to <100 mg/kg (CoEC 1994), the definition of "packaging" is general, and there are not specific references to food contact materials. The definition of packaging in this directive is:

> Packaging shall mean all products made of any materials of any nature to be used for the containment, protection, handling, delivery and presentation of goods, from raw materials to processed goods, from the producer to the user or the consumer (CoEC 1994).

World Health Organization The World Health Organization (WHO) stated in 1989 that leachable cadmium used enamel and pottery glazes are a potential source of contamination, and that cadmium should not be used as a stabilizer or a pigment in food contact plastics, however, specific regulations or actions to limit these uses were not specified in this document (WHO 1989).

Food and Drug Administration Section 6 of the *FDA Laboratory Manual* includes information on determining the level of heavy metals such as lead and cadmium in food, foodware, and housewares (FDA 2013c). Section 6.1.2.1 on *Lead and Cadmium in Foods*, states that there are "no regulatory limits (tolerances) set for toxic elements in foods" (FDA 2013c). Sections on cadmium in foodware and housewares are primarily concerned with contamination from the use of cadmium in glazes on ceramics and enamelware, and from hollowware.

Cadmium can leach from glazes or decorations used in the manufacture of ceramics, especially when acidic liquids (e.g., vinegar or juices) are in contact with the ceramics, and from metal-plated utensils and hollowware. The FDA has issued guidelines for the level of leachable cadmium from imported and domestic foodware used for preparing, serving, or storing food (FDA 2009). The following are levels of cadmium that should not be exceeded per milliliter (ml) of leaching solution from ceramics and utensils as follows: flatware, 0.5 μg/ml; small hollowware, 0.5 μg/ml; and large hollowware, 0.25 μg/ml.

No specific regulatory limits or listings of cadmium compounds, including cadmium pigments, were located in FDA packaging and food contact substances databases at http://www.fda.gov/Food/IngredientsPackagingLabeling/PackagingFCS/default.htm (e.g., *EAFUS*, *Inventory of Effective Food Contact Substance Notifications, List of Indirect Additives Used in Food Contact Substances*, and *Cumulative Estimated Dietary Intake* databases).

Limits are listed in Title 21 of the *U.S. Code of Federal Regulations* for levels of cadmium as an impurity in: (1) *color additives* that are exempt from certification—cadmium must be not more than 15 ppm in bronze powder, copper powder, zinc oxide, and luminescent zinc sulfide; and (2) *food additives* permitted for direct addition to food for human consumption—cadmium must be less 0.1 ppm in baker's yeast protein, not more than 0.05 ppm in zinc methionine sulfate, and less than 0.13 ppm in bakers yeast glycan (FDA 2013a).

The FDA requires manufacturers using recycled plastic (called post-consumer recycled plastic) in food contact-applications demonstrate that the recycling process removes possible contaminants, and guidance on specific testing approaches to be used is provided by the FDA (FDA 2014b). While no specific limits were listed for cadmium in recycled food contact plastics, if the *estimated dietary intake* of a contaminant is less than 1.5 μg per person per day, the FDA considers this exposure to be of negligible risk (FDA 2006).

10.4.4 Summary and Further Research Needs on Cadmium

While non-occupational exposure to cadmium includes dietary sources and use of tobacco products, there is evidence that consumer products also are a source of cadmium exposure to the general public. This includes detection of cadmium in jewelry, in glazes and decorations used in ceramics and glassware, some evidence of detection in food contact plastics, and in some types of household plastics, including plastic housings of electronics. As with many other toxicants, the first evidence of cadmium's toxicity in humans was observed in those with occupational exposure, and these studies provided evidence of lung cancer with inhalation exposure. The lung, kidney, bone, and immune system are the primary target organs for cadmium toxicity in mammals. There is emerging evidence from NHANES studies that there is an association between cadmium levels in urine and other health endpoints in the general U.S. population, including evidence of increased risk of cardiovascular and peripheral artery disease. There is limited evidence that some cadmium compounds may be endocrine disruptors affecting estrogenic pathways and thyroid hormone levels.

While there is legislation in the U.S. and in the E.U. regulating levels of leachable cadmium from ceramic glazes and decorations into cooking, eating, and storage vessels, and legislation restricting levels in recycled plastics, restrictions on levels in household plastics in the EU do not restrict levels of cadmium in all types of plastics (e.g., current REACH legislation does not limit levels of cadmium in HDPE or ABS plastic). Moreover, there is a lack of published studies on the extent to which cadmium leaches from cadmium-pigmented articles (painted surfaces or plastics), and the extent to which migration is affected by the type of pigment, type of plastic, age and wear, and time and temperature. Data is needed both on the ability of cadmium to leach from household plastics when handled by vulnerable populations (e.g., children and pregnant women), and the extent to which food contact plastics may be a source of cadmium in the diet.

References

Agarwal S, Zaman T, Tuzcu EM et al (2011) Heavy metals and cardiovascular disease: results from the National Health and Nutrition Examination Survey (NHANES) 1999–2006. Angiology 62(5):422–429

ATSDR (2005) Toxicological profile for naphthalene, 1-methylnaphthalene, and 2-methylnaphthalene, TP 65, August 2005. U.S. Department of Health and Human Services, Public Health Service, Agency for Toxic Substances and Disease Registry. http://www.atsdr.cdc.gov/toxprofiles/tp.asp?id=240&tid=43. Accessed 3 March 2014

ATSDR (2012) Toxicological profile for cadmium, CAS#: 7440-43-9, September 2012. U.S. Department of Health and Human Services, Public Health Service, Agency for Toxic Substances and Disease Registry. http://www.atsdr.cdc.gov/toxprofiles/tp.asp?id=48&tid=15. Accessed 5 April 2014

Batzer H (1983) Use and possibilities for substitution of cadmium stabilizers. Ecotoxicol Environ Saf 7(1):117–121

Bertin G, Averbeck D (2006) Cadmium: cellular effects, modifications of biomolecules, modulation of DNA repair and genotoxic consequences (a review). Biochimie 88(11):1549–1559

Beyersmann D, Hartwig A (2008) Carcinogenic metal compounds: recent insight into molecular and cellular mechanisms. Arch Toxicol 82(8):493–512

Bode P, De Bruin M, Aalbers TG et al (1990) Plastics from household waste as a source of heavy metal pollution. An inventory study using INAA as the analytical technique. Biol Trace Elem Res 26–27:377–383

Boffetta P (1993) Carcinogenicity of trace elements with reference to evaluations made by the International Agency for Research on Cancer. Scand J Work Environ Health 19(1):67–70

Borchers A, Teuber SS, Keen CL et al (2010) Food safety. Clin Rev Allergy Immunol 39(2):95–141

Castle L, Damant AP, Honeybone CA et al (1997) Migration studies from paper and board food packaging materials. Part 2. Survey for residues of dialkylamino benzophenone UV-cure ink photoinitiators. Food Addit Contam 14(1):45–52

Choe SY, Kim SJ, Kim HG et al (2003) Evaluation of estrogenicity of major heavy metals. Sci Total Environ 312(1–3):15–21

Council of European Communities (CoEC) (1991) Council Directive 91/338/EEC of 18 June 1991 amending for the 10th time Directive 76/769/EEC on the approximation of the laws, regulations and administrative provisions of the Member States relating to restrictions on the marketing and use of certain dangerous substances and preparations. J Euro Commun L 186(7):59–63. http://eur-lex.europa.eu/legal-content/EN/TXT/HTML/?uri=CELEX:31991L0338&rid=1. Accessed 24 April 2014

Council of European Communities (CoEC) (1994) European Parliment and Council Directive 94/62/EEC of 20 December 1994 on packaging and packaging waste. J Euro Commun L 365 37:10–23. The Council of European Communities. http://eur-lex.europa.eu/legal-content/EN/TXT/?uri=CELEX:31994L0062. Accessed 24 April 2014

Council of European Communities (CoEC) (2007) Council Directive of 15 October 1984 on the approximation of the laws of the Member States relating to ceramic articles intended to come into contact with foodstuffs (84/500/EEC). The Council of the European Communities. http://eur-lex.europa.eu/LexUriServ/LexUriServ.do?uri=CONSLEG:1984L0500:20050520:EN:PDF. Accessed 24 April 2014

Council of European Communities (CoEC) (2008) Commission Regulation (EC) No. 283/2008 of 27 March 2008 on recycled plastic materials and articles intended to come into contact with foods and amending Regulation (EC) No. 2023/2006. J Eur Union, L 86/9. The Commission of the European Communities. http://eur-lex.europa.eu/legal-content/EN/TXT/HTML/?uri=CELEX:32008R0282&from=EN. Accessed 24 April 2014

Consumer Product Safety Commission (CPSC) (1997a) CPSC releases lead and cadmium test results on vinyl products, October 9, 1997, release number: 98008. U.S. Consumer Product Safety Commission. http://www.cpsc.gov/en/Newsroom/News-Releases/1998/CPSC-Releases-Lead-and-Cadmium-Test-Results-on-Vinyl-Products/. Accessed 4 April 2014

Consumer Product Safety Commission (CPSC) (1997b) CPSC Staff Report on lead and cadmium in children's polyvinyl chloride (PVC) products, 21 November 1997. U.S. Consumer Product Safety Commission. http://www.cpsc.gov//PageFiles/95613/Pbcdtoys.pdf. Accessed 5 April 2014

Consumer Product Safety Commission (CPSC) (2010a) Cadmium in children's metal jewelry, October 2010. U.S. Consumer Product Safety Commission. http://www.cpsc.gov/PageFiles/115615/cadmiumjewelry.pdf. Accessed 5 April 2014

Consumer Product Safety Commission (CPSC) (2010b) McDonald's recalls movie themed drinking glasses due to potential cadmium risk, June 4, 2010, Release #10-257. U.S. Consumer Product Safety Commission. http://www.cpsc.gov/en/Recalls/2010/McDonalds-Recalls-Movie-Themed-Drinking-Glasses-Due-to-Potential-Cadmium-Risk/. Accessed 4 April 2014

Consumer Product Safety Commission (CPSC) (2014) Consumer Product Safety Commission recalls and news releases; cadmium. U.S. Consumer Product Safety Commission. http://www.cpsc.gov/en/Search/?query=cadmium&filters=recalls&language=en&sort=relevance&date=all. Accessed 4 April 2014

CURIA (2013) Case T-456/11, Judgment of the general court (seventh chamber) 14 November 2013 pertaining to: REACH—Transitional measures concerning the restrictions on the manufacture, marketing and use of cadmium and its compounds—Annex XVII to Regulation (EC) No 1907/2006—Restrictions on the use of cadmium pigments in plastic materials—Manifest error of assessment—Risk analysis. InfoCuria—Case-law of the Court of Justice. http://curia.europa.eu/juris/document/document.jsf?text=&docid=144481&pageIndex=0&doclang=EN&mode=lst& dir=&occ=first&part=1&cid=152956. Accessed 5 April 2014

Das M, Dixit S, Khanna SK (2005) Justifying the need to prescribe limits for toxic metal contaminants in food-grade silver foils. Food Addit Contam 22(12):1219–1223

European Chemicals Agency (ECHA) (2012) Call for evidence on the use of cadmium and cadmium compounds in plastics. European Chemicals Agency. http://echa.europa.eu/documents/10162/13641/background_document_call_for_evidence_cd_en.pdf. Accessed 4 April 2014

European Parliament and the Council of the European Union (EPCEU) (1994) European Parliment and Council Directive 94/62/EEC of 20 December 1994 on packaging and packaging waste. J Eur Commun L 365 37:10–23. European Parliament and the Council of the European Union. http://eur-lex.europa.eu/legal-content/EN/TXT/?uri=CELEX:31994L0062. Accessed 24 April 2014

European Union (EU) (2012) Commission Regulation (EU) No 835/2012 of 18 September 2012 amending Regulation (EC) No 19-7/2006 of the European Parliament and of the Council on the Registration, Evaluation, Authorisation and Restriction of Chemicals (REACH) as regards Annex XVII (cadmium). J Eur Union L 253, Eur Comm Eur Union. http://eur-lex.europa.eu/legal-content/EN/TXT/PDF/?uri=CELEX:32012R0835&from=EN. Accessed 5 April 2014

Food and Drug Administration (FDA) (2006) Section IV. Exposure to chemical contaminants. In: Guidance for industry: use of recycled plastics in food packaging: chemistry considerations. U.S. Food and Drug Administration. http://www.fda.gov/Food/GuidanceRegulation/GuidanceDocumentsRegulatoryInformation/IngredientsAdditivesGRASPackaging/ucm120762.htm. Accessed 10 April 2014

Food and Drug Administration (FDA) (2009) CPG Sec. 545.00 pottery (ceramics); import and domestic-cadmium contaminations. U.S. Food and Drug Administration. http://www.fda.gov/ICECI/ComplianceManuals/CompliancePolicyGuidanceManual/ucm074515.htm. Accessed 9 April 2014

Food and Drug Administration (FDA) (2010) Kellogg Company voluntarily recalls select packages of Kellogg's® Corn Pops®, Kellogg's® Honey Smacks®, Kellogg's® Froot Loops® and Kellogg's® Apple Jacks®, June 5, 2010. U.S. Food and Drug Administration. http://www.fda.gov/safety/recalls/ucm217338.htm. Accessed 9 March 2014

Food and Drug Administration (FDA) (2011) List of indirect additives used in food contact substances; methylnaphthalene search. U.S. Food and Drug Administration. http://www.accessdata.fda.gov/scripts/fcn/fcnNavigation.cfm?filter=methylnaphthalene&sortColumn=&rpt=iaListing. Accessed 27 April 2014

Food and Drug Administration (FDA) (2013a) CRF—Code of Federal Regulations Title 21, search on cadmium; Listing of color additives exempt from certification, Regnums 73.1646—bronze powder, 73.1647—copper powder, 73.1991—zinc oxide, 73.2995—luminescent zinc sulfide; Food additives permitted for direct addtion to food for human consumption 172.325—bakers yeast protein, 172.399—zinc methionine sulfate, 172.898—bakers yeast glycan, and Direct food substances affirmed as generally regarded as safe, Regnum 184.1983. Code of Federal Regulations, U.S. Food and Drug Administration. http://www.accessdata.fda.gov/scripts/cdrh/cfdocs/cfcfr/CFRSearch.cfm. Accessed 10 April 2014

Food and Drug Administration (FDA) (2013b) Everything added to food in the United States; methylnaphthalene, Doc. No. 0995, CAS RN 90-12-0. U.S. Food and Drug Administration.

http://www.accessdata.fda.gov/scripts/fcn/fcnNavigation.cfm?filter=methylnaphthalene&sortColumn=&rpt=eafusListing. Accessed 10 March 2014

Food and Drug Administration (FDA) (2013c) FDA Laboratory Manual, Secton 6—Elemental Analysis; Section 6.1.1 FDA Center for Food Safety, Section 6.1.2 Lead and cadmium in foods and foodware, Section 6.1.2.1 Lead and cadmium in foods, Section 6.1.2.2 Lead and cadmium in housewares. U.S. Food and Drug Administration. http://www.fda.gov/scienceresearch/fieldscience/laboratorymanual/ucm172150.htm. Accessed 10 April 2014

Food and Drug Administration (FDA) (2014a) Generally recognized as safe (GRAS) inventory. U.S. Food and Drug Administration. http://www.fda.gov/Food/IngredientsPackagingLabeling/GRAS/. Accessed 10 March 2014

Food and Drug Administration (FDA) (2014b) Recycled plastics in food packaging. U.S. Food and Drug Administration. http://www.fda.gov/Food/IngredientsPackagingLabeling/PackagingFCS/RecycledPlastics/default.htm. Accessed 10 April 2014

Harrington R (2010) Kellogg packaging problem resolved—says FDA, August 6, 2010. FoodProductiondaily.com. http://www.foodproductiondaily.com/Packaging/Kellogg-packaging-problem-resolved-says-FDA. Accessed 10 April 2014

Hosogai T, Ito S, Tada Y et al (1992) Migration test of lead and cadmium from plastic wares in contact with food. Eisei Shikenjo Hokoku 110:83–85

International Agency for Research on Cancer (IARC) (2012) IARC Monographs on the evaluation of carcinogenic risks to humans, Vol. 100 C, Cadmium and cadmium compounds, In: A review of human carcinogens: arsenic, metals, fibers, and dust, pp. 121–146. International Agency for Research on Cancer, Lyon, France. http://monographs.iarc.fr/ENG/Monographs/vol100C/. Accessed 5 April 2014

Isidori M, Cangiano M, Palermo FA et al (2010) E-screen and vitellogenin assay for the detection of the estrogenic activity of alkylphenols and trace elements. Comp Biochem Physiol C: Toxicol Pharmacol 152(1):51–56

JustPaint (2014) Will cadmium always be on the palette? Alternative pigments are becoming available. Golden Artist Colors. http://www.goldenpaints.com/justpaint/jp4article2.php. Accessed 11 March 2014

Kawamura Y, Kawasaki C, Mine S et al (2006) Contents of eight harmful elements in baby toys and their migration tests. Shokuhin Eiseigaku Zasshi 47(2):51–57

Kawamura Y, Mutsuga M, Yamauchi T et al (2009) Migration tests of cadmium and lead from paint film of baby toys. Shokuhin Eiseigaku Zasshi 50(2):93–96

Kellogg (2010) Kellogg Company voluntarily recalls select packages of Kellogg's® Corn Pops®, Kellogg's® Honey Smacks®, Kellogg's® Froot Loops® and Kellogg's® Apple Jacks®, News Release, June 25, 2010. Kellogg's Company. http://newsroom.kelloggcompany.com/index.php?s=27529&item=76331. Accessed 9 March 2014

Korsak Z, Majcherek W, Rydzynski K (1998) Toxic effects of acute inhalation exposure to 1-methylnaphthalene and 2-methylnaphthalene in experimental animals. Int J Occup Med Environ Health 11(4):335–342

Korzun EA, Heck HH (1990) Sources and fates of lead and cadmium in municipal solid waste. J Air Waste Manage Assoc 40(9):1220–1226

Kubwabo C, Kosarac I, Stewart B et al (2009) Migration of bisphenol A from plastic baby bottles, baby bottle liners and reusable polycarbonate drinking bottles. Food Addit Contam Part A Chem Anal Control Expo Risk Assess 26(6):928–937

Layton L (2010) U.S. regulators lack data on health risks of most chemicals, August 2, 2010; A01. The Washington Post. http://www.washingtonpost.com/wp-dyn/content/article/2010/08/01/AR2010080103469.html. Accessed 9 March 2014

Martinis EM, Olsina RA, Altamirano JC et al (2009) On-line ionic liquid-based preconcentration system coupled to flame atomic absorption spectrometry for trace cadmium determination in plastic food packaging materials. Talanta 78(3):857–862

Murata Y, Denda A, Maruyama H et al (1993) Chronic toxicity and carcinogenicity studies of 1-methylnaphthalene in $B6C3F_1$ mice. Fundam Appl Toxicol 21(1):44–51

Murata Y, Denda A, Maruyama H et al (1997) Chronic toxicity and carcinogenicity studies of 2-methylnaphthalene in B6C3F_1 mice. Fundam Appl Toxicol 36(1):90–93
Murata Y, Emi Y, Denda A et al (1992) Ultrastructural analysis of pulmonary alveolar proteinosis induced by methylnaphthalene in mice. Exp Toxicol Pathol 44(1):47–54
Navas-Acien A, Silbergeld EK, Sharrett R et al (2005) Metals in urine and peripheral arterial disease. Environ Health Perspect 113(2):164–169
Nnorom IC, Osibanjo O (2009) Toxicity characterization of waste mobile phone plastics. J Hazard Mater 161(1):183–188
National Toxicology Program (NTP) (2011) Cadmium and cadmium compounds. In: Report on Carcinogens, Twelfth Edition. U.S. Department of Health and Human Services, Public Health Service, National Toxicology Program, Research Triangle Park, NC, pp. 80–83
Padilla MA, Elobeid M, Ruden DM et al (2010) An examination of the association of selected toxic metals with total and central obesity indices: NHANES 99-02. Int J Environ Res Public Health 7(9):3332–3347
Perring L, Alonso MI, Andrey D et al (2001) An evaluation of analytical techniques for determination of lead, cadmium, chromium, and mercury in food-packaging materials. Fresenius J Anal Chem 370(1):76–81
Preda N, Popa L, Ariesan M (1983) The possibility of food contamination with cadmium by means of coloured plastics. J Appl Toxicol 3(3):139–142
Santos MC, Nobrega JA, Baccan N et al (2010) Determination of toxic elements in plastics from waste electrical and electronic equipment by slurry sampling electrothermal atomic absorption spectrometry. Talanta 81(4–5):1781–1787
Schroeder E (2010) Compound in Kellogg cereal packaging identified, July 12, 2010. Food Business News. http://www.foodbusinessnews.net/News/News%20Home/Food%20Safety%20News/2010/7/Compound%20in%20Kellogg%20cereal%20packaging%20identified.aspx?cck=1. Accessed 9 March 2014
Shirai S, Suzuki Y, Yoshinaga J et al (2010) Maternal exposure to low-level heavy metals during pregnancy and birth size. J Environ Sci Health A Tox Hazard Subst Environ Eng 45(11):1468–1474
Swiercz R, Wasowicz W, Majcherek W (2010) The toxicokinetics of 2-methylnaphthalene in rats. Int J Occup Med Environ Health 23(4):385–389
Underwriters Laboratory (UL) (2012) UL 126, Standard for sustainability for plastic film products. Underwriters Laboratory. http://www.ul.com/global/eng/pages/solutions/standards/accessstandards/catalogofstandards/standard/?id=126_1. Accessed 4 April 2014
United States Environmental Protection Agency (USEPA) (2003) Toxicological review of 2-methylnaphthalene (CAS No. 91-57-6), EPA 635/R-03/010. U.S. Environmental Protection Agency. http://www.epa.gov/iris/toxreviews/1006tr.pdf. Accessed 9 March 2014
World Health Organization (WHO) (1989) Evaluation of certain food additives and contaminants, thirty-third report of the Joint FAO/WHO Expert Committee on Food Additives, World Health Organization Technical Report Series, No. 776. World Health Organization. http://whqlibdoc.who.int/trs/WHO_TRS_776.pdf?ua=1. Accessed 22 April 2014
Wilson DC, Young PJ, Hudson BC et al (1982) Leaching of cadmium from pigmented plastics in a landfill site. Environ Sci Technol 16(9):560–566
Yorita Christensen KL (2013) Metals in blood and urine, and thyroid function among adults in the United States 2007-2008. Int J Hyg Environ Health 216(6):624–632

Chapter 11
Food Contact Materials: Practices, Agencies and Challenges

Jane Muncke

Abstract Foods and beverages are often processed and packaged before we consume them. Any material that intentionally comes into contact with foodstuffs is called *food contact material* (FCM). Many FCMs are plastics or are made of synthetic polymeric materials, like coatings and adhesives. Individual chemicals used for the manufacture of FCMs are called *food contact substances* (FCS). Specific regulations aim at limiting the migration of FCS into the food, thereby reducing risks of chronic chemical exposure to human health. However, FCMs are an under-recognized source of chemical food contamination. Currently, around 4,000 substances are used in FCMs. The challenge of determining which FCSs are present in food and beverages by chemical analysis is further increased by *non-intentionally added substances* (NIASs) that are impurities and breakdown products, or formed as reaction by-products of polymerization processes. Over the last few decades, scientific research has increased our understanding of risks linked to chronic chemical exposures. Of particular concern are endocrine disrupting chemicals that affect our body's hormone systems, mixture effects of chemicals present at individual no-effect levels, transgenerational inheritance of epigenetic effects, and the importance of protecting the developing fetus and infant from harmful chemical exposures. Taken together, these research findings offer important opportunities for prevention of chronic disease. This chapter summarizes current use, regulation and risk assessment of FCMs in the United States of America (U.S.) and the European Union (E.U.). Challenges to the risk assessment process arising from recent scientific insights are discussed, and recommendations how to address these challenges are made.

Keywords Food packaging · Food contact materials · Food contact substances · Migration · Non-intentionally added substance · Chemical risk assessment · Toxicological testing requirements

J. Muncke (✉)
Food Packaging Forum Foundation, Zurich, Switzerland
e-mail: jane.muncke@fp-forum.org

S. M. Snedeker (ed.), *Toxicants in Food Packaging and Household Plastics*,
Molecular and Integrative Toxicology, DOI: 10.1007/978-1-4471-6500-2_11,

11.1 Key Take Home Points

- Food contact materials include all food packaging, processing and storage equipment.
- Most food contact materials are made of different kinds of plastics.
- Food contact plastics are complex chemical mixtures, containing known intentionally added food contact substances, as well as non-intentionally added substances that may be known or unknown.
- Food contact materials can interact with food: chemicals can migrate from the food contact material to the foodstuff.
- Chemical risk assessment characterizes food contact substances for their potential harm to human health on a substance by substance approach.
- Insights from research offer new opportunities for prevention of chronic diseases, based on scientific understanding of low-dose effects, non-monotonic dose responses, mixture toxicity, developmental origins of health and disease, and epigenetics.
- Current regulatory efforts could be strengthened to better protect public health by incorporating new research findings.

11.2 Overview of the Use, Safety Issues, and Human Exposure to Food Contact Materials

11.2.1 Food Contact Materials, Migration and Non-intentionally Added Substances

Our globalized economy has led to the development of complex food production and distribution systems. Food is no longer only a typically local or regional product. It is also not necessarily a simple product but can be processed for enhancement of taste, storage time, or convenience. During all the different production stages, processing steps, and storage periods, foods (including beverages) come into contact with food contact materials (FCMs). These FCMs are, for example, the conveyor belt onto which freshly harvested potatoes are placed for initial washing, or the storage tank in which wines are fermented. The most visible FCMs are our food's packaging, such as single use plastic water bottles, plastic wrap, tuna fish cans, glass jars of tomato sauce, paperboard cereal cartons, and kitchen and tableware. Most FCMs are complex substances and are not chemically inert. When they come into contact with foods they transfer small but detectable amounts of chemicals into the food, a process known as *migration*. Migration describes the diffusion of chemicals from FCMs into food. This process is not only limited to leaching (i.e., migration into liquids); chemicals have been shown to transfer from printed paperboard cartons via the gas phase into dry foods, like breakfast cereals, rice or cakes. Currently, around 4,000 different chemicals are used in the manufacture of FCMs (Neltner et al. 2011), not including printing inks.[1] Chemicals that are intentionally added to FCMs are called *food contact substances* (FCS). They are monomers, catalysts or additives added to the FCM for specific material properties. Monomers and catalysts may also be present in the final FCM, for example antimony trioxide is used as a catalyst in the manufacturing of single use plastic beverage bottles.

Analytical chemistry has developed powerful tools for assessing food contaminants, but for each contaminant a new method needs to be developed, and existing methods need to be established in enforcement labs (Gallart-Ayala et al. 2013). Both method development and new method establishment take a month to years to develop. Therefore, the regular testing of marketed food for all of the known 4,000 FCS is de facto not achievable, and explains the focus on a selection of substances of concern.

Another issue is reaction products of the FCM migrants with the food; testing only for the known migrant will underestimate overall chemical migration if such reaction products are formed (Coulier et al. 2010). The manufacturer of a foodstuff, therefore, needs to ensure that marketed products are safe and comply with the regulations by thoroughly understanding the underlying chemistry of both their food and the FCM. Food producers are responsible for all aspects of food safety, including chemical contamination by migration from FCMs. However, the transfer of relevant chemical information throughout the supply chain is hampered by business interests aiming at the protection of trade secrets (Grob et al. 2009).

Adding to this challenge is the issue of *non-intentionally added substances* (NIAS). These are FCM manufacture by-products, impurities of starting materials, FCM-additive breakdown products or contaminants. For example, different types of plastic used as FCMs have been shown to contain a number of NIAS when extracted and analyzed (Bradley and Coulier 2007). Sometimes the chemical identity of NIAS remains unknown, especially if their concentration in the FCM and their migration into food is assumed to be low, i.e., below 10 micrograms per kilogram (μg/kg) of food (Nerin et al. 2013). Another example are recycled materials that can be sources of contaminants (like non-food grade mineral oil-based printing inks).

Migration from FCMs into food is increased at higher temperatures, during longer storage or contact time, and it depends on the chemical and physical properties of the FCM and the food. Packaging size can also be a relevant factor. Migration levels can be proportionally higher from smaller volume packaging with larger surface-to-volume ratio compared to large-volume packaged foods.

[1] For printing inks an estimated 5,000 compounds are used, whereby several ink components are also used in FCMs. The Swiss ordinance list is the most comprehensive publicly available printing ink list for FCMs (Swiss Federal Department of Home Affairs 2011).

11.2.2 The Global Food Packaging Market

In 2010, the global food and beverage packaging market was around 280 billion U.S. dollars (Rexam 2011). The most important types of food and beverage packaging materials are rigid and flexible plastic, paper and board, and glass and metal (Fig. 11.1). Packaging makes up around 20 to 50 % of retail food costs.

While paperboard and metal containers constitute a large proportion of food packaging materials (Rexam 2011), most FCMs used in food packaging are made of rigid or flexible plastic (Fig. 11.1). The entire FCM may be plastic, or the layer in direct contact with the food consists of plastic or a synthetic polymeric coating. An example is multi-layered beverage cartons (e.g., juice boxes), with the innermost, food contacting layer made of polyethylene (PE) plastic. Today PE is the most abundant type of plastic. PE is used in milk bottles [as high-density PE (HDPE)] or plastic wrap [as low-density PE (LDPE)]. Synthetic FCMs also are used in the closures of glass bottles and jars. The closures can either be made of coated metal with a polyvinyl chloride (PVC) plastic gasket or of thermoplastic elastomers (TPE). Other widely used types of plastic in FCMs are polypropylene (PP), polyethylene terephthalate (PET), polystyrene (PS), polyamide (PA), polyurethane (PU), and acrylonitrile butadiene styrene (ABS). Often, plastics are blended or layered so that the actual packaging may be a mixture of different polymers and/or plastic types that enhance performance of the final article, e.g., improving gas barrier properties.

11.2.3 Food Contact Materials and Recent Chemical Food Safety Issues

The following are recent examples of FCSs migrating from FCM into food. In many cases, changes in practices, the food packaging itself, or regulations have been necessary to protect the consumer from exposure to the FCS.

Changes in Printing Technology Due to ever increasing analytical sensitivity, more and more compounds are being detected in food, often due to migration from FCMs (Gallart-Ayala et al. 2013). But contamination can also arise due to changes in FCM chemistry or manufacturing. For example, the printing technology of photoinitiator-based ultraviolet (UV) curing has enabled rapid printing on the outside of multi-layered beverage cartons and other types of paperboard packaging. Large sheets of rolled carton packaging are printed, UV-cured to rapidly dry the printed layer, and rolled up again for storage until further use. During this storage, *set-off migration* can occur since the printed layer is in direct contact with the inner, food-contacting plastic layer. In 2005, infant formula, milk, juices and other products were found to be contaminated with the photoinitiator isopropylthioxanthone (ITX) which is used in curing printing inks (Fig. 11.2) (Morlock and Schwack 2006). ITX is consumed during the curing process, so one solution to

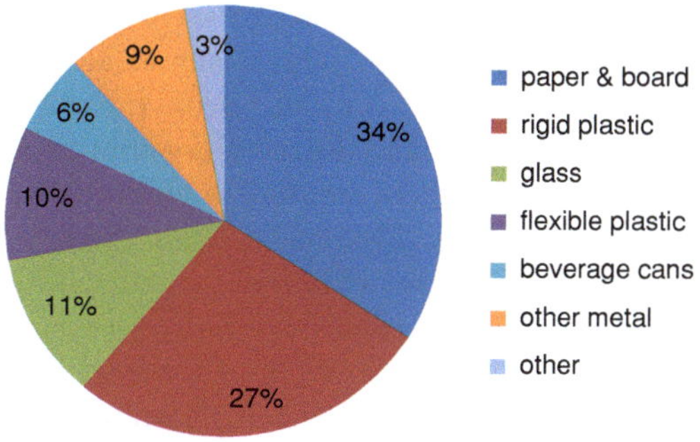

Fig. 11.1 2010 global packaging market (all types of consumer packaging, expressed as a percentage (%). Food and beverage packaging had a share of 69 % of consumer packaging in 2010 (Rexam 2011)

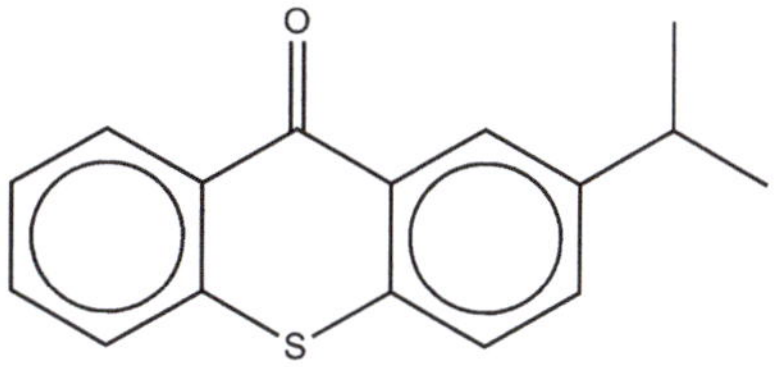

Fig. 11.2 Chemical structure of 2-isopropylthioxanthone (ITX) that is used as a photoinitiator in UV-cured printing inks

reducing ITX set-off migration was to extend curing time. Other UV-initiators used in printing paperboard cartons, like benzophenone and 4-methylbenzophenone, are not always fully consumed in the printing process, and can migrate through the porous layers of paperboard to packaged foods such as cereals (see Chap. 6 for more information on UV-initiators used in printing inks and migration to food).

Mineral Oil in Recycled Paperboard Chemical contamination may also occur due to the use of contaminated raw materials, such as recycled paperboard. This ecofriendly and economic raw material source is used to manufacture paperboard cartons for food packaging. The migration of mineral oil from recycled paperboard FCM into dry foods was detected by the 1990s in Europe, but only recently has the issue been discussed more broadly (Biedermann and Grob 2010). Printing inks used for non-food contact products like newspapers and magazines are usually based on mineral oil. During the paper recycling process these mineral oils are not removed. Due to their volatility these substances can migrate through the porous, recycled paperboard into the packaged food. The highest contamination occurs in foodstuffs with a high specific surface area, like rice or flour. These mineral oil-hydrocarbons are adsorbed to the food, sometimes in the range of hundreds of mgs per kg of food. While some mineral oil components are associated with carcinogenicity, a complete toxicological evaluation has not been performed on these compounds. A solution to this problem is the use of virgin paperboard. However, this evades the principles of sustainable development and the need for a resource efficient circular economy. Ideally, a hazard-limiting source control approach would be used for all printing inks, regardless of their use for food or non-food applications. While use of barrier technology (barrier layer between the recycled paperboard and food, or use of secondary packaging liner bag) can offer solutions to prevent migration (Biedermann et al. 2013), appropriate materials need to be chosen and tested to ensure effective barrier properties.

Cling Wraps and Films Very thin transparent plastic film is used in commercial or household settings to wrap foods, preserve their freshness and prevent loss of moisture. Originally made from PVC and the related polyvinylidene chloride with higher chlorine content, plastic wrap also contained high levels of phthalate plasticizers. These additives give plastic wrap its flexibility, but they also migrate into foods, especially if the food is fatty (McNeal et al. 2000). In 2004 the U.S.-based producer of Saran^{TM} wrap introduced a reformulation of its product and launched a PVC and phthalate-free version based on LDPE. Not all plastic wrap on the market is PVC-free. The most commonly used plasticizer in PVC films today is diethylhexyladipate.

Polycarbonate Plastics and Bisphenol-A Baby bottles were predominately made of glass until the introduction of plastics, for example those made of polycarbonate (PC). PC is made with the monomer bisphenol A (BPA), and the polymer molecule is a repeat of the BPA unit with up to 10,000 monomers covalently linked to each other. During the manufacture of PC not all BPA monomer is polymerized. The free BPA can migrate out of the plastic article during the first few uses, and migration rates increase with higher temperatures from heating or microwaving. Over time, when the plastic becomes scratched from the use, the polymer starts breaking down into its monomer and BPA is released (Le et al. 2008). Canada banned the use of PC in baby bottles in 2008 (Bailin et al. 2008), and many companies and retail establishments voluntarily ended sales of PC baby bottles and PC sports/water bottles (Bailin et al. 2008). Since 2011, PC baby bottles are banned in Europe. The U.S. Food and Drug Administration (FDA) followed with a ban of PC baby bottles and sippy training cups in July 2012 (Federal Register 2012). Replacement products made of plastic are not always BPA-free, even if advertised as such (Simoneau et al. 2012). In the wake of health concerns over PC plastic, an U.S. chemical company developed a BPA-free plastic with the same properties as PC (i.e., heat resistant, transparent, and shatter proof). Marketed under the trade name Tritan^{TM}, the new polymer is made of di-methylterephthalate, 1,4-cyclohexanedimethanol, and 2,2,4,4-tetramethyl-1,3-cyclobutanediol. Research commissioned by the company has shown that its product does not contain estrogenic or androgenic chemicals (Osimitz et al. 2012). While this finding has been confirmed by an independent study (Guart et al. 2013), Tritan^{TM} was shown to be estrogenic in another test system (Yang et al. 2011).

Closures of Glass Jars A last example for how the food packaging market has adapted to chemical concerns are the closures of glass jars. Foods that are packaged in glass and intended for long-term storage undergo sterilization, where the sealed jar is heated above 100 °C for a short period of time. Since the jar must be absolutely air tight to protect the food from micro-biological spoilage, the glass jar rim is sealed with a flexible gasket on the inside of the lid. The gasket's material is usually made of PVC plastic containing epoxidized soy bean oil as plasticizer (Fig. 11.3). Because plasticizer migration from PVC-sealed glass jar closures can

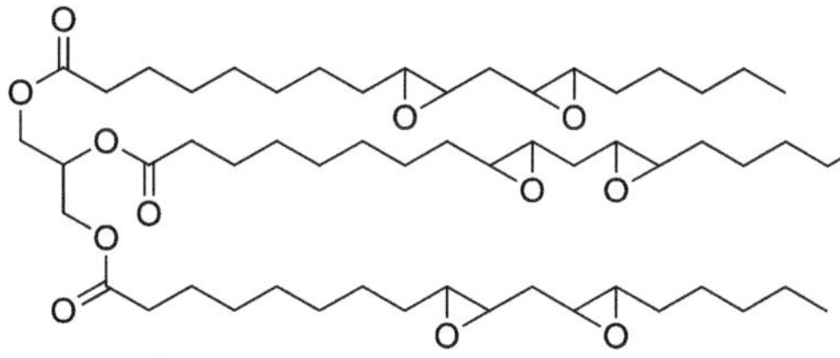

Fig. 11.3 Chemical structure of epoxidized soy bean oil, a plasticizer used in the polyvinyl chloride (PVC) gaskets of glass jar closures intended for heat sterilization

be considerable (Pedersen et al. 2008), a German company developed a PVC-free glass jar closure based on TPE, requiring less plasticizer (Pano 2011). With this closure, plasticizer migration due to heat sterilization is significantly reduced.

11.2.4 Human Exposure to Controversial Chemicals from Food Contact Materials

FCMs are a largely underestimated source of chemical food contamination (Grob et al. 2006). People are exposed to FCM-derived contaminants on a daily basis from the foods they eat and drink. Few studies have assessed actual human exposure to the many chemicals of concern arising from food packaging. Several recent studies have begun to examine whether changes in packaged food consumption can affect urinary levels of BPA and certain phthalates. In a 2011 dietary intervention study, scientists were interested in how changing dietary habits can reduce personal exposure to the FCM-associated chemical compounds BPA and phthalates (Rudel et al. 2011). Study participants that changed for several days from a common diet to eating only unpackaged, locally sourced fresh organic foods, prepared in plastic-free FCMs, had significantly reduced urinary levels of BPA and phthalates. In a similar study, participants were given canned soup during 1 week and freshly made soup during another; a marked reduction in urinary levels of BPA was observed the week when unpackaged, home-cooked food was consumed (Carwile et al. 2011). These studies did not test for actual chemical levels in the foods given to study participants. However, it is plausible that decreased dietary exposure to BPA-containing FCMs was responsible for the observed decreased urinary levels of BPA. Additional studies are needed to assess chronic human exposure to low levels of other chemicals via FCMs. There is a need for carefully designed research assessing migration of FCS from FCMs into foods, human exposure to FCS, and the impact on human health at different life-stages.

11.3 Regulation of Food Contact Materials: A Complex Challenge

11.3.1 The Food and Drug Administration

The FDA is responsible for assessing and managing the risk of FCS in the U.S. Unlike for its European counterparts, the tasks of science-based risk assessment and interest-balancing risk management are historically united within one agency.

The FDA was founded in 1927, based on the *Pure Food and Drug Act of 1906*. The *1938 Food, Drug and Cosmetic Act* introduced an absolute restriction of all poisonous substances from food. This approach was deemed unpractical and led to the application of the *de minimis* principle: poisonous chemicals are permitted in food if their concentrations are minimal, based on the work of famously misquoted Swiss physician Paracelsus ("the dose makes the poison").[2]

The *Food Additive Amendment of 1958* launched the premarket safety assessment for all food additives, including those that were considered *indirect* (United States Statutes at Large 1958). Migrants from FCMs were not intended to become part of packaged foods, but their migration was known to be inevitable due to most FCMs not being chemically inert. Migrating substances used in FCMs were termed *indirect food additives* and required clearance from FDA. Exempted from safety assessment are substances known to be *generally recognized as safe* (GRAS) and all compounds used before 1958. For all other indirect food additives a chemical risk assessment has been required since 1958, but requirements for FCS have changed since then. In this context, *safety* has been defined since 1958 as:

> ...reasonable certainty in the minds of competent scientists (U.S. Senate Committee on Labor and Public Welfare 1958)

that a compound is not harmful under the intended conditions of use.

Chemical risk is determined by a substance's exposure level and its inherent toxic effect (risk = exposure x effect). A chemical's risk is considered negligible if the *estimated daily intake* (EDI) is below the *acceptable daily intake*[3] (ADI) (Alger et al. 2013). The *de minimis* principle set legally relevant exposure levels equal to technically detectable exposure levels. However, the ability to detect low levels of exposure has improved over the last six decades due to ground breaking advances in analytical chemistry. The new problem this posed was addressed in the 1990s with the introduction of new rules: the *Threshold of Regulation* (*TOR*) *Exemption* in 1995 (FDA 2012a; Shanklin 2008) and the *Food Contact Substance Notification*

---

[2] The actual quote is "What is there that is not poison, all things are poison and nothing is without poison. Solely the dose determines that a thing is not a poison" (Deichmann et al. 1986). Notably, Paracelsus studied acute toxicity.

[3] The term ADI is used for intentionally added substances, i.e., food additives. For chemical food contaminants in the U.S. the *Reference Dose* (RfD) is common. Though FCSs are not intended to become part of the food, they are considered to be *indirect additives*, and the ADI concept is used.

(FCSN) in 1997 (FDA 2013a) with the *Food and Drug Administration Modernization Act of 1997* (FDA 2009). An important exception is made for genotoxic carcinogens [the *Delaney Clause* (FDA 2002)], known to damage DNA by mutation or chromosome aberration; for such substances the risk is considered higher due to their inherent hazard, and permitted exposure levels are lower.

The TOR is set by the FDA at 0.5 μg/kg of food (see Table 11.1 and Fig. 11.4). Chemicals that migrate from FCMs below the TOR can be exempted, because dietary exposure (assumed to be below 1.5 parts per billion (ppb) per person per day, since a person is supposed to consume 3 kg of food per day) is thought to have a negligible exposure. No toxicological testing is hence required for TOR exemptions but their chemical structures must be free from functional groups related to causing genotoxicity.

The FCSN Program allows companies to petition the FDA for approval of new FCSs (Shanklin and Sánchez 2013). FCSNs need to include toxicity data if people are exposed to the respective substance at levels between 0.5 ppb and 1.0 part per million (ppm) (Table 11.1). For these exposure levels in vitro genotoxicity and /or subchronic toxicity testing (90 days whole animal assay) are recommended (FDA 2002). Other tests can be recommended. FCSNs are specifically tied to the notifying company. When FDA receives a FCSN it has 180 days to review the submitted data, and respond either with a request for more information or a final *Letter of No Objections*. Notifications are not approved, but are only acknowledged. Only once a FCSN has been processed it is made public (FDA 2013a), but certain aspects can be treated confidentially.

If a chemical migrates into food at levels above 1 milligram (mg)/kg food (equal to 1 ppm) it is considered an *indirect food additive* and requires a dedicated petition (Table 11.1). Toxicological testing guidelines are detailed in FDA's *Redbook* (FDA 2007a). FDA recommends companies to contact them individually for discussing what toxicological tests would be relevant for specific substances.

Chemical safety of FCSs can be self-determined as GRAS by the company using them. The usual requirement for establishing a substance's GRAS status is reached by common agreement among experts. A study published in the scientific, peer-reviewed literature showing a chemical's safety under intended conditions of use is sufficient for establishing GRAS status. Alternatively, manufacturers can convene expert panels to determine GRAS status. FDA can be, but does not require to be, informed about GRAS substance use (FDA 2013b). Around 40 % of all food additives (including FCSs) in current use are GRAS self-determinations set by industry (Neltner et al. 2011).

Finally, FDA is responsible for risk management of FCM and FCS, though substances that have been in use in FCMs prior to 1958 received no formal risk assessment by the FDA. As a consequence, the use of certain chemical carcinogens in FCMs remains legal (Vogel 2012). FDA's regulatory oversight is limited to certain FCMs like packaging, but for kitchen- and tableware articles manufacturers are not required to notify or petition FDA for new FCSs (FDA 2011a). FDA currently does not require periodical reassessment of FCS safety decisions (Neltner et al. 2011). Market data used to estimate human exposure to FCS and

Table 11.1 Toxicological testing requirements for food contact substances authorization in the United States (U.S.). Adapted from (Muncke 2009)

Agency	U.S. Food and Drug Administration (FDA)			
Authorization based on	CEDI[1,2] [μg/person per day] of food contact substance			
Authorization threshold	CEDI ≤ 1.5[3]	1.5 < CEDI ≤ 150[4]	150 < CEDI < 3,000	CEDI ≥ 3,000 (for biocides: 600)
Concentration in food	≤0.5 ppb	0.5–50 ppb	50 ppb–1 ppm	>1 ppm
Specific applicable regulation	21CFR170.39 Threshold of Regulation since 1995	21CFR170.101 Food Contact Notification since 1997	21CFR170.101 Food Contact Notification since 1997	21CFR171.1 Indirect Food Additive Petition since 1958
Toxicological testing recommendations		• Gene mutations (bacteria) • Mammalian in vitro cytogenicity assay or tk+ assay	• Gene mutations (bacteria) • Mammalian in vitro cytogenicity assay or tk+ assay • Chromosomal damage in rodent hematopoietic cells in vivo • Two subchronic oral toxicity tests in vivo (rodent and non-rodent species) (90 days) • Further testing (chronic exposure) with further endpoints can be recommended (metabolism studies, teratogenicity, reproductive toxicity, neurotoxicity, immunotoxicity studies) based on FDA's case by case decision	• Gene mutations (bacteria) • Mammalian in vitro cytogenicity assay or tk+ assay • Chromosomal damage in rodent hematopoietic cells in vivo • Two subchronic oral toxicity tests in vivo (rodent and non-rodent species) (90 days) • Chronic toxicity and carcinogenicity in two rodent species (2 years), one study including in utero phase • Two generation reproductive toxicity study (in rats) • Further testing with further endpoints can be recommended based on FDA's case by case decision

(continued)

Table 11.1 (continued)

Agency	U.S. Food and Drug Administration (FDA)			
Further information requirements	• Structure analysis (carcinogen) • Chemical and physical properties • Migration into foods (as basis for setting the EDI[5] • Literature review; risk assessment if a constituent is carcinogenic	• Structure analysis (carcinogen) • Chemical and physical properties • Migration into foods (as basis for setting the EDI) • Literature review	• Structure analysis (carcinogen) • Chemical and physical properties • Migration into foods (as basis for setting the EDI) • Literature review	• Structure analysis (carcinogen) • Chemical and physical properties • Migration into foods (as basis for setting the EDI) • Literature review
FCS listing	FDA (2012a, b)	FDA (2013a)	FDA (2013a)	FDA (2011b)
	CEDI/ADI[6] database (FDA 2012b)			
Guidance	Guidance documents are available for TOR exemptions, FCN, indirect food additive petitions and GRAS notifications (FDA 2002, 2004, 2007a, b, 2011a)			

[1] Cumulative Estimated Daily Intake (CEDI)
[2] If no migration data is available for a substance a default CEDI of 7 ppb is assumed
[3] The Threshold of Regulation (TOR) is not based on the EDI but applicable for substances at or less than 0.5 ppb in the diet. The underlying assumption for setting the EDI is that 3 kg solid and liquid food is consumed per person per day which would set the EDI at 1.5 μg/person per day or below in the case of TOR substances
[4] For CEDI < 150 ppb (dietary concentration in food < 50 ppb) no ADI calculated if substance is not of toxicological concern)
[5] Estimated Daily Intake (EDI)
[6] Acceptable Daily Intake (ADI)

Credit Reprinted (adapted) from Exposure to endocrine disrupting compounds via the food chain: Is packaging a relevant source, Muncke (2009), Copyright 2009, with permission from Elsevier

determine FCS safety is partially received from industry. There is no mechanism for full public review of these data that constitute a central element of FDA's chemical risk assessment and management (Alger et al. 2013).

11.3.2 The European Food Safety Authority

The E.U.'s European Food Safety Authority (EFSA) is tasked with providing scientific opinions regarding feed and food to the European Commission (EC) and other European legal entities. Risk assessment for FCSs falls within EFSA's responsibility. Similarly to the U.S., the European FCM regulation requires pre-market safety assessments of FCS. However, unlike in the U.S., not all FCSs are

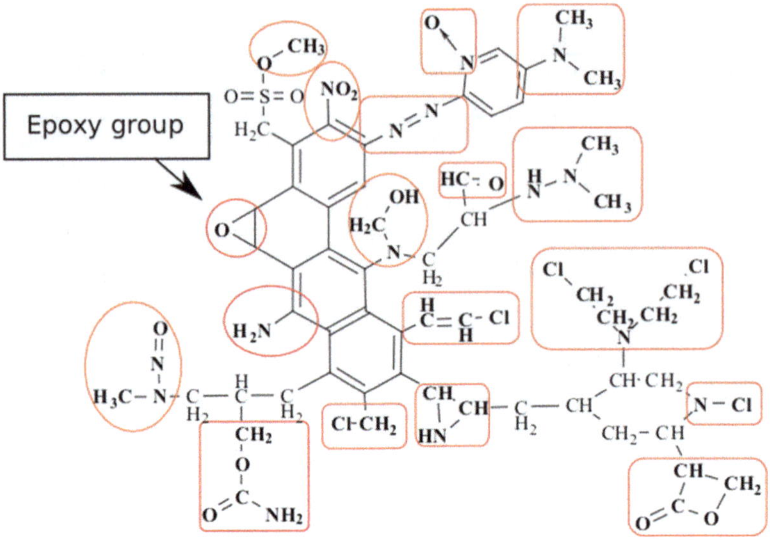

Fig. 11.4 Ashby's Structural Alerts (adaptation from original drawing by (Ashby and Tennant 1988) appearing in (Benigni and Bossa 2006). *Credit* Republished with permission from Bentham Science Publishers, from Benigni and Bossa (2006, Fig. 1), Copyright 2006

subject to specific risk assessments. Dedicated approval is given by EFSA only for plastic FCM additives (European Union 2011), plastic FCM starting materials (like monomers, catalysts and processing aids), biocides (European Union 2012), and substances used in regenerated cellulose films (European Union 2007).

EFSA was established in 2002 in response to several food crises of the 1990s (e.g., bovine spongiform encephalopathy—mad cow disease). It is intended as an independent source of scientific opinion in the area of food safety, and consults with the EC in its risk management decisions. EFSA works together with external experts who make up their thematically organized scientific panels. The *Panel on Food Contact Materials, Enzymes, Flavourings and Processing Aids* (CEF) reviews FCS applications and publishes publicly accessible scientific opinions on each substance. EFSA's scientific opinions are either based on requests by Member States, the EC, the E.U. Parliament, or are self-mandated.

Based on EFSA's scientific opinions, the EC is responsible for risk management measures. The EC places individual substances on *positive lists*. These are FCSs that have been authorized based on scientific opinions. The EC sets limits for migration levels of a chemical from FCMs to food, called the *specific migration limits* (SML) (European Commission 2013). SMLs are usually set for compounds where a *Tolerable Daily Intake* (TDI) is available. The positive listing of a substance in the *Plastics Regulation* implies that the compound may be used in any type of plastic FCM, whereby conditions of use may be specified. Where no SML is given, the *Overall Migration Limit* (OML) of 60 mg/kg food applies. Migration limits are enforced in the food product, however migration testing is commonly performed using *food simulants* (e.g., distilled water, 3 % acetic acid, 10 % ethanol, 50 % ethanol, vegetable oils, or dry food simulants like TENAX^{TM}).

Safety testing requirements for FCSs in Europe are similar to the U.S. (Table 11.2). Two important differences to the U.S. regulation exist. First, all chemicals (i.e., additives, starting materials) used in plastic FCMs in Europe need to be positive listed and must have in vitro genotoxicity testing data as a minimum requirement. However, the *Functional Barrier* concept permits the use of unapproved compounds behind layers that reduce migration below 10 μg/kg food (10 ppb). This threshold historically was arbitrarily set based on achievable chemical analytical detection limits. Second, the *Fat Reduction Consumption Factor* (FRF) in Europe takes into account that lipophilic substances will migrate to fatty foods at higher levels. The European exposure model assumes that a person (70 kg) eats 1 kg of any packaged food per day, with a packaging surface of 6 square decimeters (dm^2)/kg food. However, it is assumed that consumers are unlikely to eat large amounts (1 kg per day) of fatty foods with >20 % fat content. The FRF may be used to correct for the expected reduced consumption. Other than the FRF, the European FCM regulation currently does not include exposure-based risk management, and centers all of its FCM risk-management on migration limits. Ongoing efforts aim to create a public exposure database that would permit shifting risk management practices away from worst-case migration assumptions to an actual exposure data-based approach (JRC 2012). The effectiveness of such a change for protecting public health will depend on the quality and completeness of underlying exposure and hazard data.

11.3.3 Other Regulatory Approaches by Individual Countries

FCM regulation varies from country to country. While regulations in Switzerland are largely harmonized with the E.U., a positive list for printing inks used in food contact applications was introduced in 2010. This became a unique reference for the 5,000 substances commonly used in inks (Swiss Federal Department of Home Affairs 2011), though not all of these listed substances have undergone toxicological testing.

In Japan, the overarching *Food Sanitation Law of 1947* sets the legal framework for food safety, including FCMs. Specific migration limits from plastic FCM are subject to industry self-regulation. The Japanese plastic packaging industry is organized in material-specific associations that have adopted similar approaches to the E.U., with positive lists used for starting substances and additives (Mori 2010). An overall migration limit is set at 30 mg/kg food, half of the E.U.'s limit. Substances of concern are sometimes specifically managed, like the substitution of antimony trioxide catalyst in PET manufacture by germanium or titanium-based catalysts, or in epoxy can coatings where an additional polymer layer reduces migration of BPA.

Table 11.2 Toxicological testing requirements for food contact substances authorization in the European Union (EU), adapted from (Muncke 2009)

Agency	European Food Safety Authority (EFSA)		
Authorization based on	Migration (M) [µg/kg food] of food contact substance[1]		
Authorization threshold	M < 50	M 50–5,000	M > 5,000 (max. 60,000)
Concentration in food	<50 ppb	50 ppb–5 ppm	>5 ppm
Specific applicable regulation	EC[2] 10/2011: plastics regulation	EC 10/2011: plastics regulation	EC 10/2011: plastics regulation
Toxicological testing requirements	• Gene mutations (bacteria) • Gene mutations in mammalian cells in vitro (tk+ assay) • Chromosomal aberrations in mammalian cells in vitro	• Gene mutations (bacteria) • Gene mutations in mammalian cells in vitro (tk+ assay) • Chromosomal aberrations in mammalian cells in vitro • Two subchronic oral toxicity tests in vivo (rodent and non-rodent species) (90 days)	• Gene mutations (bacteria) • Gene mutations in mammalian cells in vitro (tk+ assay) • Chromosomal aberrations in mammalian cells in vitro • Two subchronic oral toxicity tests in vivo (rodent and non-rodent species) (90 days) • ADME[3] study in vivo • Reproduction study (one species), developmental toxicity (in two species) • Chronic toxicity and carcinogenicity in two species (2 years)
Further information requirements	• Chemical, physical and microbial properties • Migration, residual levels in food contact materials • Literature review	• Log K_{OW} data; if Log $K_{OW} > 3$ an ADME[3] study may be required • Microbial properties • Migration, residual levels in food contact materials • Literature review	• Microbial properties • Migration, residual levels in food contact materials • Literature review

(continued)

Table 11.2 (continued)

Agency	European Food Safety Authority (EFSA)
Substance listing	EU Commission, Directorate General for Health and Consumers, food contact materials database (European Commission 2013)
Guidance	Guideline of the EFSA CEF[4] scientific panel[5] (EFSA 2008)

[1] For authorized FCS, specific migration limits (SMLs) may be defined, based on the Tolerable Daily Intake (TDI), if available
[2] *EC* European Commission
[3] *ADME* Absorption, distribution, metabolism, and excretion
[4] EFSA Scientific Panel on food contact materials, enzymes, flavourings and processing aids (CEF)
[5] EFSA's CEF Panel is currently updating the guidelines; draft publication is expected in December 2013
Credit Reprinted (adapted) from Exposure to endocrine disrupting compounds via the food chain: Is packaging a relevant source, Muncke (2009), Copyright 2009, with permission from Elsevier

11.3.4 Guidance for Regulatory Food Safety by the United Nations

The United Nations (UN) agencies Food and Agriculture Organization (FAO) and World Health Organization (WHO) established a Joint FAO/WHO Expert Committee on Food Additives (JECFA). JECFA is an international expert board convened by FAO and WHO in 1955. Since 1956, JECFA meetings gather specialists from different disciplines to discuss food safety matters, including chemical contaminants from FCMs. JECFA publishes reports that are publicly available and used by many national agencies to inform their risk assessment (JECFA 2013).

11.4 Chemical Risk Assessment: The Scientific Basis of Chemical Risk Management

11.4.1 Overview of Risk Assessment and Risk Management Terminology

Risk management is:

> ...a decision-making process that entails weighing political, social, economic, and engineering information [and it] entails risk-related information to develop, analyse and compare regulatory options and select the appropriate regulatory response to a potential health or environmental hazard (van Leeuwen 2007).

Risk assessment is a central pillar of risk management for FCSs, and is the scientific basis for the decision-making process. JECFA has published technical guidance for chemical risk assessment of food additives (IPCS 2009). Risk assessment relies on unbiased, science-based information, and the process of risk management also relies on value-based information.

A chemical's risk is the product of a substance's toxicological effect (called its *hazard*) and its exposure levels. Risk may be considered unacceptable if an estimated exposure exceeds the estimated human *no effect level* (NEL). Risk can also be too large if a substance has inherent hazard properties that are considered inacceptable, like mutagenicity.

Human health risk assessment has been structured into four steps: hazard identification, hazard characterization (also called effect assessment or dose-response assessment), exposure assessment, and risk characterization (USEPA 2012). Usually, chemical risk assessment for FCMs is performed for single compounds.

11.4.2 Hazard Identification

This first step in chemical risk assessment is concerned with a chemical's inherent toxicological properties. For example, an epoxide group would be a known structural alert for mutagenicity (Fig. 11.4). Hazard identification is increasingly being performed in silico using computer-based approaches like quantitative structure activity relationships, three dimensional protein-binding assays, and others. Many in vitro assays exist for hazard identification screening. Screening assays need to be sufficiently sensitive, i.e., correctly identify as many substances as possible for an assay's given endpoint. Developments in high-throughput screening assays such as Tox21 (see Sect. 11.5.6) will likely increase the efficiency and expand the comprehensiveness of hazard identification for FCSs in the near future.

FDA uses a tiered approach for hazard identification. When a substance's exposure is estimated to be below the TOR (1.5 μg/person per day), in silico analysis is considered sufficient, while for higher exposures in vitro genotoxicity (mutagenicity, chromosome aberration) assays are commonly required. In Europe, any FCS that requires authorization needs to be screened for genotoxicity. A default requirement for endocrine disruption screening is currently not in place in the U.S. or the E.U.

11.4.3 Exposure Assessment

Exposure assessment collects information about the use of a chemical, human exposure routes and environmental exposure sources. For new FCSs this information is based on estimates, derived from migration testing or modeling, and standard assumptions that EFSA and FDA use in their risk assessments.

FDA's model for risk assessment assumes that a total of 3 kg of food and beverages are consumed per person and day, with 1 kg of foodstuff packaged in 6.45 dm^2 (10 g food per square inch packaging). FDA publishes consumption factors and food type distribution factors that permit an estimate of FCM type frequency in the average daily diet, for aqueous, acidic, alcoholic and fatty foods (FDA 2007b). The exact use of a FCS is often confidential (what type of FCM an additive will be used in). Together with the substance's specific migration data from a given FCM, an estimate for human exposure can be calculated, known as the *Cumulative Estimated Daily Intake* (CEDI). FDA makes CEDI data for selected FCS publicly available (FDA 2012b). Depending on the CEDI, tiered hazard assessment requirements exist (Table 11.1). FDA's exposure assessment procedures are reviewed in detail by Alger et al. (2013).

The EFSA's risk assessment model differs notably from FDA's approach. EFSA assumes that a person[4] consumes 1 kg of any packaged foodstuff per day

[4] EFSA recently changed its assumption of person weight from 60 to 70 kg (EFSA 2012a).

with a surface area 6 dm^2 (11 g food/in^2 packaging) and does not include any further estimates of exposure. While this approach is conservative, because high consumers most likely will be sufficiently protected, it can be considered a worst-case assumption and too high of a burden for industry. The FACET project (***F**lavourings, **A**dditives and food **C**ontact materials **E**xposure **T**ask*) in Europe is currently compiling a publicly accessible database for providing actual data on consumption factors for potential use in EFSA's risk assessment process (JRC 2012). However, the legal basis for this change in risk assessment still needs to be provided. Requirements for hazard assessments are tiered in Europe, based on migration data from testing or modeling (Table 11.2) (EFSA 2008).

11.4.4 Hazard Characterization

During this step of risk assessment, a chemical's dose-dependent toxicological properties are of interest. A substance is studied at different concentrations (doses) in biological assays (e.g., oral dosing over 90 days in rodents to assess sub-chronic toxicity) for determining the *lowest observed effect concentration* (LOEC). This information can then be used to derive a *no observed effect concentration* (NOEC), a level that is of interest for human health risk assessment. Effects observed in laboratory animal studies may vary depending on the dose, life stage, duration and route of exposure, and sex and strain of the test animal.

Not all substances used in FCMs undergo hazard characterization, and actual testing depends on the estimated exposure level of a FCS. If a FCS's estimated exposure level remains below the U.S. TOR (1.5 μg/person per day) and its chemical structure has no structural alerts for genotoxicity, no toxicological data is required (Table 11.1). For estimated exposures between 1.5 and 150 μg/person per day, in vitro mutagenicity testing is usually required, and if levels are between 150 and 3,000 μg/person per day additional in vitro testing for genotoxicity, and 90 day subchronic toxicity in two different species are required. Further toxicological testing, according to recommendations in FDA's Redbook, may be performed but is not required by default (FDA 2007a). Hence, developmental toxicity testing will not be routinely required for FCS with estimated exposures below 3,000 μg/person per day. FDA's hazard characterization procedures have been reviewed (Maffini et al. 2011), and recommendations on how they can be improved based on current scientific understanding have been made (Maffini et al. 2013).

In Europe, the *Functional Barrier* concept permits the use of untested substances behind a migration-reducing barrier, provided their migration does not exceed 10 ppb and they are neither carcinogenic, mutagenic or toxic to reproduction (CMR). In addition, no nanoparticles may be used behind a Functional Barrier (European Union 2011).

11.4.5 Risk Characterization and Safety Factors

The final step in risk assessment integrates all collected information on hazard and exposure. Further assumptions are made regarding the predicted NEL for humans. NOEC data are derived from animal experiments, and safety factors are used to extrapolate these data to humans (Falk-Filipsson et al. 2007). A safety factor of 100 is used to extrapolate from animal data to the estimated human NEL (i.e., the NOEC is divided by 100): a factor of 10 allows for uncertainty in the extrapolation from animal data to humans (interspecies), and another factor of 10 accounts for differences in sensitivity (differences in chemical metabolism) within the human population (Renwick 1991). Further information on the derivation of safety factors and their limitations has been reviewed by Martin et al. (2013). Additional safety factors are used to extrapolate from in vitro data or from sub-chronic toxicity data to human life time exposure. Sensitive population groups like children and infants may be protected by applying additional safety factors (Renwick 2004). Finally, the *hazard quotient* (HQ) is calculated based on the ratio of estimated human exposure concentration/estimated human NEL. A HQ above 1.0 implies that the estimated human exposure is higher than the estimated no effect concentration. Hence, for HQs of 1.0 and above, risk management may result in risk reduction measures, including not granting authorization for a substance's intended use, or limiting its permitted migration into food. An alternative approach is to use the *Margin of Exposure* (MoE), where the HQ ratio is simply inversed and a larger MoE denotes a higher level of safety.

11.5 Current Challenges in Chemical Risk Assessment of Food Contact Materials

Our society has become increasingly dependent on chemicals in everyday products, including food packaging. Chemistry-enabled technical innovations have become part of our modern lives, with the chemical risk assessment process an integral part of these developments. At the same time, we have improved our tools and methods for measuring chemicals in packaging, food, and people, and with these new approaches, new questions regarding the safety of chemicals are being asked.

11.5.1 Non-intentionally Added Substances

When FCM plastics are extracted using organic solvents, these extracts can be analyzed with the aim of identifying their components and determining the composition of the plastic(s). The resulting chromatograms, showing a signal for

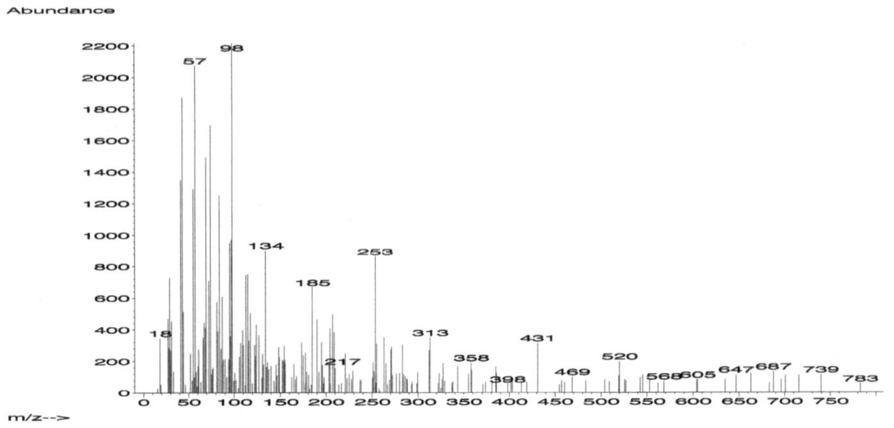

Fig. 11.5 The Forrest of Peaks. Ethanol and isooctane extracts of polypropylene (PP) measured using GC-MS for an unknown compound. From (Bradley and Coulier 2007). *Credit* Reprinted with permission from the Food Standards Agency, from a final research report entitled: An investigation into the reaction and breakdown products from starting substances used to produce food contact plastics, Report FD 07/01, Project Number A03054, Figure PP429, page 421, Central Science Laboratory, London, http://www.foodbase.org.uk//admintools/reportdocuments/518-1-911_A03054_reaction_and_breakdown_products_final_report.pdf, Bradley and Coulier (2007), Copyright 2007

each chemical in the extract, are dubbed *forest of peaks* (Fig. 11.5), illustrating the complexity of multiple signals detected. Plastics are composed of starting materials (like monomers, catalysts and processing aids), additives, impurities and reaction by-products. While the chemical identity of intentionally added substances (i.e., monomers, additives, processing aids) is known, there can be other substances present in the extract that were not intentionally added (the NIASs) and may be of unknown identity. The presence and identification of NIASs are a significant challenge for establishing FCM safety (Nerin et al. 2013).

In 2007, British scientists were tasked with a project aimed at clarifying how well all extractable plastic components, including NIAS, could be measured and identified (Bradley and Coulier 2007). Five commodity plastics were custom made under controlled laboratory conditions, extracted and chemically analyzed. All plastics contained substances in the extract with unknown chemical identity. The authors concluded that anyone currently is

> …unlikely to be capable of detecting and identifying every non-intentionally added substance in food contact plastics (Bradley and Coulier 2007).

Risk assessment and risk management continues to be a challenge for both identified and unidentified NIASs (Nerin et al. 2013), and this has led to the development of auxiliary approaches using generic thresholds.

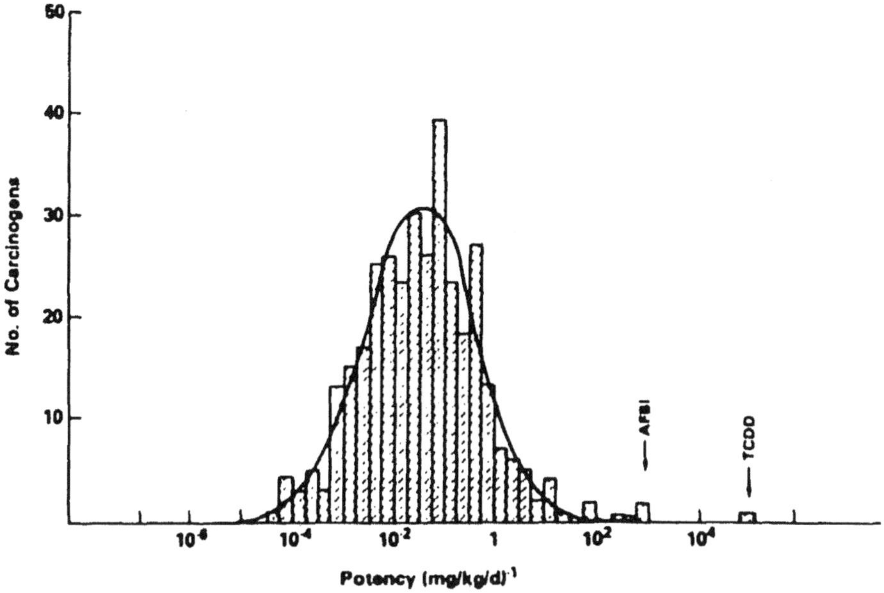

Fig. 11.6 Histogram of 343 substances analyzed for their tumorigenic potency over lifetime exposure showing a normal distribution. From this analysis it was assumed that a probability can be calculated for most compounds to cause cancer. From Rulis (1987). *Credit* Republished with permission of Taylor and Francis Group LLC Books, from Chap. 2: *De Minimis* and the Threshold of Regulation, Fig. 1: Histogram and nonlinear least squares best fit Gaussian to the potencies of Gold et al., authored by Rulis (1987), ISBN 978-087371-047-3. Copyright 1987; permission conveyed through the Copyright Clearance Center, Inc

11.5.2 Generic Thresholds as Part of Chemical Risk Assessment

Most of the approximately 4,000 substances knowingly used in FCMs, and the majority of the identified NIAS, have little or no hazard characterization data that is publically available (Neltner et al. 2013). In the absence of toxicological data, risk assessors have developed concepts using generic thresholds, based on toxicity data of known substances.

The TOR is an FDA rule that permits FCS without toxicological data to be present in foods at levels below 0.5 ppb (corresponding to 0.5 μg/kg food and 1.5 μg/person per day estimated exposure) (Shanklin 2008). The TOR does not apply to substances with chemical structures related to mutagenicity (FDA 2012a). It was developed in the 1980s by FDA scientists, based on an analysis of in vivo carcinogenicity data of 343 substances (Rulis 1987). The substances had been tested for their tumorigenic potency in test animals. The probability of a substance causing a tumor in 50 % of animals was found to be normally distributed, with a narrow dose range where most of the substances caused the effect (Fig. 11.6).

Based on this analysis it was assumed that for all substances a similar distribution exists, and the assumedly safe threshold of 0.5 ppb for non-genotoxic carcinogens in food was derived. Currently, FDA uses a tiered approach with three levels (1.5, 15, and 45 μg/person per day) reflecting structural differences of chemicals (Cheeseman et al. 1999).

The *Threshold of Toxicological Concern* (TTC) assigns a threshold based on a chemical's two-dimensional structure. The different threshold groups are derived from an analysis of chronic toxicity data (carcinogenicity, neurotoxicity, developmental toxicity, immunotoxicity) for a variety of different compounds according to their chemical structures (Kroes et al. 2004; Munro et al. 1996). The TTC concept is used for prioritization purposes by EFSA (EFSA 2011) and in the U.S. by industry for GRAS self-determinations. The TTC has been suggested for use as hazard identification tool (Dewhurst and Renwick 2012), though this concept is scientifically controversial (Nordic Council of Ministers 2005).

11.5.3 Compatibility of Low-Dose Effects with Thresholds

The assumption of low-dose linearity for genotoxic carcinogens has been widely accepted because a NOEC cannot be established for such compounds. This assumption is reflected in the minimum pre-market testing requirements for FCS in Europe and the U.S. (Table 11.1). The scientific justification of thresholds for genotoxicants is complex and has been reviewed extensively (Neumann 2009).

Currently, there is a similar discussion ongoing regarding thresholds for endocrine disrupting chemicals (EDCs). EDCs are natural or synthetic compounds that interact with the hormone system. For example, EDCs may be estrogenic if they mimic the female sex hormone estrogen (like BPA). Other types of endocrine disruption may affect the male sex hormone testosterone, or the thyroid hormone system. EDCs have been linked to a number of health endpoints, such as cancer, reproduction, development, behavior and cognition, immune system, diabetes and cardiovascular disease. In 2013 the UN Environmental Programme (UNEP) and WHO published a comprehensive report on EDCs, collecting the currently available scientific knowledge on this diverse group of compounds (UNEP/WHO 2013). In the context of EDCs, *low dose* can have several different definitions. It can be referred to as doses below an established NOEC or as the assumed or measured average human exposure level from biomonitoring studies.

EDCs have been an issue in chemical risk assessment for several decades. The FDA's *Select Committee on GRAS Substances* recommended in 1982 that FDA take hormone receptor binding at low dose levels into account for risk assessment (Maffini et al. 2013). In 1996, the U.S. Congress passed the *1996 Food Quality Protection Act and the Safe Drinking Water Act Amendments*, requiring the U.S. Environmental Protection Agency (USEPA) to screen high volume chemicals and certain pesticides for endocrine disrupting effects (USEPA 2013a). This has resulted in the *Endocrine Disruptor Screening Program* that is in the process of developing testing methods to screen chemicals for estrogenic-, androgenic- and thyroid-hormone-like action (USEPA 2013a). Chemicals on the current priority list for screening include some of the phthalates and several other FCSs (USEPA 2009, 2013b). This screening program is still being validated. Dedicated EDC screening is currently not routinely required in FDA's, or EFSA's hazard characterization step.[5]

While several known or suspected EDCs are legally used as FCS in the U.S. and Europe (Muncke 2009), several endocrine scientists have stated that for EDCs no safe thresholds can be assumed (Zoeller et al. 2012).

11.5.4 Non-monotonic Dose Responses

In a comprehensive review, Laura Vandenberg and colleagues (Vandenberg et al. 2012) published the currently available scientific evidence for *non-monotonic dose responses* (NMDRs). Hormone-like chemicals may exert different, or opposite, effects at a lower dose range compared to the high dose range, resulting in a dose response curve that has a U-shape or inverted U-shape. This phenomenon is currently not being taken into account in FCS chemical risk assessments. In the regulatory context, toxicological data is commonly analyzed *top–down;* if no adverse effects are detected at the highest tested dose, lower dose data are not further considered (Maffini et al. 2011). The risk assessment process is not designed to consider NMDRs and new approaches taking these into account are needed. For newly designed chemicals, a comprehensive *Tiered Protocol for Endocrine Disruption* (TiPED) testing has been established that helps identification of NMRDs during development and serves as independent guidance for industry (Schug et al. 2013).

11.5.5 Mixture Toxicity

An additional challenge to assessing the risk of FCMs is that people are exposed to low doses of many different chemical substances at the same time. Natural and synthetic chemicals can act additively at low levels to cause effects that single substances would not induce at the same levels. This phenomenon was studied for estrogenic compounds in vitro, and has since been called "Something from Nothing" (Silva et al. 2002). Mixtures of anti-androgenic chemicals have been shown to act additively in vivo (Christiansen et al. 2008; Rider et al. 2008). First suggestions have been made how mixture toxicity can be integrated into chemical

[5] EFSA is currently in the process of revising its toxicology testing guidelines for FCS, the draft is expected for December 2013.

risk assessment, but a central prerequisite is the availability of high-quality exposure data (Kortenkamp and Faust 2010). Hence, the challenge of mixture toxicity is also a challenge of information transparency on chemical use, a data gap that has also been highlighted for FCS (Alger et al. 2013).

11.5.6 Chronic Toxicity Testing in Chemical Risk Assessment

Chemical risk assessment makes use of single substance toxicity data collected in vivo using laboratory animal models and in vitro cell-based assays. This includes the use of cancer cell lines derived from human or animal tumors, and bacteria for genotoxicity studies (e.g., the Ames test). Commonly the testing guidelines issued by the Organization for Economic Co-operation and Development (OECD) are followed (OECD 2013). Single substances are tested over a range of doses to derive a NOEC.

This practice implies two assumptions: (1) a toxicological threshold exists for the tested chemicals and (2) the NOEC is a reliable enough reflection of the human NEL. The NOEC is usually defined as the lowest tested dose that did not cause the effect of interest in the test system (usually rodents). Its value will consequently depend on the number and range of doses, the sample size per test group, and test duration. Animal cancer bioassays require a long duration because of the long latency period for most cancers and because certain types of tumors may develop late in the test animal's life. It has been argued that using NOECs may encourage study designs resulting in higher safe values (Falk-Filipsson et al. 2007). Most in vivo toxicity testing is carried out in highly inbred laboratory rodent strains. This can facilitate testing, because statistical significance of toxic effects may be more easily established, but some strains of laboratory animals may be overall less sensitive to specific effects of certain chemicals, due to their specific genetic traits. Including positive controls for selected endpoints of interest, for example those addressing endocrine disruption, has been suggested (vom Saal et al. 2010).

New developments in the area of high-throughput testing offer opportunities for reducing or even eliminating animal testing. The U.S. is pioneering these efforts with the programs *ToxCast*, carried out at the USEPA's National Center for Computational Toxicology (USEPA 2013c) and the interagency program *Tox21* (USEPA 2013d). Tox21 is examining the biological pathways of approximately 8,200 chemicals using computational biology to better predict how these chemicals will affect toxicological and health endpoints. The FDA, along with the USEPA and the National Institute of Environmental Health Sciences, are among the agencies participating in this effort that is using robotics and knowledge of biological and toxicological pathways to not only screen chemicals, but to prioritize which chemicals require more extensive evaluation of their toxicological risk (USEPA 2013d). Results of this large chemical library will be available to risk assessors and regulators. These new approaches address the need to assess toxicity for thousands of substances already intentionally used or present in FCM as impurities, break down or polymerization by-products that have no available toxicological data. Further, science-related issues like low-dose effects, mixture toxicity and variability between humans can be addressed using these approaches.

11.5.7 Scientific Basis for Safety Factors

Alternatives have been suggested to the traditional, but arbitrarily set safety factors. Species-specific allometric scaling factors have been proposed to account for differences in toxicokinetics between animal and human species (Falk-Filipsson et al. 2007). Other uncertainties include a lack of data showing the extent of inter-individual sensitivity in humans (genetic polymorphisms), or individual sensitivity and susceptibility to chronic diseases. Because of these uncertainties, use of chemical or effect-specific factors in chemical risk assessment has been suggested as a more appropriate approach resulting in higher confidence of sufficient protection (Martin et al. 2013).

11.5.8 Developmental Origins of Health and Disease

The developing fetus is especially sensitive to certain chemicals, as the tragic cases of thalidomide and diethylstilbestrol have shown. Known as the *Developmental Origins of Health and Disease*, the developmental vulnerability has been linked with other chronic diseases that become apparent later in life, such as metabolic disease, hypertension and cardiovascular disease. Common industrial chemicals, like certain FCS (e.g., BPA and some of the phthalates), have been shown to cause relevant effects in laboratory animal studies. Evaluating these hypotheses on human populations is more challenging, because of ubiquitous exposure to FCS and subsequent lack of a non-exposed human reference population. In addition, clinical health effect(s) may take many years or decades to manifest in exposed humans. Fetal development is a sensitive window for chemical exposures and acknowledging its importance offers the possibility for prevention of chronic diseases (Balbus et al. 2013). Regulatory risk assessment in the U.S. and E.U. require two-generation developmental toxicity tests for FCS migrating above 1 and 5 ppm, respectively (Tables 11.1 and 11.2).

11.5.9 Transgenerational Inheritance of Chronic Disease

Inheritance of phenotypes, like disease susceptibility, has traditionally been associated with genetic information stored in DNA's genetic code. During the last decade, however, research has shown that further, non-genetic traits called *epigenetic factors*, play a role in phenotype inheritance.

Epigenetics is defined as:

> …the study of mitotically and/or meiotically heritable changes in gene function that cannot be explained by changes in DNA sequence (Russo et al. 1997).

These epigenetic factors have been shown to regulate gene expression, including the timing, level of expression, and tissue specific expression. While they are diverse, the best-studied epigenetic factor to date is DNA methylation. Feil and Fraga (2012) summarized environmental factors associated with epigenetic changes, including chemical exposures. For example, BPA-exposure during embryonic development changed DNA methylation at a specific DNA location in the Agouti mouse model, a strain used to study epigenetic effects (Dolinoy et al. 2007). Mixtures of plastic-associated chemicals have been shown to induce effects that persist four generations after the developmental exposure (Manikkam et al. 2013). This transgenerational epigenetic inheritance of disease phenotypes that are caused by chemical exposures are a new challenge for chemical risk assessment. Current developmental toxicity testing is carried out in two-generation in vivo studies. With an improved understanding of underlying molecular mechanisms, epigenetic effects may become part of chemical hazard identification in future (Rasoulpour et al. 2011).

11.6 Summary

FCMs are all materials that come into direct contact with food and beverages. The term collectively describes food packaging, food processing and handling equipment, storage containers, kitchenware and tableware. Depending on their chemical and physical properties, FCMs can transfer their chemical components into the food and beverages they contact. People are exposed to intentionally added FCS and NIAS at low levels on a daily (chronic) basis. Regulatory risk assessment characterizes the known exposures and estimates resulting human health harm, where data are available, for single substances. Chemical risk assessment has been based on the principle that low exposure generally implies low risk. This paradigm has been challenged by important advances in our understanding of how chemicals affect health. Notably, EDCs can mimic natural hormones and may interfere with development and homeostasis. Further, EDCs and other chemicals may act in mixtures to cause adverse health effects, even if the individual mixture components are present below their single substance effect levels. Especially during the period of fetal development these aspects are of concern, since these chemical exposures may be associated with chronic disease later in life. Finally, research into transgenerational inheritance has shown that epigenetic imprinting can be affected by chemical exposures. Implementing these important insights from academic research into regulatory risk assessment of chemicals is a challenge requiring new approaches. High-throughput screening assays such at ToxCast and Tox21 present promising approaches to providing toxicological information on many chemicals used or present unintentionally in FCM for which there is currently little or no chemical risk information. Industry has already addressed some of the related concerns and changed product formulations accordingly.

11.7 Data Gaps and Recommendations

Using Biomonitoring Data In order to correctly assess the human population risk from chemicals, reliable, transparent and up-to-date exposure data for FCS needs to be available. Risk assessment for chemicals present in FCMs must also include other uses and exposure routes for these compounds. One possible approach to risk assessment of known FCS is to use human biomonitoring data. In the U.S., the Centers for Disease Control and Prevention (CDC) collect representative information in their National Health and Nutrition Examination Survey (NHANES), including blood and urinary levels of over 300 environmental chemicals (CDC 2012). FDA already makes use of the NHANES food consumption database for estimating cumulative dietary exposure (Alger et al. 2013). The benefit of using CDC's biomonitoring data is that all human exposure routes for many chemicals, not only exposure from FCMs, are taken into consideration.

Identifying Non-intentionally Added Substances NIAS are a major challenge for establishing the safety of FCMs. Improved identification by chemical analysis, but also data-based hazard characterization is necessary. A promising approach is the combination of appropriate in vitro bioassays with chemical fractionation analysis. Such tools acknowledge the chemical complexity of plastics, and take mixture toxicity into account. It will be of importance to select sensitive and meaningful bioassays for relevant endpoints, for example hormone binding for endocrine disruption and chromosome aberration for genotoxicity.

Screening for Endocrine Disruption Assessing endocrine disruption properties for new FCS in all premarket assessments is a minimum step to ensuring FCM safety, based on recent scientific findings. Suitable in vitro screening assays for different types of endocrine disruption are available (Schug et al. 2013). New chemicals should only be permitted for use in FCMs if they have been tested for these hazards, similarly to genotoxicants. Thereby, the development of reliable in silico tools for identifying EDCs in current (postmarket assessment) and for new FCS is a priority.

Translating Scientific Findings for Risk Assessment and Regulation One major challenge for regulatory agencies across the world is updating legal requirements with the latest scientific findings. Society pays for scientific research into health effects of chronic chemical exposures, because there is a moral need for better understanding how chronic diseases can be prevented. This research has delivered important results over the last decades. On the other hand, businesses need to operate in a reliable, predictable and stable market environment. Regularly changing and updating legal requirements according to new research results may imply an increased operational risk for industry. Society needs to define a constructive mechanism how to address this dichotomy of needs, where legal requirements can be based on the most current scientific information available with the aim of best protecting public health. A possible structural adaptation in the U.S. may be a clear separation between the risk assessing agency and a democratically elected body responsible for risk management.

Acknowledgements All statements made in this text reflect the author's personal professional opinion and are not an expression of the Food Packaging Forum Foundation's views.

References

Alger HM, Maffini MV, Kulkarni NR et al (2013) Perspectives on how FDA assesses exposure to food additives when evaluating their safety: workshop proceedings. Compr Rev Food Sci F 12(1):90–119. doi:10.1111/j.1541-4337.2012.00216.x

Ashby J, Tennant RW (1988) Chemical structure, *Salmonella* mutagenicity and extent of carcinogenicity as indicators of genotoxic carcinogenesis among 222 chemicals tested in rodents by the U.S. NCI/NTP. Mutat Res,Genet Tox 204(1):17–115. doi:10.1016/0165-1218(88)90114-0

Bailin PS, Byrne M, Lewis S et al (2008) Public awareness drives market for safer alternatives. Bisphenol A market analysis report. Investor Environmental Health Network. http://www.iehn.org/documents/BPA%20market%20report%20Final.pdf. Accessed 2 July 2013

Balbus JM, Barouki R, Birnbaum LS et al (2013) Early-life prevention of non-communicable diseases. Lancet 381(9860):3–4. doi:10.1016/S0140-6736(12)61609-2

Benigni R, Bossa C (2006) Structural alerts of mutagens and carcinogens. Curr Comput Aided Drug Des 2(2):169–176

Biedermann M, Grob K (2010) Is recycled newspaper suitable for food contact materials? Technical grade mineral oils from printing inks. Eur Food Res Technol 230(5):785–796

Biedermann M, Ingenhoff J-E, Zurfluh M et al (2013) Migration of mineral oil, photoinitiators and plasticizers from recycled paperboard into dry foods: a study under controlled conditions. Food Addit Contam Part A 30(5):885–898. doi:10.1080/19440049.2013.786189

Bradley E, Coulier L (2007) An investigation into the reaction and breakdown products from starting substances used to produce food contact plastics. Central Science Laboratory, London. http://www.foodbase.org.uk//admintools/reportdocuments/518-1-911_A03054_reaction_and_breakdown_products_final_report.pdf. Accessed 2 July 2013

Carwile JL, Ye X, Zhou X et al (2011) Canned soup consumption and urinary bisphenol A: a randomized crossover trial. JAMA 306(20):2218–2220. doi:10.1001/jama.2011.1721

Centers for Disease Control and Prevention (CDC) (2012) National biomonitoring program. Environmental chemicals. http://www.cdc.gov/biomonitoring/environmental_chemicals.html. Accessed 2 July 2013

Cheeseman MA, Machuga EJ, Bailey AB (1999) A tiered approach to threshold of regulation. Food Chem Toxicol 37(4):387–412. doi:10.1016/S0278-6915(99)00024-1

Christiansen S, Scholze M, Axelstad M et al (2008) Combined exposure to anti-androgens causes markedly increased frequencies of hypospadias in the rat. Int J Androl 31(2):241–248. doi:10.1111/j.1365-2605.2008.00866.x
Coulier L, Bradley EL, Bas RC et al (2010) Analysis of reaction products of food contaminants and ingredients: bisphenol A diglycidyl ether (BADGE) in canned foods. J Agric Food Chem 58(8):4873–4882. doi:10.1021/jf904160a
Deichmann WB, Henschler D, Holmsted B et al (1986) What is there that is not poison? A study of the Third Defense by Paracelsus. Arch Toxicol 58(4):207–213. doi:10.1007/BF00297107
Dewhurst I, Renwick AG (2012) Evaluation of the threshold of toxicological concern (TTC)—challenges and approaches. Regul Toxicol Pharmacol 65(1):168–177. doi:10.1016/j.yrtph.2012.03.007
Dolinoy DC, Huang D, Jirtle RL (2007) Maternal nutrient supplementation counteracts bisphenol A-induced DNA hypomethylation in early development. Proc Natl Acad Sci USA 104(32):13056–13061
European Commission (2013) Food contact materials. https://webgate.ec.europa.eu/sanco_foods/main/?event=display. Accessed 16 June 2013
European Food Safety Authority (EFSA) (2008) Guidance document on the submission of a dossier on a substance to be used in food contact materials for evaluation by EFSA by the panel on additives, flavourings, processing aids and materials in contact with food. EFSA J. doi:10.2903/j.efsa.2008.21r
European Food Safety Authority (EFSA) (2011) Report of ESCO WG on non-plastic food contact materials. European Food Safety Authority. http://www.efsa.europa.eu/en/supporting/pub/139e.htm. Accessed 16 June 2013
European Food Safety Authority (EFSA) (2012a) Guidance on selected default values to be used by the EFSA scientific committee, scientific panels and units in the absence of actual measured data. EFSA J 10(3):2579–2611. doi:10.2903/j.efsa.2012.2579
European Food Safety Authority (EFSA) (2012b) Scientific opinion on exploring options for providing advice about possible human health risks based on the concept of threshold of toxicological concern (TTC). EFSA J 10(7):2750–2853. doi:10.2903/j.efsa.2012.2750
European Union (2007) Commission directive 2007/42/EC of 29 June 2007 relating to materials and articles made of regenerated cellulose film intended to come into contact with foodstuffs. http://eur-lex.europa.eu/LexUriServ/LexUriServ.do?uri=OJ:L:2007:172:0071:0082:EN:PDF. Accessed 18 July 2013
European Union (2011) Commission regulation (E.U.) No 10/2011 of 14 January 2011 on plastic materials and articles intended to come into contact with food. http://eur-lex.europa.eu/LexUriServ/LexUriServ.do?uri=OJ:L:2011:012:0001:0089:EN:PDF. Accessed 18 July 2013
European Union (2012) Regulation (E.U.) No 528/2012 of the European Parliament and of the Council of 22 May 2012 concerning the making available on the market and use of biocidal products. http://eur-lex.europa.eu/LexUriServ/LexUriServ.do?uri=OJ:L:2012:167:0001:0123:EN:PDF. Accessed 18 July 2013
Falk-Filipsson A, Hanberg A, Victorin K et al (2007) Assessment factors: applications in health risk assessment of chemicals. Environ Res 104(1):108–127
Food and Drug Administration (FDA) (2002) Guidance for industry: preparation of food contact notifications for food contact substances: toxicology recommendations. http://www.fda.gov/Food/GuidanceRegulation/GuidanceDocumentsRegulatoryInformation/IngredientsAdditivesGRASPackaging/ucm081825.htm. Accessed 17 June 2013
Food and Drug Administration (FDA) (2004) Guidance for industry: frequently asked questions about GRAS. http://www.fda.gov/Food/GuidanceRegulation/GuidanceDocumentsRegulatoryInformation/IngredientsAdditivesGRASPackaging/ucm061846.htm. Accessed 4 July 2013
Food and Drug Administration (FDA) (2007a) Guidance for industry and other stakeholders: toxicological principles for the safety assessment of food ingredients. Redbook 2000. http://www.fda.gov/Food/GuidanceRegulation/GuidanceDocumentsRegulatoryInformation/IngredientsAdditivesGRASPackaging/ucm2006826.htm. Accessed 16 July 2013

Food and Drug Administration (FDA) (2007b) Guidance for industry: preparation of premarket submissions for food contact substances: chemistry recommendations. http://www.fda.gov/Food/GuidanceRegulation/GuidanceDocumentsRegulatoryInformation/IngredientsAdditivesGRASPackaging/ucm081818.htm#aivti. Accessed 14 June 2013

Food and Drug Administration (FDA) (2009) Food and Drug Administration Modernization Act (FDAMA) of 1997. http://www.fda.gov/RegulatoryInformation/Legislation/FederalFoodDrugandCosmeticActFDCAct/SignificantAmendmentstotheFDCAct/FDAMA/default.htm. Accessed 2 July 2013

Food and Drug Administration (FDA) (2011a) Guidance for industry: submitting requests under 21 CFR 170.39 Threshold of regulation for substances used in food-contact articles. Exemptions for houseware articles. http://www.fda.gov/Food/GuidanceRegulation/GuidanceDocumentsRegulatoryInformation/IngredientsAdditivesGRASPackaging/ucm081833.htm#_ftn2. Accessed 14 June 2013

Food and Drug Administration (FDA) (2011b) List of indirect additives used in food contact substances. http://www.accessdata.fda.gov/scripts/fcn/fcnNavigation.cfm?rpt=iaListing. Accessed 4 July 2013

Food and Drug Administration (FDA) (2012a) Threshold of regulations exemptions. List of issued exemptions based on the threshold of regulation, since 1996 (updated December 2012). http://www.fda.gov/Food/IngredientsPackagingLabeling/PackagingFCS/ThresholdRegulationExemptions/ucm093685.htm. Accessed 14 June 2013

Food and Drug Administration (FDA) (2012b) Cumulative estimated daily intake. http://www.accessdata.fda.gov/scripts/sda/sdnavigation.cfm?sd=edisrev. Accessed 4 July 2013

Food and Drug Administration (FDA) (2013a) Inventory of effective food contact substance (FCS) notifications. http://www.accessdata.fda.gov/scripts/fcn/fcnnavigation.cfm?rpt=fcslisting. Accessed 14 June 2013

Food and Drug Administration (FDA) (2013b) GRAS notice inventory. http://www.accessdata.fda.gov/scripts/fcn/fcnNavigation.cfm?rpt=grasListing. Accessed 14 June 2013

Federal Register (2012) Indirect food additives: polymers. Polycarbonate resins. 21 CFR 177.1580. https://federalregister.gov/a/2012-17366. Accessed 17 July 2013

Feil R, Fraga MF (2012) Epigenetics and the environment: emerging patterns and implications. Nat Rev Genet 13(2):97–109. doi:10.1038/nrg3142

Gallart-Ayala H, Núñez O, Lucci P (2013) Recent advances in LC-MS analysis of food-packaging contaminants. TrAC 42:99–124. http://dx.doi.org/10.1016/j.trac.2012.09.017

Grob K, Biedermann M, Scherbaum E et al (2006) Food contamination with organic materials in perspective: packaging materials as the largest and least controlled source? A view focusing on the European situation. Crit Rev Food Sci Nutr 46(7):529–535

Grob K, Stocker J, Colwell R (2009) Assurance of compliance within the production chain of food contact materials by good manufacturing practice and documentation—part 1: legal background in Europe and compliance challenges. Food Control 20:476–482

Guart A, Wagner M, Mezquida A et al (2013) Migration of plasticizers from Tritan™ and polycarbonate bottles and toxicological evaluation. Food Chem 141(1):373–380. doi:10.1016/j.foodchem.2013.02.129

IPCS (2009) Principles and methods for the risk assessment of chemicals in food, vol 240. Environmental Health Criteria. WHO, Geneva. http://www.who.int/foodsafety/chem/principles/en/index1.html. Accessed 2 July 2013

Joint FAO/WHO Expert Committee on Food Additives (JECFA) (2013) Joint FAO/WHO expert committee on food additives publications. http://www.who.int/foodsafety/chem/jecfa/publications/en/index. Accessed 14 June 2013

Joint Research Centre (JRC) (2012) The FACET Project: flavorings, additives and food contact materials exposure task. http://ihcp.jrc.ec.europa.eu/our_activities/food-cons-prod/chemicals_in_food/FACET. Accessed 14 June 2013

Kortenkamp A, Faust M (2010) Combined exposures to anti-androgenic chemicals: steps towards cumulative risk assessment. Int J Androl 33(2):463–474. doi:10.1111/j.1365-2605.2009.01047.x

Kroes R, Renwick AG, Cheeseman M et al (2004) Structure-based thresholds of toxicological concern (TTC): guidance for application to substances present at low levels in the diet. Food Chem Toxicol 42(1):65–83

Le HH, Carlson EM, Chua JP et al (2008) Bisphenol A is released from polycarbonate drinking bottles and mimics the neurotoxic actions of estrogen in developing cerebellar neurons. Toxicol Lett 176(2):149–156

Maffini MV, Alger HM, Bongard ED et al (2011) Enhancing FDA's evaluation of science to ensure chemicals added to human food are safe: workshop proceedings. Compr Rev Food Sci F 10(6):321–341. doi:10.1111/j.1541-4337.2011.00165.x

Maffini MV, Alger HM, Olson ED et al (2013) Looking back to look forward: a review of FDA's food additives safety assessment and recommendations for modernizing its program. Compr Rev Food Sci F 12(4):439–453. doi:10.1111/1541-4337.12020

Manikkam M, Tracey R, Guerrero-Bosagna C et al (2013) Plastics derived endocrine disruptors (BPA, DEHP and DBP) induce epigenetic transgenerational inheritance of obesity, reproductive disease and sperm epimutations. PLoS ONE 8(1):e55387. doi:10.1371/journal.pone.0055387

Martin OV, Scholze M, Kortenkamp A (2013) Dispelling urban myths about default uncertainty factors in chemical risk assessment—sufficient protection against mixture effects? Environ Health 12(1):53. doi:10.1186/1476-069X-12-53

McNeal TP, Biles JE, Begley TH et al (2000) Determination of suspected endocrine disruptors in foods and food packaging. In: Keith LH, Jones-Lepp TL, Needham LL (eds) Analysis of environmental endocrine disruptors, vol 747., ACS symposium series. American Chemical Society, Washington, DC, pp 33–52

Mori Y (2010) Rules on food contact materials and articles in Japan. In: Rijk R, Veraart R (eds) Global legislation for food packaging materials. Wiley, Weinheim, pp 291–318

Morlock G, Schwack W (2006) Determination of isopropylthioxanthone (ITX) in milk, yoghurt and fat by HPTLC-FLD, HPTLC-ESI/MS and HPTLC-DART/MS. Anal Bioanal Chem 385(3):586–595

Muncke J (2009) Exposure to endocrine disrupting compounds via the food chain: is packaging a relevant source? Sci Total Environ 407(16):4549–4559

Munro IC, Ford RA, Kennepohl E et al (1996) Correlation of structural class with no-observed-effect levels: a proposal for establishing a threshold of concern. Food Chem Toxicol 34(9):829–867

Neltner TG, Kulkarni NR, Alger HM et al (2011) Navigating the U.S. Food Additive Regulatory Program. Compr Rev Food Sci F 10(6):342–368. doi:10.1111/j.1541-4337.2011.00166.x

Neltner TG, Alger HM, Leonard JE et al (2013) Data gaps in toxicity testing of chemicals allowed in food in the United States. Repro Tox. doi:10.1016/j.reprotox.2013.07.023

Nerin C, Alfaro P, Aznar M et al (2013) The challenge of identifying non-intentionally added substances from food packaging materials: a review. Anal Chim Acta 775:14–24 (2013). doi:10.1016/j.aca.2013.02.028

Neumann HG (2009) Risk assessment of chemical carcinogens and thresholds. Crit Rev Toxicol 39(6):449–461

Nordic Council of Ministers (2005) Threshold of toxicological concern (TTC): Literature review and applicability, Nordic Council. http://www.norden.org/da/publikationer/publikationer/2005-559/at_download/publicationfile. Accessed 2 July 2013

Organization for Economic Co-operation and Development (OECD) (2013) OECD guidelines for the testing of chemicals. http://www.oecd-ilibrary.org/environment/oecd-guidelines-for-the-testing-of-chemicals_chem_guide_pkg-en;jsessionid=jj1gcntzpbmu.x-oecd-live-02. Accessed 13 June 2013

Osimitz TG, Eldridge ML, Sloter E et al (2012) Lack of androgenicity and estrogenicity of the three monomers used in Eastman's Tritan™ copolyesters. Food Chem Toxicol 50(6):2196–2205. doi:10.1016/j.fct.2012.02.010

Pano (2011) Pano blueseal. http://www.pano.de/en/products/metal-packaging/pano-blueseal. Accessed 14 June 2013

Pedersen GA, Jensen LK, Fankhauser A et al (2008) Migration of epoxidized soybean oil (ESBO) and phthalates from twist closures into food and enforcement of the overall migration limit. Food Addit Contam Part A Chem Anal Control Expo Risk Assess 25(4):503–510

Rasoulpour RJ, LeBaron MJ, Ellis-Hutchings RG et al (2011) Epigenetic screening in product safety assessment: are we there yet? Toxicol Mech Methods 21(4):298–311. doi:10.3109/15376516.2011.557883

Renwick AG (1991) Safety factors and establishment of acceptable daily intakes. Food Addit Contam 8(2):135–149

Renwick AG (2004) Risk characterisation of chemicals in food. Toxicol Lett 149(1–3):163–176. doi:10.1016/j.toxlet.2003.12.063

Rexam (2011) Consumer packaging report 2011–2012. http://www.rexam.com/files/pdf/packaging_unwrapped_2011.pdf. Accessed 2 July 2013

Rider CV, Furr J, Wilson VS et al (2008) A mixture of seven antiandrogens induces reproductive malformations in rats. Int J Androl 31(2):249–262

Rudel RA, Gray JM, Engel CL et al (2011) Food packaging and bisphenol A and bis(2-ethyhexyl) phthalate exposure: findings from a dietary intervention. Environ Health Perspect 119(7):914–920

Rulis AM (1987) *De Minimis* and the threshold of regulation. In: Felix CW (ed) Food protection technology: "current and projected technologies for food protection—recommendations and implementation". Proceedings of the 1986 conference for food protection. Lewis, Chelsea, MI, pp 29–37

Russo VEA, Martienssen RA, Riggs AD (1997) Epigenetic mechanisms of gene regulation. Cold Spring Harbor Laboratory Press, New York

Schug TT, Abagyan R, Blumberg B et al (2013) Designing endocrine disruption out of the next generation of chemicals. Green Chem 15(1):181–198. doi:10.1039/C2GC35055F

Shanklin AP (2008) How FDA's threshold of regulation program works. Food safety magazine. http://www.foodsafetymagazine.com/magazine-archive1/december-2008january-2009/how-fdas-threshold-of-regulation-program-works/. Accessed 2 July 2013

Shanklin AP, Sánchez ER (2013) Regulatory report: FDA's food contact substance notification program. http://www.fda.gov/Food/IngredientsPackagingLabeling/PackagingFCS/ucm064161.htm. Accessed 2 July 2013

Silva E, Rajapakse N, Kortenkamp A (2002) Something from "nothing"—eight weak estrogenic chemicals combined at concentrations below NOECs produce significant mixture effects. Environ Sci Technol 36(8):1751–1756

Simoneau C, Van den Eede L, Valzacchi S (2012) Identification and quantification of the migration of chemicals from plastic baby bottles used as substitutes for polycarbonate. Food Addit Contam Part A Chem Anal Control Expo Risk Assess 29(3):469–480. doi:10.1080/19440049.2011.644588

Swiss Federal Department of Home Affairs (2011) Ordinance of the federal department of home affairs on materials and articles (817.023.21). Section 8b: Packaging inks. http://www.bag.admin.ch/themen/lebensmittel/04867/10015/index.html?lang=en. Accessed 14 June 2013

United Nations Environment Programme/World Health Organization (UNEP/WHO) (2013) State of the science of endocrine disrupting chemicals 2012. In: Bergman A, Heindel JJ, Jobling S, Kidd KA, Zoeller RT (eds), United Nations Environmental Programme and the World Health Organization. http://www.who.int/ceh/publications/endocrine/en/ Accessed 2 July 2013

United States Statutes at Large (1958) An Act to protect the public health by amending the Federal Food, Drug and Cosmetic Act to prohibit the use in food of additives which have not been adequately tested to establish their safety. http://www.govtrack.us/congress/bills/85/hr13254. Accessed 2 July 2013

U.S. Environmental Protection Agency (USEPA) (2009) Final list of chemicals for initial Tier 1 screening. http://www.epa.gov/endo/pubs/prioritysetting/finallist.html. Accessed 2 July 2013

U.S. Environmental Protection Agency (USEPA) (2012) Human health risk assessment. http://www.epa.gov/risk_assessment/health-risk.htm. Accessed 14 June 2013

U.S. Environmental Protection Agency (USEPA) (2013a) Endocrine disruptor screening program (EDSP). http://www.epa.gov/endo/. Accessed 2 July 2013
U.S. Environmental Protection Agency (USEPA) (2013b) Overview of the second list of chemicals for Tier 1 screening. http://www.epa.gov/endo/pubs/prioritysetting/list2facts.htm. Accessed 2 July 2013
U.S. Environmental Protection Agency (USEPA) (2013c) ToxCast. Screening chemicals to predict toxicity faster and better. http://www.epa.gov/ncct/toxcast/. Accessed 13 June 2013
U.S. Environmental Protection Agency (USEPA) (2013d) Tox21. http://epa.gov/ncct/Tox21/. Accessed 14 June 2013
U.S. Senate Committee on Labor and Public Welfare (1958) Food additives amendment of 1958 committee report—Senate report No. 2422. Congressional Record, Washington, D.C
van Leeuwen CJ (2007) Toxicity testing for human health risk assessment. General introduction. In: van Leeuwen CJ, Vermeire TG (eds) Risk assessment of chemicals: an introduction. Springer, Dordrecht, pp 1–36
Vandenberg LN, Colborn T, Hayes TB et al (2012) Hormones and endocrine-disrupting chemicals: low-dose effects and nonmonotonic dose responses. Endocr Rev 33(3):378-455. doi:10.1210/er.2011-1050
Vogel SA (2012) Chapter 1: Plastic food. Defining safety in the food additives act. In: Is it safe? BPA and the struggle to define the safety of chemicals. University of California Press, Berkeley, p 35
vom Saal FS, Akingbemi BT, Belcher SM et al (2010) Flawed experimental design reveals the need for guidelines requiring appropriate positive controls in endocrine disruption research. Toxicol Sci 115(2):612–613. doi:10.1093/toxsci/kfq048
Yang CZ, Yaniger SI, Jordan VC et al (2011) Most plastic products release estrogenic chemicals: a potential health problem that can be solved. Environ Health Perspect 119(7):989–996
Zoeller RT, Brown TR, Doan LL et al (2012) Endocrine-disrupting chemicals and public health protection: a statement of principles from the endocrine society. Endocrinology 153(9):4097–4110. doi:10.1210/en.2012-1422

Editor Biography

Suzanne M. Snedeker is an independent consultant working as a science writer and editor in Ithaca, NY. Her interests include evaluating the relationships between exposures to endocrine disrupting chemicals found in the home and workplace and the development of chronic diseases. She developed an interest in toxicants found in food packaging while a Visiting Fellow in the Cornell University Department of Food Science. She previously served as Associate Director for the Cornell University Program on Breast Cancer and Environmental Risk Factors (BCERF), a translational research program that included a public health outreach component. At the National Institute of Environmental Health Sciences (NIEHS) her work as a Staff Fellow focused on the role of peptide growth factors in the development of the mammary gland, and mechanisms of metal-induced kidney neoplasia. As a Biologist and Project Officer in the National Toxicology Program, she designed large-scale, multigenerational animal studies to assess the effects of low-level exposure to environmental endocrine disrupting chemicals on reproductive and cancer endpoints. She has served on numerous state and federal advisory panels and groups, including the New York State (NYS) Task Force on Flame Retardants (NYS Department of Health), the Working Group for the Breast Cancer and Environmental Research Centers [NIEHS and National Cancer Institute (NCI)], the Advisory Panel for the Sister Study (NIEHS), and the Agricultural Health Study Risk Communication Working Group (NCI). She also served the co-chair for the Program Work Team on Environmental Health Risks in Communities (Cornell Cooperative Extension). She is an experienced public speaker and has given numerous presentations to scientific, medical, workplace, cancer survivor, and general consumer audiences, and has developed technical reports, websites, consumer factsheets, occupational brochures, and YouTube videos on the health effects of environmental chemicals. She received her Ph.D. from the University of Wisconsin, Madison, WI, and her B.S. from Cornell University, Ithaca, NY.

S. M. Snedeker (ed.), *Toxicants in Food Packaging and Household Plastics*,
Molecular and Integrative Toxicology, DOI: 10.1007/978-1-4471-6500-2,

Index

S. M. Snedeker (ed.), *Toxicants in Food Packaging and Household Plastics*, Molecular and Integrative Toxicology, DOI: 10.1007/978-1-4471-6500-2,

E

F

G

H

V

W

Z

MIX
Papier aus verantwortungsvollen Quellen
Paper from responsible sources
FSC® C105338

If you have any concerns about our products,
you can contact us on
ProductSafety@springernature.com

In case Publisher is established outside the EU,
the EU authorized representative is:
Springer Nature Customer Service Center GmbH
Europaplatz 3, 69115 Heidelberg, Germany

Printed by Libri Plureos GmbH
in Hamburg, Germany